AF412183

Contemporary Clinical Neuroscience

Series editor

Mario Manto, Unité d'Etude du Mouvement (UEM), FNRS, Neurologie
ULB-Erasme, Bruxelles, Belgium

More information about this series at http://www.springer.com/series/7678

Christophe Habas

Editor

The Neuroimaging of Brain Diseases

Structural and Functional Advances

 Springer

Editor
Christophe Habas
Service de NeuroImagerie CHNO des
15/20
Paris, France

Contemporary Clinical Neuroscience
ISBN 978-3-319-78924-8   ISBN 978-3-319-78926-2   (eBook)
https://doi.org/10.1007/978-3-319-78926-2

Library of Congress Control Number: 2018944279

This Springer imprint is published by the registered company Springer International Publishing AG part of Springer Nature.
The registered company address is: Gewerbestrasse 11, 6330 Cham, Switzerland

Foreword

With its key contributions, the book edited by Doctor Christophe Habas is a timely contribution for the readership of *Contemporary Clinical Neurosciences*. Indeed, there is a now a consensus that modern neuroimaging has revolutionized not only traditional clinical branches, such as neurology, psychiatry or neurosurgery, but also fundamental neurosciences. Novel techniques undergo continuous refinements, especially for the numerous techniques tackling the imaging of the brain. No one could consider anymore modern neuroimaging as an accessory discipline for the advancement of neurosciences and even clinical care.

This book reviews the most recent developments in neuroimaging tools, highlighting translational research in neuroimaging for the multiple disciplines addressing brain disorders. The joint effort of the contributors under the careful guidance of the Guest Editor provides detailed information on the biology of the brain, from anatomical to functional and metabolic aspects. Habas' book is by essence multidisciplinary by the way it addresses important brain disorders which contribute to the very high burden.

The chapters are highly accessible to the readers. Starting from the basic mechanisms underlying the MRI technique, the reader is taken by the hand to understand the multimodal approaches currently used in many laboratories and stepping in the departments of radiology or neuroradiology at a world level. Thanks to novel hybrid systems, our view of the brain is now evolving even more quickly. The advances in sensitivity of techniques, the increasing speed to perform examinations and the successful efforts to reduce motion artefacts will not only allow an earlier detection of brain disorders but also improve the follow-up of their progression or provide relevant and reliable information in terms of accurate evaluation of therapies. Indeed, the aim to identify unambiguous radiological biomarkers is now clearly at the horizon of this century. Furthermore, the assessment of brain integrity has become an important field of research. A typical example of disorders whose pathogenesis is being unravelled at a high speed thanks to neuroimaging techniques is the heterogeneous group of psychiatric diseases, such as bipolar disorders or schizophrenia. In only three decades, neuroimaging has allowed the extraction of critical and so far unsuspected defects in circuitry, in particular at the level of brain networks. The large amount of information requires a critical and updated overview.

The various chapters are clearly written and complementary, with practical examples. The style is concise and the size of the book ideal for gaining novel knowledge with a broad perspective in mind. It will be an excellent platform of learning for the neuroscientists, clinicians and trainees who are willing to jump in the field of neuroimaging. It will also serve as a springboard for the students discovering how one technique has evolved fastly as a must for brain appraisal. Training the next generation of students and talented neuroscientists is a goal of our book series. This book is also tailored to this objective.

Mons, Belgium

Mario Manto

Preface

Recent neuroimaging developments allow a better diagnosis, follow-up and prognosis of neurological, neuropsychological and psychiatric diseases. If part of these improvements concern high spatial and temporal resolution obtained, for instance, with high-field magnetic resonance imaging (3 T–7 T) or specific metabolic tracers in nuclear medicine, the major part of them are due to clinical application of research imaging and post-processing techniques. For instance, voxel-based morphometry and cortical thickness measurements can detect subtle regional volumetric variations of white and grey matter. Spectroscopy gives access to the metabolic glial, neuronal and tumoural composition and alteration, as well as radiotracer in nuclear medicine. Diffusion imaging and tractography analyse anatomical, inter-areal fascicular connectivity and provide biomarkers of tissue integrity. T1- and T2*-relaxometry can also give information about chemical composition of brain tissues. Stimulation and resting-state functional imaging, using blood oxygenation dependent information or arterial spin labelling, furnishes functional brain mapping in relation with task-related circuits, and with intrinsically connected networks, such as the default-mode network. For functional magnetic resonance (fMRI), several mathematical methods can be further applied to determine causal effects between co-activated brain areas, and to discriminate subtle interregional topological arrangement or disease-related perturbations. Moreover, all the data gathered with all this techniques and constituting potential biomarkers can also be treated by advanced multi-variate approaches which can identify correlated and complex multidimensional patterns, and which can be implemented by artificial neural networks (deep learning). Finally, functional brain imaging, such as neurofeedback, can also be used to help patients to self-regulate some symptoms, such as pain.

Therefore, besides the classical imaging methods studying the brain gross anatomy, advanced neuroimaging coupled with performant statistical and deep-learning-based techniques now offers a richer and deeper view of the brain (micro-) architecture, physiology and functioning. The present book aims at illustrating the growing clinical applications of the above-mentioned powerful techniques to a wide field of brain diseases, and the gain obtained in terms of diagnosis and prognosis. The first two chapters cover briefly the physical basis of morphological and functional magnetic resonance imaging, and the main classical and advanced statistical methods applied to the raw data in order to generate the final clinical images, as well as functional and metabolic mappings. All the remaining chapters deal with specific application to some neurologic (ataxias, multiple sclerosis, epilepsy, vision impairments and rehabilitation, pain, brain tumours, stroke, Parkinson disease and related syndromes), neuropsychological (autism spectrum disorders) and psychiatric (dementia, bipolar disorder, schizophrenia, obsessive-compulsive disorders) affections.

Undoubtedly, recent neuroimaging advances, at least, in terms of MRI sequence, algorithms and artificial intelligence will likely supply more exhaustive, sensitive, specific, robust and predictive information about structural and functional brain alteration and recovery, and will thus be more and more incorporated into clinical routine imaging.

Paris, France   C. Habas

Contents

Contributors

Sara Regina Meira Almeida Department of Neurology, School of Medical Sciences, University of Campinas (UNICAMP), Campinas, SP, Brazil

Brazilian Institute of Neuroscience and Neurotechnology (BRAINN), Campinas, SP, Brazil

Amir Amedi Department of Medical Neurobiology, The Institute for Medical Research Israel-Canada, Faculty of Medicine, The Hebrew University of Jerusalem, Jerusalem, Israel

The Edmond and Lily Safra Centre for Brain Sciences (ELSC), The Hebrew University of Jerusalem, Jerusalem, Israel

Cognitive Sciences Program, The Hebrew University of Jerusalem, Jerusalem, Israel

Alexander Barnett Krembil Research Institute, University Health Network, Toronto, ON, Canada

Deparment of Psychology, University of Toronto, Toronto, ON, Canada

Nicolaas I. Bohnen Department of Radiology, University of Michigan, Ann Arbor, MI, USA

Department of Neurology, University of Michigan, Ann Arbor, MI, USA

Morris K. Udall Center of Excellence for Parkinson's Disease Research, University of Michigan, Ann Arbor, MI, USA

Hugo Botha Department of Neurology, Mayo Clinic, Rochester, MN, USA

Gabriela Castellano Neurophysics Group, Institute of Physics Gleb Wataghin, University of Campinas (UNICAMP), Campinas, SP, Brazil

Brazilian Institute of Neuroscience and Neurotechnology (BRAINN), Campinas, SP, Brazil

Francine Chassoux Department of Neurosurgery, Centre de Psychiatrie et Neurosciences, INSERM U894, Paris, France

Daniel-Robert Chebat Visual and Cognitive Neuroscience Laboratory (VCN lab), Department of Behavioral Sciences, Faculty of Social Sciences and Humanities, Ariel University, Ariel, Israel

A. D'Abreu Neuroimaging Laboratory, School of Medical Sciences, State University of Campinas, Campinas, Brazil

Neurology Department, School of Medical Sciences, State University of Campinas, Campinas, Brazil

G. de Marco Laboratoire CeRSM (EA-2931), Equipe Analyse du Mouvement en Biomécanique, Physiologie et Imagerie, Université Paris Nanterre, Nanterre, France

Bertrand Devaux Department of Neurosurgery, Centre de Psychiatrie et Neurosciences, INSERM U894, Paris, France

Myriam Edjlali Department of Radiology, Centre Hospitalier Sainte-Anne, Université Paris Descartes Sorbonne, Paris, France

S. Espinoza Service de NeuroImagerie, CHNO des 15-20, Paris, France

Remy Guillevin Clinical Neuroimaging & Research team DACTIM-MIS/LMA CNRS 7348, CHU and University of Poitiers, Poitiers, France

C. Habas Service de NeuroImagerie, CHNO des 15-20, Paris, France

Laura Hatchondo University Hospital of Poitiers, Poitiers, France

DACTIM-MIS team LMA/ CNRS 7348, Poitiers University, Poitiers, France

Benedetta Heimler Department of Medical Neurobiology, The Institute for Medical Research Israel-Canada, Faculty of Medicine, The Hebrew University of Jerusalem, Jerusalem, Israel

The Edmond and Lily Safra Centre for Brain Sciences (ELSC), The Hebrew University of Jerusalem, Jerusalem, Israel

Arjan Hillebrand Department of Clinical Neurophysiology and MEG Center, VU University Medical Center, Amsterdam, The Netherlands

Shir Hofsetter Department of Medical Neurobiology, The Institute for Medical Research Israel-Canada, Faculty of Medicine, The Hebrew University of Jerusalem, Jerusalem, Israel

The Edmond and Lily Safra Centre for Brain Sciences (ELSC), The Hebrew University of Jerusalem, Jerusalem, Israel

David T. Jones Department of Neurology, Mayo Clinic, Rochester, MN, USA

Hirotaka Kosaka Research Center for Child Mental Development, University of Fukui, Fukui, Japan

Laurence Legrand Department of Radiology, Centre Hospitalier Sainte-Anne, Université Paris Descartes Sorbonne, Paris, France

Mary Pat McAndrews Krembil Research Institute, University Health Network, Toronto, ON, Canada

Deparment of Psychology, University of Toronto, Toronto, ON, Canada

Neuropsychology Clinic, Toronto Western Hospital, Toronto, ON, Canada

Jean-François Meder Department of Radiology, Centre Hospitalier Sainte-Anne, Université Paris Descartes Sorbonne, Paris, France

Charles Mellerio Department of Radiology, Centre Hospitalier Sainte-Anne, Université Paris Descartes Sorbonne, Paris, France

Department of imaging, Centre cardiologique du Nord, Saint-Denis, France

Li Li Min Department of Neurology, School of Medical Sciences, University of Campinas (UNICAMP), Campinas, SP, Brazil

Brazilian Institute of Neuroscience and Neurotechnology (BRAINN), Campinas, SP, Brazil

Martijn L. T. M. Müller Department of Radiology, University of Michigan, Ann Arbor, MI, USA

Morris K. Udall Center of Excellence for Parkinson's Disease Research, University of Michigan, Ann Arbor, MI, USA

Yuko Okamoto Research Center for Child Mental Development, University of Fukui, Fukui, Japan

Catherine Oppenheim Department of Radiology, Centre Hospitalier Sainte-Anne, Université Paris Descartes Sorbonne, Paris, France

I. Peretti Laboratoire CeRSM (EA-2931), Equipe Analyse du Mouvement en Biomécanique, Physiologie et Imagerie, Université Paris Nanterre, Nanterre, France

C. C. Piccinin Neuroimaging Laboratory, School of Medical Sciences, University of Campinas, Campinas, SP, Brazil

Menno Schoonheim Department of Anatomy & Neurosciences, VU University Medical Center, Amsterdam, The Netherlands

Prejaas Tewarie Department of Clinical Neurophysiology and MEG Center, VU University Medical Center, Amsterdam, The Netherlands

Jessica Vicentini Department of Neurology, School of Medical Sciences, University of Campinas (UNICAMP), Campinas, SP, Brazil

Brazilian Institute of Neuroscience and Neurotechnology (BRAINN), Campinas, SP, Brazil

Chapter 1
Magnetic Resonance Imaging: Basic Principles and Applications

G. de Marco and I. Peretti

Nuclear Magnetic Resonance

All substances in nature have, at the microscopic level (atoms or nuclei), more or less significant magnetic properties. They are identical to magnets and characterized by a more or less intense magnetism. The direction of the magnetism is usually represented by a north-south pole, as in the case of the magnetic compass needle. The magnetic properties are represented by a vector μ called "elementary magnetic moment," whose length and direction correspond to its intensity and its direction, respectively.

In MRI and MRS, the magnetism that is measured is nuclear [1, 2]. Atomic nuclei are in fact formed from two types of particles called protons, which are positively charged, and neutrons. Both types of particle have a magnetic moment. Atomic nuclei with an even number of protons and neutrons have no magnetic property. On the other hand, those with an odd number of protons or neutrons can have a detectable magnetism. Elements such as hydrogen 1, carbon 13, and phosphorus 31 may be the cause of a nuclear magnetization.

The value of the magnetic moment μ is proportional to a characteristic of the particle to rotate on itself and which is called spin angular momentum or simply "spin." Due to its magnetic and kinetic spin properties, the hydrogen nucleus consisting only of a proton is the basis of most medical applications of MRI. Hydrogen is an essential component of water, organic liquid, and fats. It is therefore very abundant in the human body.

In a small element of volume of biological tissue, different hydrogen nuclei (also called "protons") have randomly oriented individual magnetic moments μ (Fig. 1.1). Their sum corresponds to zero total magnetization M. So that this is not zero, we

G. de Marco (✉) · I. Peretti
Laboratoire CeRSM (EA-2931), Equipe Analyse du Mouvement en Biomécanique, Physiologie et Imagerie, Université Paris Nanterre, F92000 Nanterre, France

© Springer International Publishing AG, part of Springer Nature 2018
C. Habas (ed.), *The Neuroimaging of Brain Diseases*, Contemporary Clinical Neuroscience, https://doi.org/10.1007/978-3-319-78926-2_1

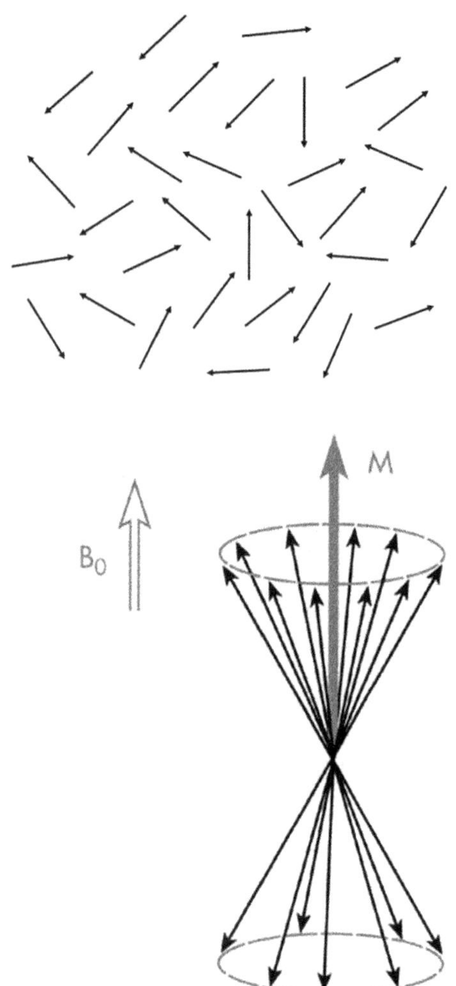

Fig. 1.1 Magnetic moments μ of different hydrogen nuclei of an element of volume of biological tissue placed in a natural environment. These magnetic moments are randomly oriented, and their sum corresponds to zero magnetization M

Fig. 1.2 Individual magnetic moments μ of the protons of a biological sample placed in an intense B_o magnetic field

place the patient in a strong magnetic field and uniform B_0. The intensity of this field usually varies between 0.02 and 3 Tesla.

Subjected to the B_0 field, the individual magnetic moments μ of the hydrogen nuclei of a given sample will polarize, i.e., move in the direction of B_0 field. They then distribute along two energy levels and turn on two cones, called precession cones around the axis of this B_0 field. The B_0 "parallel" spins find themselves on the level of lowest energy (or basic level), and the B_0 "antiparallel" spins find themselves on a higher energy level (or excited level) (Fig. 1.2). They therefore have two different energy states, and the hydrogen nuclei are found in the state of thermodynamic equilibrium. Due to better stability in the basic state, the spins find themselves slightly in excess of this level. This slight excess of spins on the basic state is responsible for the detectable nuclear magnetization in NMR.

In equilibrium, the total nuclear magnetization M is parallel to B_o. The total nuclear magnetization M, describing the sum of the magnetic moments μ of all the hydrogen nuclei, is then defined from two components. The first, called "longitudinal magnetization" and labeled M_L, matches the projection M on the xy plane perpendicular to B_o. In equilibrium, the random orientation of the elementary magnetic moments μ of the precession cones leads to a resulting zero transverse magnetization M_T and a non-zero and maximal longitudinal magnetization M_L.

In order to detect both M_T and M_L components of the magnetization of a fabric, it is then necessary to place the system out of its equilibrium position. This requires providing additional energy. The sample is therefore subjected to the action of an electromagnetic radiation similar to that used in radio. The transfer of energy of the radio wave to the hydrogen nuclei only occurs, however, if the frequency of the radio wave is determined such that its energy is equal to the difference between the two energy states of the hydrogen cores.

We then say that there is resonance. This phenomenon is similar to that which exists in many physical situations. There is a maximum energy transfer between two physical systems if the frequencies characteristic of the two systems are equal. It is well known, for example, that a person making a sound can break a crystal glass remotely if the frequency of the wave is equal to the frequency of vibration of the glass.

In nuclear magnetic resonance, the frequency of resonance f of the radio wave is proportional to the intensity of the field B_o; $f = (\gamma/2\pi)\ B_o$ (the constant γ, called gyromagnetic ratio, is characteristic of a given core). For a field of 0.5 Tesla, the frequency of resonance f (also called Larmor frequency) of the hydrogen nuclei is equal to 21.29 MHz (21.29 million cycles per second).

Radio waves used in magnetic resonance imaging have very short durations, to the tune of several milliseconds. They are "pulses." The energy transfer of the radio wave to the cores results in a tilting of the total magnetization M with respect to its initial position. The value of the tilting angle θ is a function of the amplitude and the duration of the excitation pulse. We call the pulse of 30°, 90°, or 180°, a radio wave of intensity and duration such that, immediately after the pulse, the magnetization M makes an angle θ of 30°, 90°, or 180° with the B_o field. We most often use pulses of 90° or 180°. Some rapid imaging techniques, however, rely on a pulse angle that is less than 90°.

Return to Equilibrium of Magnetization

After the end of excitation of the sample by the radio wave, the nuclear magnetization returns to its position of equilibrium parallel to B_o. This return to equilibrium, called "relaxation," is not instantaneous. Its evolution over time is characterized by two time durations T1 and T2, which vary according to the normal and pathological state of the tissue.

Immediately after stopping a 90° pulse, the total magnetization M is perpendicular to the z-axis with B_o. It then gradually joins this z-axis, with the end of the magnetization vector delineating a helical path, of decreasing radius. The values of the two components of the magnetization therefore vary over time. This return to equilibrium in fact corresponds to the disappearance of the transverse magnetization and the recovery of the longitudinal magnetization. However, it is necessary to separate these two processes because the magnetization does not keep a constant module during this relaxation. In biological tissues, in particular, the reduction in transverse magnetization is more rapid than the increase in longitudinal magnetization.

Relaxation Times

The "longitudinal relaxation time" T1 characterizes the regrowth of the longitudinal magnetization M_L over time by the following Eq. 1.1:

$$M_L(t) = M_{eq} \cdot \left(1 - e^{-\frac{t}{T_1}}\right) \tag{1.1}$$

After stopping a 90° pulse, it corresponds to the time taken by M_L to reach 63% of its equilibrium value (Fig. 1.3).

The numerical value of T1 time durations of the biological tissues depends strongly on the intensity of the B_0 magnetic field used. The T1 value is also based on the microviscosity of the medium. It also depends on the mass and size of the molecules that make up the tissue. In pure water, movements of rotation and translation of the molecules are very fast, and they have frequency characteristics that are much larger than the Larmor frequency. There is therefore little energy

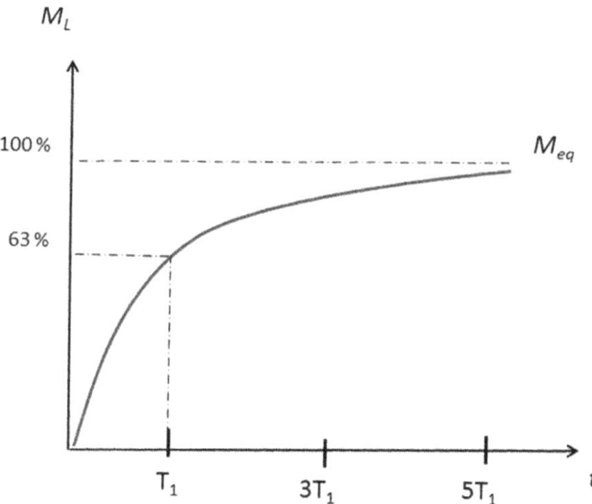

Fig. 1.3 The evolution of the longitudinal increase M_L after a 90° pulse. After a time interval equal to T1 (longitudinal relaxation time), the longitudinal magnetization recovered 63% of its value at equilibrium M_{eq}

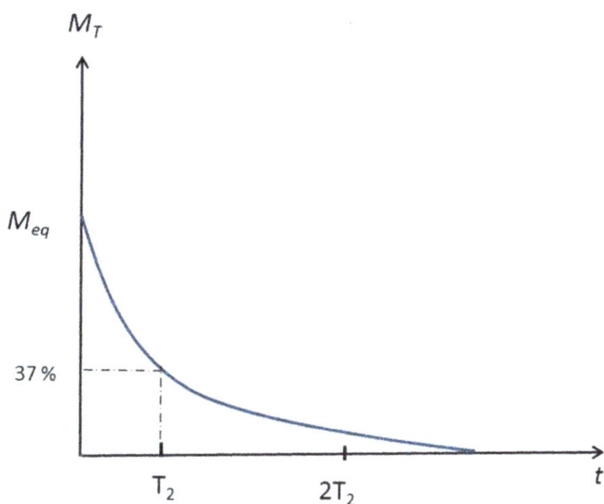

Fig. 1.4 Decrease in transverse magnetization M_T after a 90° pulse. After a time interval T2 (transverse relaxation time), the transverse magnetization decreased to 63% and reached 37% of its initial value

exchange with the stimulated hydrogen nuclei, and the longitudinal relaxation time is long. If the test medium contains medium-sized molecules such as lipids, the molecular movements are slower. They are most effective; therefore, the T1 relaxation time of protons is shorter.

For fluid media such as cerebrospinal fluid, the T1 values are greater than the second. For more structured tissues (muscle, liver, gray matter, white matter, etc.), the T1 values are to the tune of a few hundred milliseconds. The possible presence of ions or paramagnetic molecules also changes the value of the relaxation time. This property can be used to change the contrast imaging.

The decrease in transverse magnetization M_T over time is characterized by the "transverse relaxation time" T2 and corresponds to a reduction of 63% of M_T by the following Eq. 1.2:

$$M_T(t) = M_{eq} \cdot e^{-\frac{t}{T_2}} \tag{1.2}$$

After a 90° pulse, the value of the transverse magnetization decreases gradually to a zero value which corresponds to a random spread of all the magnetizations of protons (Fig. 1.4).

For pure water the T2 value is 3 s, and the T1/T2 ratio is equal to one. In biological tissues, the numerical values of T2 are generally lower than T1, the T2 time duration being all the longer the more fluid the sample is.

Nuclear Magnetic Resonance Signal

Detecting the nuclear magnetization is carried out by placing a "sensor coil" or "antenna," called a measuring plane, in the plane perpendicular to B_0. The transverse magnetization rotational movement creates an induced electric current in the

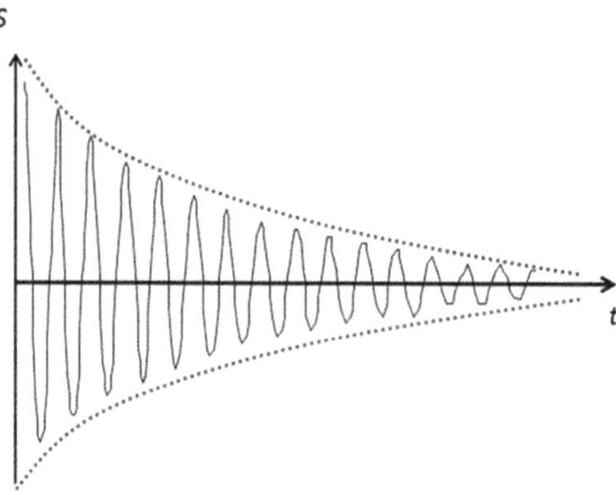

Fig. 1.5 show the appearance of the NMR signal obtained after a radio pulse of 90°, for a homogeneous sample placed in a uniform magnetic field B_0. This signal appears in the form of damped oscillations because the transverse magnetization, which is the only one to induce a current, decreases over time. The signal is positive or negative according to whether the end of the magnetization vector is directed toward the antenna or in the opposite direction. The measurement of this induced current, also known as "free precession signal," allows the determination of parameters characteristic of the NMR signal: frequency, duration, amplitude, and possibly its phase

antenna, that can be measured after amplification, and which constitutes the NMR signal (Fig. 1.5). It is well known that if a bar magnet in rotation is placed in front of a wire coil, an electric current appears in the coil which is more intense the faster the movement.

Accurate measurement of the frequency provides information on the molecular structure. It is the basis of spectroscopic techniques. In imaging, the determination of the frequency allows localization of the signal. With the value of resonance frequencies high, the measurement is carried out by comparing the signal to a reference oscillation. The shape of the NMR signal corresponds to a damped oscillation. It is therefore characterized by its envelope, in other words by the evolution over time of its lower and upper limits. The height of the envelope, immediately after the end of the excitation pulse, is proportional to the amplitude of the magnetization that can then be measured.

This measurement allows imaging to evaluate the relative amount ρ of hydrogen nuclei per unit of volume in the sample. It is also possible to indirectly derive the relaxation times T1 and T2 by using different sequences of pulses to change, before the 90° pulse, the intensity of the longitudinal magnetization according to the relaxation time. The decrease in the signal over time is due to the exponential decrease of the transverse magnetization. The rate of decrease is directly related to the transverse relaxation time value T2 if the magnetic field B_0 is perfectly uniform. However some physical factors can cause a faster decrease of the signal. In particular, if the magnetic field B_0 is not strictly uniform throughout the sample, the resonance frequencies of the different elements of volume of this sample are not quite identical. The decrease of the signal is then characterized by the parameter T2* ("T2 star"), which is smaller than the T2 parameter of the sample. The existence of flow (blood or cerebrospinal) in the sample can also be a cause of more rapid decrease of the signal.

The phase of the signal is represented by the temporal position of the maxima and minima within the envelope of the NMR signal. It provides, at every moment, the angular position of the transverse magnetization of the sample.

Due to its very low intensity, the magnetic resonance signal is highly dependent on the "noise" of the antenna and the noise associated with the sample. One method for improving the signal to noise ratio is to acquire each signal n times (usually 2 or 4), with the accumulation of signals followed by an averaging performed by the computer. The signal to noise ratio is then multiplied by the square root of n.

Classical Sequences

The determination of "tissue parameters" ρ (density of hydrogen nuclei), T1 (longitudinal relaxation time), and T2 (transverse relaxation time) for each small element of volume (or voxel) of biological tissue studied is not carried out from a single radio pulse but from radio frequency pulse sequences. There are different types of sequences according to the duration, the repetition time of the pulses, and the magnetization of the tilt angle.

Spin Echo Sequence

This is the base sequence in imaging by magnetic resonance. In spectroscopy, it corresponds to the T2 measurement sequences. The elementary sequence is comprised of two pulses separated by a time interval, which is denoted usually by TE/2.

The first pulse is a radio frequency pulse of $90°$, which swings the total magnetization M in the measuring plane. Immediately after this pulse, the longitudinal magnetization M_L is then zero. The second pulse is a $180°$ pulse which will cause a progressive rephasing of the different individual magnetic moments. It is then possible to record an "echo" of the original signal at the time of TE. The TE time interval is called "echo time." Each elementary sequence of two pulses can be repeated after a time interval TR called "repetition time" (Fig. 1.6). Both TE and TR parameters, called acquisition parameters, are selected to favor a signal depending preferentially on a given tissue parameter, in other words either T1 or T2.

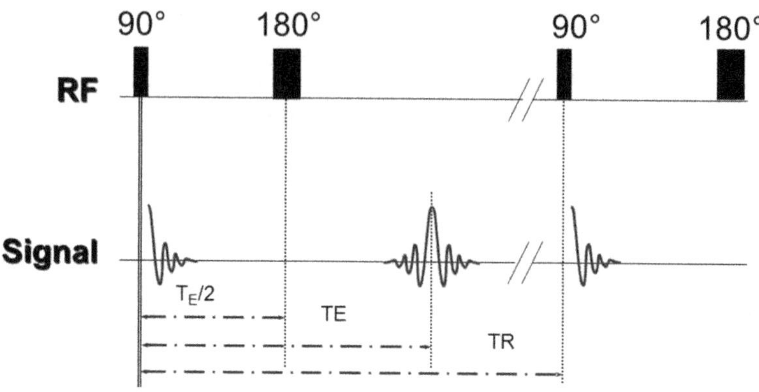

Fig. 1.6 The elementary sequence of spin echo (SE) is composed of a 90° pulse, followed, after a time interval of TE/2, by a 180° pulse. It is possible to register an echo of the original signal at the TE time (echo time). Each elementary sequence of two pulses is repeated after a time interval TR (repetition time)

Weighted Signal ρ (or Proton Density)

If the repetition time TR is "long," much higher than the T1 values of the examined tissues, the intensity of the measurement signal is practically independent of T1. With a "short" echo time TE chosen well below the T2 values of the tissues, the intensity of the measured signal is practically independent of T2 as well. We then say the signal is weighted in ρ.

Weighted Signal in T2

By contrast, with a "long" TE, chosen from the same order of magnitude as the T2 times of the tissues, associated with a "long" TR, the intensity of the signal depends on the values of ρ and T2. We then say that the signal is weighted in T2.

The sequences used enable the acquisition of several echoes. Both types of signals in ρ and T2 are then obtained in a single sequence, called "long sequence." In practice, the TR value is selected above 1500 or 2000 ms depending on the intensity of the magnetic field and the TE value between 5 and 20 ms for first echo weighted in ρ, between 75 and 100 ms for the echo weighted in T2.

Weighted Signal in T1

If the echo time TE is significantly lower than T2, the intensity of the signal is practically independent of the T2 time values of the tissues examined. Then a "short" repetition time TR should be selected so that the signal strength depends mainly on T1 and ρ. Such a sequence is called a "short sequence." The TR value generally ranges between 400 and 600 ms and that of TE between 5 and 20 ms.

Magnetic Resonance Imaging (MRI)

Principle of Signal Localization

It is necessary to form an image, discriminating the magnetic resonance signals from the different voxels of a section or element of examined volume.

The radio waves used in MRI cannot be directed at a specific region. They irradiate the whole body placed in the transmit coil. It is therefore impossible to directly separate the signals from different voxels. The localization of the origin of the signal will therefore be obtained by means of an experimental artifice based on the fact that the resonance frequency of a sample is, at any point, proportional to the intensity of the magnetic field. The principle of magnetic resonance imaging is to produce a variation in the intensity of the magnetic field of a voxel to another by adding a "field gradient" of very low intensity to the magnetic field B_o (to the tune of 10 milliTesla per meter).

A gradient is characterized by its amplitude and direction. The gradients used in MRI are such that the difference in intensity of the magnetic field between two points is proportional to the distance separating between these two points in the direction of the gradient. The resonance frequencies of the protons will then be different, from one voxel to another in the sample in a given direction parallel to the applied field gradient. It is thus possible to locate the origin of the signal along this axis.

The origin of the signal should, in fact, be located in the three directions x, y, and z in the space. It is therefore necessary to use three field gradients applied successively. Their simultaneous interventions lead to a single gradient applied in an oblique direction.

Imaging by Two-Dimensional Fourier Transformation

The signal localization is carried out in three successive stages. The first selects a section. It therefore has the goal of obtaining a signal that only comes from one section of the examined sample.

The method consists of applying a field gradient Gs during an excitation pulse of 90°. The orientation of the selected section is perpendicular to the direction of this gradient. Its thickness is inversely proportional to the intensity of Gs and proportional to the width of the frequency band of the radio wave. The value of the central frequency of the pulse also selects the section level. Unlike XCT, with MRI, it is possible using this method called selective excitation to select a section in any orientation plane without changing the patient's position.

The following steps were designed to locate the signal in both directions of the selected section. The first direction of the section is marked by the method "coding by frequency." It involves applying a Gf field gradient perpendicular to the slice selection gradient during signal acquisition. The protons located in the section on the line perpendicular to the direction of the gradient Gf, therefore, have the same resonant frequency. It is then necessary to discriminate all the points in the same line.

The second dimension of the cutting plane is "encoded by the phase." In this method, a Gp gradient duration of determined amplitude is applied immediately after the excitation pulse. Its effect is to gradually change the phase of the elementary magnetic moments located in the direction of this gradient.

The technique for acquiring the image by two-dimensional Fourier transformation (TF2D) therefore consists of combining coding by phase and coding by frequency (Fig. 1.7). A coding gradient is applied before each acquisition. During acquisition of the signals, a coding gradient by frequency is applied in a perpendicular direction to the coding by phase gradient. Several signals are acquired successively by repeating a basic sequence. From one signal to another, the intensity of the coding by phase gradient is changed. A signal matrix is thus obtained which, after Fourier transformation, gives the image of the section.

Each cycle allows the acquisition of a signal. From one signal to another, only the value of the coding by phase gradient is changed: G_s, selection gradient; G_p, coding by phase gradient; and G_f, coding by frequency gradient.

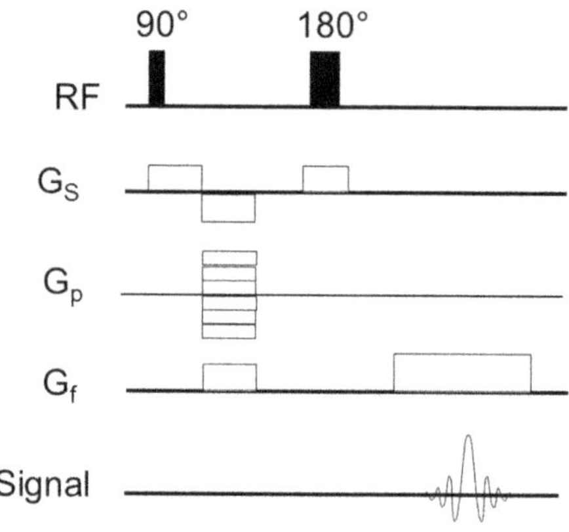

Fig. 1.7 Technique for acquiring an image by two-dimensional Fourier transformation (TF2D) in spin echo sequence

Fourier Plane (or k-Space)

The concept of the Fourier plane takes on an important role in MRI and represents an abstract mathematical concept that is necessary to assimilate and to understand how an image is manufactured. An MR image is formed in two distinct stages: the first consists of acquiring a raw data matrix (discrete signals) in the frequency domain filling the Fourier plane; the second stage is a mathematical operation (two-dimensional Fourier transform), which transforms this matrix into an interpretable image in the spatial domain. In fact, we find the same information in the Fourier plane and in the image but this information is organized differently.

In the spin echo technique, the filling of the Fourier plane is done line by line. After each selective excitation, the application of the reading gradient is used to fill a line of the Fourier plane (or k-space) in its entirety. A line is scanned in the k-space during each TR. The position of the line in the Fourier plane is determined by the value of the coding by phase gradient. The increment of this value makes it possible to move the acquisition of a line to the next. The time required to completely fill the k-space is therefore given by the relationship: N x TR, where N represents the number of phase coding lines and TR the repetition time of the sequence.

The relationship between the MRI data from a section and the image is not immediate. Indeed, there is no simple relationship between the k-space lines (points) and the lines (points) of the image. Each point of the Fourier plane will contribute to the entire image and the translation of this point in the image will depend on its position in the Fourier plane. Indeed, the points that are at the center of the Fourier plane determine the contrast of the image by providing information on the main structures contained therein; they correspond to low spatial frequencies. The points that are located on the periphery of the Fourier plane determine the spatial resolution of the image by providing information on the details of the image. They correspond to high spatial frequencies.

Rapid Imaging

In imaging by two-dimensional Fourier transformation (TF2D), the acquisition time T_{acq} for a series of images is given by the following equation: $T_{acq} = TR. N_y$, N_{acc}, or N_{acc} represents the number of accumulations (or "excitations") of the signal and N_y the number of codings by phase carried out, therefore the number of lines of the image matrix. With the classical sequences, this time is relatively long since it varies from 3 to 17 min depending on the sequences. The use of fast imaging sequences enables significant time saving, and for this reason they were first used as reference images.

The rapid imaging techniques are increasingly used because, in addition to saving time, they have a number of advantages. They make it possible to significantly improve patient comfort, to reduce artifacts related to patient movement (e.g., physiological movements such as heartbeat or breathing). They may also have economic implications by reducing the cost of MRI.

Rapid imaging paves the way for other applications such as dynamic analysis of contrast enhancement, the study of motion of a joint, the study of cardiac kinetics, and obtaining abdominal images in apnea. Rapid techniques also allow three-dimensional image acquisition. Several approaches can shorten the acquisition time.

Decrease in the Number of Collected Data

A simple method consists of carrying out only one accumulation: $N_{acc} = 1$ instead of 2 or 4. A second simple method consists of reducing the number of N_y lines of the image. For example, we can choose $N_y = 28$ instead of 192 or 256.

A partial scan of the Fourier plane is also possible. Some manufacturers offer a number of fractional accumulations: $N_{acc} = 0.5$ or 0.75, for example. This is actually a reduced number of phase codings made carried out during the acquisition, with the missing data then deduced by symmetry.

Rapid Scanning of the Fourier Plane

Currently, the most rapid method is the "echo-planar" method, usually denoted EPI for echo planar imaging. With acquisition times to the tune of a few tens of milliseconds, it offers great potential. In fact, it is possible to obtain images almost instantaneously. This method consists of scanning the entire Fourier plane into one signal ("single shot"), which implies a very rapid reversal of the gradient while reading the signal (Fig. 1.8). It requires the use of "hypergradients" characterized, on the one hand, by a maximum amplitude of high gradient (to the tune of a few tens of milliTesla per meter) and, on the other hand, by a very short rise time (to the tune of hundreds of microseconds). These technological performances make the acquisition of this technique of rapid imaging costly.

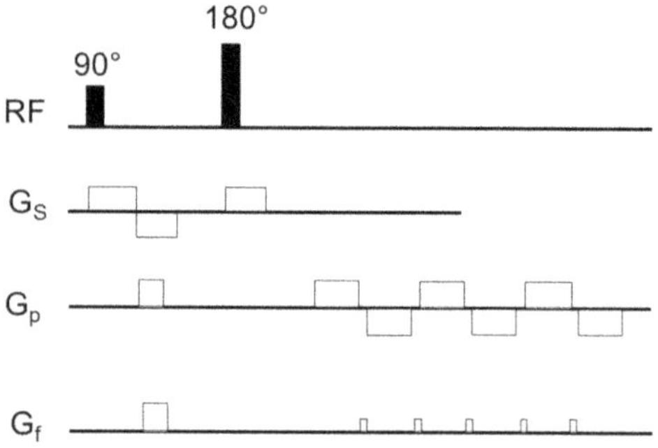

Fig. 1.8 Example of echo-planar SE (blipped) sequence. All data is recorded during the acquisition of a single signal

Several distinct echo planar sequences are available. The echo gradient EPI sequence, consisting mainly of a single 90° pulse followed by the reading of the signal, provides T2* weighted images. These images thus depend on the T2 relaxation time of the tissues, but they are also very sensitive to inhomogeneities within the magnetic field. The EPI sequence in spin echo comprises, after the 90° pulse, a 180° pulse followed by the reading of the signal. Due to the presence of the 180° pulse, the images obtained are T2-weighted. There is also a version of EPI sequence in inversion recovery comprising a 180° pulse before that of 90°, providing T1-weighted images. It should be noted that the echo planar technique is very sensitive to the chemical shift between the water molecules and lipid molecules; to avoid the corresponding artifact, manufacturers supply EPI sequences with suppression of the fat component.

The echo-planar opens up the pathway to new fields of MRI applications. In particular, it is possible to obtain, in addition to morphologic information, functional information about the tissue characterization. Thus, diffusion imaging, perfusion imaging, and imaging of brain activation are most often based on echo-planar acquisitions.

Principles of MR Diffusion-Weighted Imaging, Voxel-Based Morphometry, Cortical Thickness, and MR Spectroscopy

Introduction

Magnetic resonance imaging (MRI) techniques have undergone extraordinary progress in regard to the design of imagers as well as in terms of associated digital technologies. Thus, advances in terms of magnetic field gradients (e.g., rapid gradient rise times), the development of new echo-planar sequences (providing excellent temporal resolution), and improvements in the processing and analysis of images have paved the way for new applications in medical imaging. Imaging by diffusion magnetic resonance provides access to physiological and pathophysiological mechanisms that take place at the microscopic level in human tissues [3, 4]. The data obtained can be quantified through digital processing. Indeed, analysis software allows quantitative maps to be made and apparent diffusion coefficients to be determined in humans. Diffusion imaging can be performed with a clinical MR imager equipped for ultra-rapid imaging sequencing. Excellent temporal resolution of EPI-type sequences allows an image to be captured in 50 to 100 msec, thus permitting artifacts due to physiological movements to be reduced and integration of the diffusion technique into clinical practice.

What Is Diffusion in Imaging?

Typically, the intensity of the MRI signal is obtained from a population of protons contained in the voxel of the image, and it depends primarily on the density of the protons and the relaxation times T1, T2, and T2*. Tissue characterization, based on the mobility of water molecules, can nowadays be achieved using MRI. Indeed, water molecules act as genuine microscopic markers of tissue structure. The diffusion of water molecules across regions of inhomogeneity in the magnetic field (linked to heterogeneity of the environment) leads to a substantial degree of dispersion of the phase of the hydrogen nuclei. The microscopic molecular movements correspond with speeds of several tenths to several hundredths of a millimeter per second. In order to be able to detect these short-range movements, high-intensity pulsed fields are applied. When the water molecules cross these areas of gradient, they undergo a change in phase depending on their direction and their speed. Too much movement of the molecules causes a decrease in the amplitude of the echo (signal) and this translates into a decrease in the signal intensity on the image.

Random Movements of Molecular Diffusion

Molecular diffusion is a dynamic process that involves translational and random movements by water molecules. Water molecules are in perpetual short-range motion due to thermal excitation, and molecular diffusion is represented mathematically by a Gaussian distribution with an average of zero. Thus, the probability of molecules moving in one direction is equal to the probability of movement of the molecules in the opposite direction. The displacement r between two molecular collisions in a three-dimensional space increases with the square root of time according to the following relationship (Eq. 1.3):

$$r = \sqrt{6Dt} \tag{1.3}$$

The proportionality constant D is called the diffusion coefficient, and it characterizes the mobility of a water molecule in its environment.

The Apparent Diffusion Coefficient and Restrained Diffusion

The diffusion coefficient of water molecules in tissues is lower relative to that of pure water. The water molecules encounter all sorts of obstacles to their movement, which slows their diffusion. Thus, their displacement is hindered by membranes, cell organelles, and molecular barriers. With biological tissues one hence refers to apparent diffusion coefficients (or ADCs). Due to a higher viscosity of water in tissues, the diffusion coefficient is two to three times lower that of water on its own. Using Eq. (1.3), knowledge of the diffusion coefficient and the observation time allow the distance travelled by the water molecules between two collisions to be determined, hence providing a value for the dimensions of the restrictions that limit the movement of the water molecules. For example, during an observation time of 40 msec equivalent to the duration of the echo time (T), the water molecules in tissue diffuse freely over a distance in the order of 10 micrometers, corresponding to a diffusion distance two times lower than that travelled by pure water at 37 °C and in the same time span. The average diameter of cells is approximately equivalent to this distance. This equation hence allows assessment of the dimensions of cells, or also to document changes in the permeability in the myelin sheaths of white matter.

Anisotropy of Diffusion

Diffusion of water in gray matter is known to be isotropic; that is to say, it appears to be identical in all three dimensions of space. On the other hand, diffusion in white matter is anisotropic due to the structure of the myelin fibers. The water molecules are, in fact, more free to move along the axonal fibers than perpendicular to them, since the permeability of the myelin sheath is low. Two very different approaches allow this diffusion anisotropy to be characterized.

Tensors of Diffusion and the Trace

In order to more precisely characterize anisotropy of the environment, it is necessary to measure a tensor D of diffusion, represented by a matrix of three lines and three columns. This allows the principal axes of diffusion (eigenvectors) to be determined and to calculate the diffusion coefficients (intrinsic values) along each of these axes.

These axes accurately reflect the diffusion characteristics of the tissue. The axis associated with the highest ADC in particular is the preferred direction of diffusion of the water molecules. For example, for a voxel of white matter this direction corresponds with the orientation of the fibers. This technique can be useful for the investigation of demyelinating diseases or for studies of brain myelinization in neonates.

The various components measured of the tensor depend on the coordinate system of the magnet. By contrast, the trace Tr(D) of this tensor, obtained from the sum of the three diagonal components, does not depend on it. Pulsed sequences allow maps of the Tr(D) to be obtained directly. These sequences have the advantage of being fast while separately measuring the various components of the tensor D.

Diffusion Sequences

Stejskal and Tanner Sequences

In 1965, Stejskal and Tanner developed a diffusion imaging technique. This sequence, usually referred to as PGSE, consists of using a spin echo (SE) sequence, to which two pulsed gradients area added that are of short duration and that are placed symmetrically on one side and the other of the 180° refocalizing pulse. The first gradient defocalizes the protons, while the second rephases the protons which have been able to diffuse during the time interval that separates the application of the two gradients. In this case, rephasing of the protons is nearly complete, and consequently the amplitude of the signal will be little changed by the application of the gradients. By contrast, if the water molecules are moving fast, it results in an incomplete rephasing of the spins by the second gradient and hence a notable decrease in the signal intensity. Attenuation (Att) of the signal due to the diffusion is determined based on the following equation (Eq. 1.4):

$$Att = \exp(-b \times CDA) \tag{1.4}$$

The parameter b is a key factor depending both on the intensity G of the duration δ of the diffusion gradients and the time Δ separating the two gradients. b is expressed according to the following equation (Eq. 1.5):

$$b = \gamma^2 \delta^2 G^2 (\Delta - \delta/3) \tag{1.5}$$

where γ represents the gyromagnetic ratio of the hydrogen nucleus (42 MHz/T).

How to Obtain a Diffusion Map?

A commonly used method consists of acquiring a first image obtained without activating the diffusion gradients (b = 0), followed by three diffusion-weighted images, each taken after use of a very high diffusion gradient (b = 1000 sec/mm) applied along each direction x, y, and z. The first image serves as a reference for the following images, so as to eliminate the T1 and T2 weighting on each pixel of the image. The differences in diffusion speeds (slow and fast) are more readily discernible on the image when high values of b (intense gradients) are chosen. For elevated values of b, the signal to noise ratio decreases by about a factor of ten. For a precise calculation of the ADCs (x, y, and z), several values of b are applied successively. A subsequent digital processing allows a quantitative map of the ADC to be made. The values for the ADC are then represented as a color scale ranging from blue (low diffusion) to red (high diffusion). A map of the ADC, established pixel by pixel, is then obtained for each section, in the three dimensions of space.

How to Obtain a Tractography?

MR diffusion tractography is a method for identifying white matter pathways in the brain (Fig. 1.9). These pathways form the substrate for information transfer between remote brain regions and are therefore central to our understanding of function in both the normal and diseased brain. As mentioned in Yendiki et al. [5], several diffusion tractography methods have been proposed over the years to reconstruct white-matter pathways.

Most early methods were deterministic and followed the streamline approach, which modeled a path as a one-dimensional curve. The curve was grown from a starting point by taking steps in directions that were determined by the diffusion orientation in the underlying voxels. Other deterministic methods were volumetric, modeling the path as a volume, and allowing it to grow in three dimensions. Both streamline and volumetric approaches were local, in the sense that the algorithm considered the image data at a single location to determine how to grow the path at

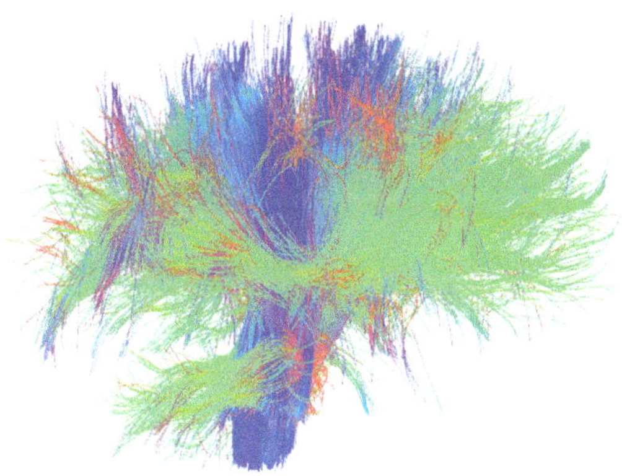

Fig. 1.9 Example of fiber tracking of the corticospinal and corticopontine tracts (blue) and inferior longitudinal fasciculus (green). (From Trakvis (0.6.1) software)

each step. Compared to deterministic approaches in which the estimated fiber orientation (e.g., direction of the maximum diffusivity for the tensor model) is assumed to represent the best estimate to propagate streamlines, probabilistic methods generate multiple solutions to reflect also the variability or uncertainty of the estimated fiber orientation. Local tractography algorithms, whether deterministic or probabilistic, are best suited for exploring all possible connections from one brain regions, which is used as the tractography seed, to any other region.

Global tractography methods were suggested as an alternative approach to address the problem of identifying specific white-matter pathways. The global approach defines both end regions where the pathway is thought to determinate and searches the space of all possible connections between these two regions for the connection that bests fits the data. Thus the entire pathway is estimated at once, rather than step-by-step. The solution is symmetric with respect to the two end regions, instead of treating one as the "seed" and the other as the "target." Since global optimization integrates along the length of the pathway, it is less sensitive to localized regions of high uncertainty (e.g., pathway crossing) than the streamline approach. A challenge with global tractography is the size of the solution space, which consists of all possible connections between two regions. To address these issues, a computational method "TRACULA" (TRActs Constrained by UnderLying Anatomy) has been developed. This method for automated reconstruction of major white-matter pathways that is based on the global probabilistic approach and utilizes prior information on the anatomy of the pathways from a set of training subjects to derive a description of the pathways in terms of the structures that they intersect and neighbor [6]. The knowledge on path anatomy that is extracted from the training set is then used to initialize a global probabilistic tractography algorithm and also to constrain its search space penalizing connections that do not match the prior anatomical knowledge. This allows the algorithm to reconstruct the pathways reliably in a novel subject with no manual intervention, facilitating the analysis of large data sets.

Another method called "TBSS" (tract-based spatial statistics) aims at modeling or quantifying changes in white matter across individuals. This method relies on the precise changes in white matter across individuals, in particular on the precise matching of anatomical locations across subjects, which is necessary in order for the experimenter to be confident that any result from the analysis is due to genuine changes in white matter microstructure rather than variations in brain structures shape, size, or position. TBSS restricts the statistical comparisons to the centers of white matter tracts after non-linear registration of different subjects into a common space. TBSS uses fractional anisotropy (FA) measurements to realign subjects and extract the centers of white-maters tracts.

Sources of Artifacts in Diffusion-Weighted MRI

The biggest hurdle in diffusion imaging involves movement artifacts that appear on the image due to macroscopic movements by the patient or physiological movements (e.g., breathing, heartbeat, and CSF). The latter can be reduced by applying a cardiac synchronization during the image acquisition. The movement of the protons across the magnetic field gradients can cause substantial shifts in the phase of the signal and can perturb the localization of the signal during the coding phase. These variations in the phase generally translate into phantom images. This artifact problem is amplified considerably in the presence of diffusion gradients. Other types of artifacts due to the substantial difference in magnetic susceptibility that exists at air-tissue and bone-tissue interfaces can result in substantial distortions in the images. This is particularly so for those done in temporal regions and in the posterior fossa. Such artifacts may be minimized by applying highly intense field gradients.

Voxel-Based Morphometry and Cortical Thickness

The morphology of cortical gray matter is commonly assessed using T1-weighted MRI together with automated computerized methods such as voxel-based morphometry (VBM) and cortical thickness (VBCT). Findings suggest that while VBM provides a mixed measure of gray matter including cortical surface area or cortical foldings, as well as cortical thickness, VBCT selectively investigates cortical thickness [7]. VBM allows for the examination of brain changes and/or group differences across the entire brain with a high-regional specificity (i.e., voxel by voxel), without requiring the a priori definition of particular ROIs [8]. VBM is an approach that enables a voxel-wise prediction of the local amount of a specific tissue. Classically, VBM is directed at examining gray matter, but it can also be used to examine white matter. In the latter case, however, the sensitivity is limited, for white matter areas are characterized by large homogeneous regions with only subtle changes in intensity [9]. For white matter, it is strongly advisable to use DTI. The reader interested in the VBM can download VBM8 toolbox at http://www.neuro.uni-jena.de/vbm/down load/ which runs with SPM8 (http://www.fil.ion.ucl.ac.uk/spm/).

Given the range of normal individual variance in cortical morphologic features, such as gyral and sulcal patterns, the use of voxel-based tools that transform and smooth individual MRI data into common coordinate spaces may remove the precise features of interest for studies investigating within-group correlations between cortical morphometry and cognitive performance (for example) and reduce the ability to specifically localize findings. Furthermore, the measure typically analyzed by voxel-based techniques (gray matter density) is difficult to interpret quantitatively with respect to a particular morphometric property of cerebral tissue (i.e., volume, thickness, surface area). To enable the study of morphometric properties of the human cerebral cortex and their relationship to cognitive function, disease state or other behavioral variables, automated methods have been developed for segmenting and measuring the cerebral cortex from MRI data [10, 11]. The validation of MRI-derived cortical thickness measurements has been performed against manual measurements derived from both in vivo and postmortem MRI brain scans [11]. Since measures of cortical thinning are sensitive and reasonably specific [12], at least in the context of neurodegenerative diseases or performance ability, these measures are a promising candidate MRI biomarker. Reconstruction of the cortical surfaces and measurements of cortical thickness were classically performed using FreeSurfer toolkit which is freely available to the research community via the Internet at https://surfer.nmr.mgh.harvard.edu/.

Magnetic Resonance Spectroscopy (MRS)

MRS can have several areas of application [13–15]. In vitro analyses of biological tissue can be carried out. The future field of application concerns in vivo and in situ explorations of small volume elements of living organs localized spatially. In vivo spectrometry can be used to complete the morphological and functional exploration obtained by magnetic resonance imaging. It provides access to information related to cell metabolism. This technique takes advantage of noninvasive MRI and includes the quantitative information of biochemistry [16].

Magnetic resonance spectroscopy may use nuclei other than hydrogen. Some of these nuclei have sufficient natural isotopic abundance in biological tissues. These are endogenous nuclei, phosphorus 31, and sodium 23, for example. Other nuclei termed exogenous, such as fluorine 19 and lithium 7, can be used as tracers.

Localized Spectroscopy and In Vivo Spectroscopic Imaging

There are two techniques for locating the spectroscopic signal.

"Monovoxel" MRS

One possible method of spatial localization of the volume element that we want to study in in vivo spectroscopy consists of choosing a single volume of interest and using a surface coil placed near the study area. The depth of the examined region is in the order of magnitude of the radius of the antenna used and the volume studied in the order of approximately 10 cm^3. This technique is applied in particular for superficial and homogeneous lesions.

Metabolic Imaging by Spectrometry

The spatial localization of spectra obtained using magnetic field gradients such as imaging (see MRI paragraph) can analyze all voxels in a cut. Post-processing provides access to parametric images of different metabolites. This technique is well suited for comparative studies between lesions and contralateral normal tissue or lesions that may be heterogeneous and extended.

Physiochemical Parameters

Chemical Shift

We have previously seen that the magnetic resonance signal appeared after Fourier transformation in the form of a peak (also called ray) centered at a frequency V_s of width Δv. The frequency of resonance of the nuclei will depend on the chemical bond involving the nucleus. Indeed, the resonance frequency of a nucleus is slightly modified by the electron cloud of adjacent nuclei, thus creating a very small local magnetic field in addition to the B_0 field. Therefore, the effective magnetic field (B_{eff}) perceived by the nucleus will be slightly different from the B_0 field. The effective resonance frequency will be written: $V_{eff} = \Upsilon B_{eff}/2\pi$. For example, if the hydrogen nucleus of a lipid resonates at 42 MHz for a field of 1 Tesla, the hydrogen nucleus of water will have a Larmor frequency which differs from 142 Hz to that of the lipid. The resonance frequency is therefore different from one nucleus to the other according to the nature of the molecule and the electronic environment.

This variation in frequency is referred to in the terms "chemical shift." This is measured by a number d independent of the value of the external magnetic field. It is expressed in p.p.m. (parts per million): $d = 10^6 (V_s - V_r)/V_r$ ou V_s represents the actual resonance frequency of the nucleus in question and V_r a reference frequency of this nucleus. In spectrometry of the proton, the resonance frequency corresponds to that of tetramethylsilane (reference molecule = 0 ppm); the chemical shift of different chemical groups thus becomes measurable, making it possible to identify different metabolites. In the previous example, the chemical shift of the lipid proton will be equal to 3.4 ppm based on the water proton.

Take the example of phosphorus 31 which is a nucleus that is especially used in MRS. The chemical shift in muscle physiology is usually calculated taking the phosphocreatine ray V_r as the reference. The normal muscle spectrum will consist of a number of V_s rays corresponding to inorganic phosphorus, ATP, and ADP, each being identified by the value of the chemical shift relative to phosphocreatine.

Spectral Characteristics

The area demarcated by the peak and the x-axis (the "surface of the ray") is proportional to the density of the resonating nuclei. The width of this peak, measured at mid-height of the maximum, is inversely proportional to the transverse relaxation time.

NMR Signal, Spectral Resolution, and Coupling

In spectrometry, two factors are crucial in the quality of a spectrum: firstly the signal to noise ratio which determines the sensitivity of the measurement and also the spectral resolution that corresponds to the separation of the different resonance rays. The spectral resolution can be enhanced with the homogeneity and intensity of the main field. Due to the abundance of water in the biological tissues, the relative signal of the nuclei contained in the metabolites of interest is embedded in the water. It is therefore essential to overcome the water signal, for example, by applying a selective pulse cancellation of the water signal. It is then possible to detect intracellular metabolites at concentrations in the order of millimoles per liter.

To improve the signal to noise ratio, it is necessary to accumulate the signal and to select a relatively large measuring volume. This results in degrading the spatial resolution of spectroscopic images to the tune of one centimeter, compared to submillimeter MR images. If the equipment has sufficient resolution, some resonance rays can appear split or multiple. This duplication comes from an interaction between neighboring nuclear spins. The interval between two rays defines a coupling constant characteristic of this interaction.

Measurement of pH

Sometimes the chemical shift of a nucleus reflects the acidity of the investigated medium. This is the case in spectrometry of phosphorus 31. There is a difference between the values of chemical shift for the two types of phosphoric acid (mono or divalent) present in a biological medium. The position of the resonance peak of the inorganic phosphorus results from a balance between these two forms and will therefore depend on the physiological pH value. Thus, a calibration will determine the pH value according to the position of the phosphor ray.

Magnetization Transfer

It is possible to analyze the evolution of a biochemical reaction between the two compounds A and B, following the variation in the intensity of their respective rays during the reaction. For example, the kinetics of an enzymatic reaction will be characterized by the rate of increase in the intensity of the ray A and the intensity of the decreasing ray B ("magnetization transfer" from A to B).

In Vivo Spectrometry

Spectrometry of Phosphorus 31 of Muscle Tissue

The first medical application of spectrometry by nuclear magnetic resonance concerns spectrometry of the phosphorus 31 nucleus in the exploration of muscle metabolism.

A stress test is usually performed. A series of spectra is recorded during the successive phases of rest, exercise, and recovery. In normal subjects, the NMR spectra show a decrease in the phosphocreatine peak and ATP level, which remains relatively stable during exercise. The breakdown of ATP is indeed compensated by its resynthesis: phosphocreatine splits into creatine and phosphate ion, and thus constitutes a first reserve of energy. Spectroscopic quantitative study makes it possible to determine the amount of energy used and the origin of this energy. In vivo NMR spectroscopy of phosphorus 31 makes it possible to study the muscle physiology and pathophysiology. Its contribution can be important in the diagnosis of certain myopathies.

Proton Spectroscopy of Healthy and Pathological Brain Tissue

As regards the study of the cerebral metabolism, the spectrometry of hydrogen nucleus can provide important information. Brain tumors are indeed a major field of application for MRS. In fact, the recorded spectra make it possible to identify metabolites such as (1) N-acetyl-aspartate which is an indicator of the density and neuronal viability, (2) choline which provides information on the membrane metabolism, and (3) creatinine which represents the energy metabolism of cells. The

Fig. 1.10 Recording spectra at a brain metastasis of a lung cancer (A) and the contralateral healthy zone (B). The identification of lipids at the tumor is related to the breakdown of cell membranes. (Adapted from J.-F. Le Bas and S. Grand)

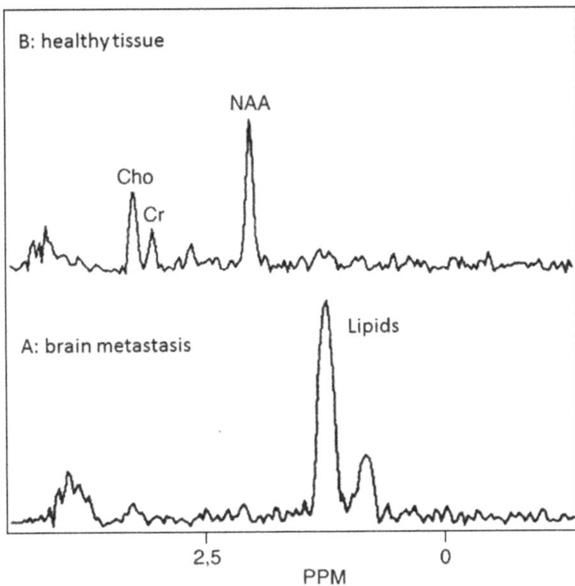

absolute quantification of metabolites is still difficult because of an often low signal to noise ratio. Thus, the relative ratio measurements of choline with respect to N-acetyl-aspartate or creatinine or indeed N-acetyl-aspartate ratios between healthy tissue and tumor tissue are typically carried out. Tumors, for example, are accompanied by an increase in lactate, a marker of anaerobic glycolysis, and a significant decrease in NAA indicating a metabolic deterioration of neurons and an elevation of choline which may reflect cell proliferation (Fig. 1.10).

Diagnostic Role of MRS and Therapeutic Follow-Up

MRS can track the effect of radiotherapy to predict the effectiveness of a treatment. Indeed, if the irradiation is effective, a reduction in choline is observed and may indicate a decrease in the number of tumor cells. MRS is also promising in many diffuse diseases such as viral and metabolic encephalopathies, related to an enzyme deficiency, for example. It also allows the study of brain maturation. Spectrometry also opens the way to the metabolic study of focal lesions, such as cerebral ischemia, which is accompanied by an early increase in lactate levels.

Imaging of Proton Chemical Shift

Several techniques are possible. The first type of method, seldom used, provides a spectrum for each volume element of the studied section. The second type of method helps to understand the composition of water and fat in each voxel. The chemical shift of the hydrogen nucleus, depending on whether it belongs to a lipid molecule or a water molecule, can indeed be utilized to obtain the spatial distribution of these two types of molecules.

The spectroscopic imaging method described by Dixon (1984) consists of successively using two spin echo sequences. The first is a normal sequence, in other words the signal is acquired at time TE where the magnetizations of water and fat are added together since they are in phase. The second is slightly modified to acquire the signal at a time TE $\pm\ \tau$ where the magnetization of water and fat are out of phase and therefore escape. This produces two images for the same cut, the first representing the sum of water + fat and the second the difference of water - fat. We then simply add the two images acquired to obtain the spatial distribution of water and subtract these two images to obtain that of fat. Chemical shift imaging thus makes it possible to obtain biochemical information in addition to morphological information and has a significant contribution in tissue characterization.

References

1. Bloch F (1953) The principle of nuclear induction. Science 118(3068):425–430
2. Purcell EM (1953) Research in nuclear magnetism. Science 118(3068):431–436
3. Le Bihan D (2014) Diffusion MRI: what water tells us about the brain. EMBO Mol Med 6 (5):569–573
4. Basser PJ, Jones DK (2002) Diffusion-tensor MRI: theory, experimental design and data analysis – a technical review. NMR Biomed 15(7–8):456–467
5. Yendiki A et al (2011) Automated probabilistic reconstruction of white-matter pathways in health and disease using an atlas of the underlying anatomy. Front Neuroinform 5:23
6. Jbabdi S et al (2007) A Bayesian framework for global tractography. NeuroImage 37 (1):116–129
7. Hutton C et al (2009) A comparison between voxel-based cortical thickness and voxel-based morphometry in normal aging. NeuroImage 48(2):371–380
8. Ashburner J, Friston KJ (2000) Voxel-based morphometry – the methods. NeuroImage 11(6 Pt 1):805–821
9. Kurth F, Gaser C, Luders E (2015) A 12-step user guide for analyzing voxel-wise gray matter asymmetries in statistical parametric mapping (SPM). Nat Protoc 10(2):293–304
10. Dale AM, Fischl B, Sereno MI (1999) Cortical surface-based analysis. I. Segmentation and surface reconstruction. NeuroImage 9(2):179–194
11. Fischl B, Dale AM (2000) Measuring the thickness of the human cerebral cortex from magnetic resonance images. Proc Natl Acad Sci U S A 97(20):11050–11055
12. Du AT et al (2007) Different regional patterns of cortical thinning in Alzheimer's disease and frontotemporal dementia. Brain 130(Pt 4):1159–1166
13. Proctor W, Yu F (1950) The dependence of a nuclear magnetic resonance frequency upon chemical compound. Phys Rev 77(5):717

14. Gutowsky H, McCall D, Slichter CP (1953) Nuclear magnetic resonance multiplets in liquids. J Chem Phys 21(2):279–292
15. Frahm J et al (1989) Localized high-resolution proton NMR spectroscopy using stimulated echoes: initial applications to human brain in vivo. Magn Reson Med 9(1):79–93
16. Frahm J et al (1989) Localized proton NMR spectroscopy in different regions of the human brain in vivo. Relaxation times and concentrations of cerebral metabolites. Magn Reson Med 11 (1):47–63

Chapter 2
Brief Overview of Functional Imaging Principles

C. Habas and G. de Marco

Physiological and Physical Bases of BOLD Response

BOLD signal roots in local and transient hyperoxygenation exceeding neuronal metabolic demand of activated neurons and is due to vasodilation partly and feed-forwardly controlled by vasoactive molecules in relation to (inter-)neuronal activity and to energetic metabolism [1] and modulated by more distant neural influences.

Neurovascular Coupling with Glutamatergic Neurons At rest, the main neuronal consumption of energy produced by aerobic glycolysis (tricarboxylic acid cycle) is linked to the maintenance of gradients of ionic concentration around the plasmic membrane and intracellular biosynthesis. During synaptic activation, neuronal metabolism and, consequently, neuronal consumption/production of energy are increased mainly in relation with the recruitment of Na+/K+ ATPase pumps involved in gradient ion restoration. A tight neuron-glial-vascular coupling allows to provide sufficient amount of oxygen and glucose to neurons (and astrocytes) during sustained activity and during restoration of their energetic reserves and to get rid of metabolic potentially noxious by-products [2]. This coupling induces (1) a strong and nonlinear augmentation of cerebral blood flow (CBF) and (2) a moderate augmentation of cerebral blood volume (CBV), while a weak cerebral metabolic rate of oxygen utilization (CMRO2) is observed [3]. This hemodynamic response is based on a vasodilation of arterioles and capillaries causing increased CBF and, in downstream veins, an increased speed of blood flow [2]. These vascular processes would rely on Ca^{2+}-dependent release of vasodilators by neurons (or interneurons)

C. Habas (✉)
Service de NeuroImagerie, CHNO des 15-20, Paris, France
e-mail: chabas@15-20.fr

G. de Marco
Laboratoire CeRSM (EA-2931), Equipe Analyse du Mouvement en Biomécanique, Physiologie et Imagerie, Université Paris Nanterre, Nanterre, France

© Springer International Publishing AG, part of Springer Nature 2018
C. Habas (ed.), *The Neuroimaging of Brain Diseases*, Contemporary Clinical Neuroscience, https://doi.org/10.1007/978-3-319-78926-2_2

and by the associated astrocytes whose metabotropic receptors are activated by the glutamate released in the synaptic cleft and whose end feet surround the microvasculature. These molecules can act on pericytes and smooth muscle cells. Several factors can contribute to the local vasodilation [4]: astrocytic extrusion of potassium, prostaglandins, and epoxyeicosatrienoic acids; neuronal liberation of vasoactive mediators such as nitric oxide (NO), prostaglandins, vasoactive intestinal peptide (VIP), adenosine, or adenine; and variation of blood pCO_2 and pH and lactate/ pyruvate ratio. It is noteworthy that this vasodilation can be potentially counterbalanced by vasoconstrictors such as norepinephrine. Two other mechanisms can also participate to this functional hyperemia. First, some vasoactive substances may be released from distant subcortical afferents such as from the cholinergic substantia innominata, serotoninergic raphe, dopaminergic ventral tegmental area, and noradrenergic locus coeruleus [2, 4]. Second, (retro-)propagation of vasodilation can occur in relation to endothelial direct action on smooth muscle cells via hyperpolarization-dependent [5] or calcium wave-dependent vasoactive factors [6]. This mechanism would explain why vasodilation can also spread to the arterioles toward the cortical surface [2].

Neurovascular Coupling with GABAergic Interneurons Stimulation of GABA-A receptors produces arteriolar dilation in the neocortex and hippocampus so that GABAergic interneurons may contribute to positive BOLD response [7]. However, several experiments have demonstrated that inhibitory GABAergic synapse could also induce arteriolar vasoconstriction and subsequent decrease in CBV and blood oxygenation, likely through neuropeptide corelease [4]. This hemodynamic response would partly explain negative BOLD response. However, interneurons, such as cerebellar basket and stellate cells, can also secrete molecules such as NO producing vasodilation, while GABA has no vasoactive effects in the cerebellum [8]. Finally, local neurovascular response represents summed and complex effects of both excitatory and inhibitory synapses.

Neuronal Metabolism of Glutamatergic Neurons Moreover, part of the glutamate delivered in the synaptic cleft is captured by contiguous astrocytes and transformed into glutamine before being released back to presynaptic neuron where it will be recycled [9, 10]. The glial sodium-glutamate cotransport stimulates anaerobic glycolysis which produces ATP necessary to extrude sodium through a sodium-potassium ATPase and lactate delivered to the neuron using monocarboxylate transporters. Lactate is then transformed into pyruvate by neuronal lactate dehydrogenase before undergoing oxidative phosphorylation, according to the lactate shuttle hypothesis. This mechanism would explain the observation that only a small fraction of blood oxygen is extracted, while a substantial uptake of glucose occurs.

Local Magnetic Field Owing to functional hyperemia, the surplus of diamagnetic oxyhemoglobin (mainly due to the increased CBV) flows in the local capillaries and venous system which become more diamagnetic [11]. Let's remind that, at rest, the difference of magnetic susceptibility between the paramagnetic deoxyhemoglobin-rich vascular compartment and the more diamagnetic surrounding tissue creates a microscopic field gradient around vessels causing variation of the precessional frequency with space and time. Moreover, randomly diffusing spins (hydrogen atoms of water molecules) due to thermal energy experience different amplitudes of the local magnetic field and exhibit subsequent accrual phase. However, this effect dominates mainly around the small vessels (capillaries and venules) as the diameter of the spin diffusion sphere on the order of 10 µm and the spatial spread of the gradient are approximatively of the same range. In larger vessels (arterioles and large venules), spins can be regarded as static in the perivascular space. It is noteworthy that dephasing also takes place in the intravascular compartment where diffusing spins undergo direct magnetic influence of deoxyhemoglobin because of their spatial closeness. Conversely, after neuronal activation, the increased concentration of oxyhemoglobin in vessels leads to decreased local field distortions and consequently lesser phase dispersion of water molecules. In summary, augmentation of CBF due to neuron-glial-vascular coupling is accompanied by a net increased oxygenation of the microvasculature, while CBV and oxygen consuming induce a counterbalancing but weaker augmentation of deoxyhemoglobin, so that the local magnetic field is significantly diminished. This change of local tissue oxygenation-related magnetism is recorded by fMRI as BOLD response.

BOLD Signal and fMRI After radio-frequency pulse, fMRI detects transversal magnetization whose decay depends upon static and fluctuating field inhomogeneities, including the deoxyhemoglobin-related field gradient, and on spin motion [11, 12]. These magnetic inhomogeneities contribute to faster loss of spin-phase coherence, and this effect is reflected in the $T2^*$ relaxation time which appears in the equation describing the MRI signal evolution in function of time S(t) with a baseline signal equal to S_0:

$$S(t) = S_0 . \exp[-t/T2^*]$$

After neuronal activation, $T2^*$ increases to $T2^{*\prime}$ because of the decrease of intravascular and perivascular paramagnetic distortions of the main field exerted by deoxyhemoglobin transiently replaced by oxyhemoglobin. Therefore, the measured fMRI signal appears stronger. It can be easily calculated that the percentage of change of the signal between the rest baseline and the activated state, which represents the BOLD signal, is approximatively proportional to $[1/T2^* - 1/T2^{*\prime}]$ TE, where TE refers to echo time of the MRI sequence. Several biophysical models have been developed to express the relaxation time $T2^*$ in terms of local field and vascular parameters. For instance, it has been shown that [12, 13]:

$$1/T2^* \approx k . ([dHb] . B_0)^\beta V$$

where k designates a constant, [dHb] the concentration of deoxyhemoglobin which determines the perivascular difference of magnetic susceptibility, B_0 the main magnetic field strength, V the blood volume fraction, and $\beta = 1.5$. More complex and accurate models integrate the dependency of [dHb] on CBF, CBV, and the rate of oxygen extraction. For instance, one possible model for the BOLD response ΔS can be approached by the following nonlinear (in CBF) equation [12]:

$$\Delta S(t) = A.(1 - 1/n - \alpha_V).(1 - CBF(t)/CBF_{baseline})$$

where the complex factor A is mainly proportional to [dHb] (and consequently to hematocrit) and O2 extraction during the baseline state, n is the fractional changes in CBF, and α_V refers to the venous volume changes. It is important to emphasize that the BOLD signal in each pixel is strongly influenced by baseline values of CBF and CBV, for instance. Therefore, BOLD signal does not only depend on neuronal activation but also on regional-specific microvascular anatomy and physiological changes related to age, medication, or concomitant pathology.

As mentioned above, negative BOLD response can also be observed. Kim and Ogawa [14] summarized several possible explanations of stimulus-induced negative BOLD signals: decreased CBF due to neuronal inhibition, or regional flow reallocation, and increased O2 consumption without augmented CBF. Furthermore, they reported Shih et al.'s study [15] showing subcortical negative BOLD response (in rats) mediated by focal release of vasoconstrictors decreasing CBV despite nociceptive-induced neuronal activation. Therefore, negative BOLD signal must be carefully interpreted and must not be regarded as exclusive signature of neuronal inhibition.

Specific sequences of MRI are applied to encode this hemodynamic response within the T2*-weighted image such as T2*-weighted gradient echo (GE) and echo-planar imaging (EPI) with usual spatial and temporal resolutions 2–4 mm^3 and 2–4 s, respectively. BOLD signal detected by EG sequence predominantly derives from venous system. Two reasons may explain this fact. First, upstream, capillaries contain more oxyhemoglobin than veins. Second, in small vessels, erratic moving spins sample all local perivascular magnetic fields, which results in narrowing the phase distribution and consequently in reduced BOLD signal [12]. However, this effect can be utilized during spin echo (SE) imaging since the refocusing 180° RF pulse eliminates only dephasing effects of static magnetic inhomogeneities, but not the phase accumulation of diffusing spins. SE sequence is therefore sensitive to smaller vessels than GE ones.

Phenomenologically, BOLD response, i.e., the percentage of change in MRI signal caused by a very brief stimulation, is characterized by a brief and inconstant initial "dip" followed by the rise of the signal to a peak (4–6 s) due to blood oxygenation increase and followed by slower fall (around 15 s) and finally by a shallow post-stimulus undershoot of variable duration before reaching the baseline (Fig. 2.1). The vascular mechanisms producing the dip and the undershoot remain still a matter of debate. It is noteworthy that BOLD response (20–25 s) lasts longer than neuronal electrical events (several milliseconds) and can exhibit interindividual and interregional variability concerning its time-course. The task-related BOLD increase is usually <5%. Finally, BOLD response can be regarded as a linear

Fig. 2.1 Curve representing schematically the time-varying BOLD response caused by a brief stimulus applied at time t = 0. The dashed line corresponds to the baseline level of BOLD signal

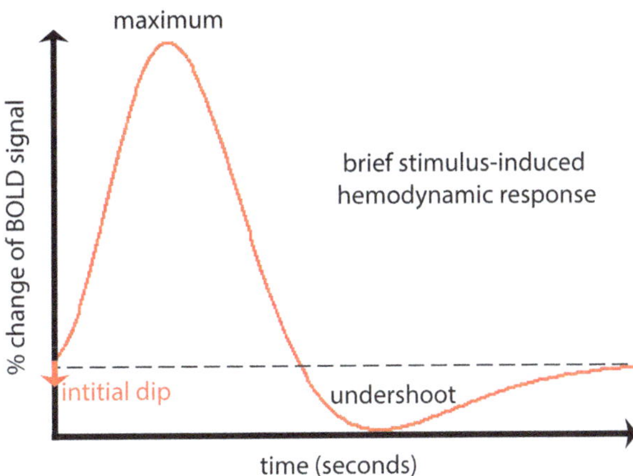

function of the underlying neuronal response in first approximation and is correlated with local potential fields which reflect afferent synaptic inputs and postsynaptic dendritic processing rather than spiking [15], although discrepant results found correlation with spiking [11].

In conclusion, positive BOLD signal mainly reflects intra- and extravascular magnetic field alterations due to focal hyperemia-related, capillaro-venous hyperoxygenation induced by tight neurovascular coupling during neuronal activation. Conversely, negative BOLD signal can be caused by neuronal inhibition. However, regional and stimulus-dependent variations can occur such as inhibitory interneuron-related positive BOLD signal in the cerebellar cortex or striatal negative BOLD signal despite neuronal activation.

CBF-Based Functional Imaging Besides BOLD fMRI, arterial spin labeling (ASL) can also be used for obtaining functional images [16], even in the brain resting state [17]. This MRI sequence usually allows generation of perfusion-weighted images and quantification of absolute perfusion. Put in a nutshell, radio-frequency pulses and gradients applied to the neck invert spin magnetization of hydrogen atoms in arterial blood. In other words, water in the blood is used as endogenous tracer. After a short delay (<5 s) necessary for the labeled spins to reach the studied tissue upstream, two-dimensional EPI images of the brain are acquired. Then, subtraction between paired images with and without spin labeling (control image) enables to suppress static tissue signal and consequently to retain perfusion-related signal. In clinical and research routines, pseudo-continuous ASL is preferentially utilized. Therefore, ASL can detect slow task-related changes of local brain perfusion only due to arteriolar or capillary dilation or contraction. This ASL-based and CBV-related functional mapping improves spatial resolution for localizing accurately activated areas, especially in regions with high susceptibility, and it excludes complex dependency of the signal to other parameters, such as CBV and $CMRO2$ modulating the BOLD signal [18]. Moreover, the ability to measure perfusion permits to evaluate not only interregional signal correlation but also local and global variations of signal amplitude [18, 19]. It is noteworthy that both ASL and BOLD signals can be recorded during a unique dual-echo EPI sequence. The BOLD signal can be extracted from the control (task-related or rest) image.

Brief Survey of Basic fMRI Design and Processing

The aim of the study is to identify brain areas specifically recruited by an experimental task (movement, perception, cognition) performed by the subject inside the MRI machine. Task-related activation must be extracted from the basal metabolism-related activation pattern of the rest of the brain. Two main experimental designs can be applied: *block design* and *event-related design*. In block design, the functional run encompasses a regular alternation of "on and off/controlled" blocks. Within each on-block, subjects are requested to carry out the same experimental activity continuously, while, within each off-block, they remain at rest. Block repetition will sample weak BOLD responses thus increasing signal-to-noise ratio, and subtraction of off-blocks from on-blocks will only retain task-specific activated areas. Alternatively, in event-related design, discrete stimuli are presented repeatedly but briefly and randomly, and they are interspaced by off epochs of variable duration. Therefore, event-related design is more prone to record transient brain activity than block design more sensitive to sustained brain activity but with a risk of habituation or task anticipation.

Raw functional images must then undergo a complex preprocessing before computing task-related "activation maps." A preliminary analysis is indeed required since, for instance, raw data are noisy due to several scanner artifacts, head motion, breathing, and heartbeat and encode not only a weak task-related BOLD signal, less than 5% of the total BOLD signal, but also BOLD variations caused by simultaneous but task-unrelated mental activity.

The main steps of preprocessing of single-subject data include (1) distortion correction, (2) motion correction and slice realignment, (3) slice-timing correction (correction of interslice delays of acquisition), (4) spatial normalization (realignment of individual anatomical data into a common framework for interindividual comparisons and group analysis), (5) spatial smoothing with a Gaussian kernel (removal of high-frequency noise), and (6) temporal filtering (removal of low-frequency noise such as the scanner drift).

Afterward, statistical modeling and inferences will determine which voxel's signal are significantly and specifically correlated with the experimental design. The most popular method utilized to achieve this goal is the general linear model (GLM). GLM will relate a dependent variable (voxel's BOLD signal time series from observed data) to one or more independent explanatory variables (voxel's predicted BOLD signal time-course or perfusion time-course using functional ASL). For example, the predicted BOLD time-course can be modeled by a boxcar, representing alternation of on-off epochs, convolved with a hemodynamic response function. Observed data $\mathbf{Y}$ are, thus, represented as a linear combination of regressors: the predicted neuronal response (design matrix) X weighted by unknown parameters β quantifying effect sizes, and of Gaussian noise (unexplained variance) ζ with a null mean and a variance σ^2, as expressed in the following matrix equation:

$$\mathbf{Y} = [\boldsymbol{\beta}].\mathbf{X} + \zeta$$

The β values can be estimated by minimizing the sum of squared residuals, i.e., differences between observed and predicted values so that the model best fits the data and, after some mathematical manipulations, can be written as:

$$[\boldsymbol{\beta}] = \left(\mathbf{X^t X}\right)^{-1}\mathbf{X^t Y}$$

where $\mathbf{X^t}$ denotes the transposition matrix and provided that $\mathbf{X^t\,X}$ is invertible necessitating that X has a full column rank.

Moreover, GLM requires that the components of the noise ζ remain temporally uncorrelated and share the same variance, which is not always the case. Therefore, after the temporal filtering, prewhitening the data could be applied to remove autocorrelation. Then, a statistical test, such as t-test or F-test, will determine whether the observed data estimated by the set of β parameters and the matrix design may result from random fluctuations of brain activity or of all other null controlled condition (H_0 hypothesis accepted) or not (H_0 rejected: "there is a significant task-related effect"). More formally, the testing relies on contrasts $\mathbf{c}$ defined as a weighted linear combination of $\boldsymbol{\beta}$ values representing effect of interest:

$$\mathbf{c} = \mathbf{c^t}[\boldsymbol{\beta}]$$

where c denotes the contribution of β to the BOLD signal at a particular TR and in a specific voxel, $\mathbf{c}$ can take $-1, 0,$ or 1 values, and then the H0 and H1 hypotheses can be expressed by:

$$H0 : \quad c = 0 \text{ versus } H1 : \quad c > 0$$

If a t-test is applied, the statistics following the t-distribution with (number of rows – number of columns in X + 1) degrees of freedom is expressed by:

$$t = \mathbf{c^T\beta}/\sigma\left[\mathbf{c^T}\left(\mathbf{X^t X}\right)^{-1}\mathbf{c}\right]^{1/2}$$

Statistical inferences, therefore, require to define the contrast (performance against rest or performance against another performance), the type of level inference (voxel versus cluster), and the statistical threshold (P-value uncorrected or, most often, corrected for multiple comparisons, e.g., P (T > t) = 0.05 corrected, where T refers to the random variable following the t-distribution having the same degrees of freedom than the computed t). Multiple comparisons is an important problem. If the probability of false positives is fixed at 0.05 for each test per voxel, then the probability of false positives for all the N studied voxels equals 0.05 N number

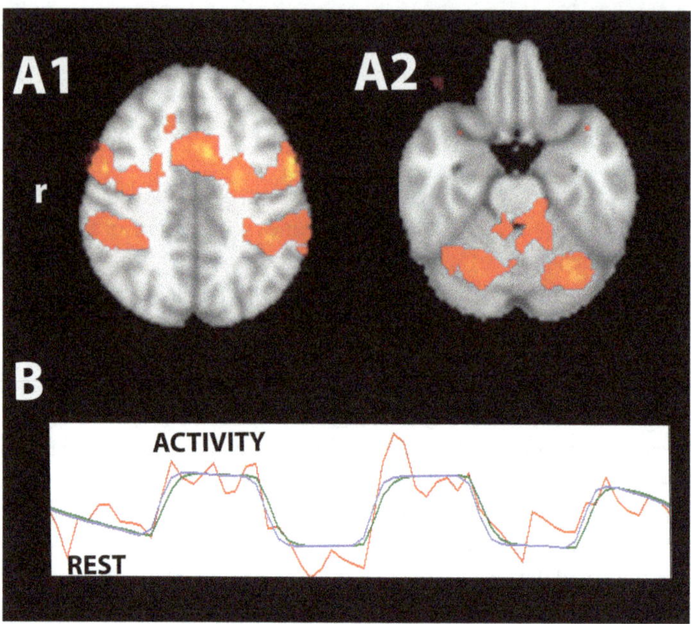

Fig. 2.2 Statistical parametric map of a subject performing alternating phases of rest and activity (motor mental imagery task) during a functional MRI scan. (**a**) Axial slices showing bilateral (pre-) frontoparietal cortical activation (A1) and cerebellar activation (A2). (**b**) The curves represent the detected time-varying BOLD signal (red) and the full model fit versus data obtained by GLM method (blue) (post-processing using analysis of FMRIB Software Library v5.0, Oxford, UK: https://fsl.fmrib.ox.ac.uk/fsl/fslwiki/FEAT)

which can be very high. Therefore, it is absolutely necessary to resort to multiple comparison corrections. Several procedures can be used such as family-wise rate correction (Gaussian random field), for false discovery rate correction controlling the false positives only among the significant voxels. The final results, the statistical parametric map, are visualized on thresholded activation map overlaid on T1 high-resolution anatomical images, where the voxel statistical significance is represented by a color gradient (Fig. 2.2).

Further higher-level statistics using single-subject statistical maps can be performed to create a single-group activation maps and to test intersession or intergroup differences in the brain activation pattern. Generally, as we want to extrapolate information from the limited group of subjects to the whole population, they belong to, random effect analyses are preferentially carried out rather than fixed effect.

Resting-State Functional Connectivity

Definition There exist tight temporal correlations between spontaneous BOLD signal fluctuations at low frequency [0.01–0.1 Hz] and at rest, between spatially remote but functionally related brain areas [20, 21]. The spontaneous brain activity at rest thus results from complex dynamical interactions between well-segregated but partially overlapping, large-scale neural networks involved in motor, perceptual, and cognitive functions [22]. rsfMRI is devoted to identify these resting-state networks (RNs), reminiscent of task-related networks, and their interactions (correlation/ decorrelation and anticorrelation) across time. This internally driven brain activity is assumed to reflect, at least, not only the concomitant free conscious experience referred as "mind wandering" but also spontaneous neural events constrained by experience-dependent neuroplastic changes and network structural and/or functional topological architecture. Let's mention that ASL-based rsfMRI can also be used [17].

Neurophysiological Basis The neurophysiological mechanisms underlying these interregional synchronizations are poorly understood and still a matter of debate. First, interregional functional coherence only partially reflects the anatomical wiring. This fact is illustrated by left and right BA 17 regions, which belong to the same primary visual RN although they are not anatomically interconnected, and by the dorsal attentional network and the visual RNs which are functionally separated although they anatomically linked [23]. Notwithstanding, callosotomy markedly reduces interhemispheric connectivity. Second, there exist transient and complex correlations between BOLD fluctuations and alpha, beta, gamma, delta, and theta EEG rhythms [24]. For instance, the default mode network and the dorsal attentional network exhibit positive and negative correlations with beta and alpha rhythms, respectively, whereas the salience networks are active in the gamma range. MEG recordings have also highlighted relationships between BOLD fluctuations and amplitude envelope of band-limited rhythms [25]. Third, in monkey, spontaneous BOLD fluctuations were consistently correlated with neural spiking activity and gamma-range band-limited power [26]. In human, correlations were also found between spontaneous BOLD and the slow cortical potentials. Fourth, several other putative factors have been assumed to shape BOLD spontaneous fluctuations [23, 27]: cellular metabolism (redox variation associated with energetic metabolism), biophysics (up-down state transition of the membrane voltage, subthreshold oscillations, microstates), biochemistry (quantal exocytosis), network constrains (time delay, noisy transmission, nonlinear attractor dynamics [28]), anatomical connections (thalamus), or neuromodulation. Therefore, spontaneous BOLD fluctuations seem to be spatially and temporally multiscale organized from the cell to the network.

MRI Sequence Inside the scan and during the whole experiment, subjects are requested to remain still and eyes closed and to focus their attention upon nothing in particular. Then, the same EPI sequences than in classical fMRI are applied to measure the resting-state brain activity. For example, around 200 volumes covering the whole brain are acquired with a TR = 2 s. Utilization of accelerated sequences, such as multiband (simultaneous acquisition of multiple slices), can notably diminish acquisition time and TR value (<1 s) and/or to increase spatial resolution. It is noteworthy that the raw data will be preprocessed as in GLM method.

Statistical Analysis The two main techniques for performing functional connectivity analysis are region-of-interest (ROI) correlations and independent component analysis (ICA) [29]. The first method focusing on a specific brain area is based on calculating correlation (Pearson score) of the BOLD signal between a predefined ROI and the rest of the brain. Alternatively, ICA is a multivariate, data-driven method allowing the identification of multiple coexisting whole-brain networks and subnetworks by voxel-to-voxel analyses. To achieve this goal, ICA solves the *blind source separation problem*, which requires distinction between a set of unknown sources underlying the observed data. ICA assumes that these sources are statistically independent, non-Gaussianly distributed, and linearly mixed. For functional imaging and more formally, spatial ICA seeks to separate all the *hidden* distinct neural networks $\mathbf{X}$ and, in probabilistic ICA model, possible source of normal noise ξ contributing altogether and simultaneously to the observable whole-brain BOLD signal $\mathbf{Y}$ [30]:

$$\mathbf{Y} = \mathbf{M}.\mathbf{X} \, (+\xi)$$

where $\mathbf{M}$ denotes the *unknown mixing* matrix. Sometimes, the raw data have been beforehand prewhitened by principal component analysis in order to reduce the dimensionality of data spaces and to remove Gaussian noise. ICA algorithm, such as maximization of negentropy or maximum likelihood, for example, then aims to determine the *unmixing* matrix in order to recover the unknown sources $\mathbf{X}$ from $\mathbf{Y}$. This algorithm computes a set of statistically independent components using, for instance, maximization of non-Gaussianity (using higher-order statistics). Each neural network and artifactual sources $\mathbf{X}$ correspond to a specific ICA component (spatial maps) and are associated with a specific BOLD time-course. The number of components remains a free parameter which can be fixed a priori (around 15–20) or estimated by the algorithm (around 30–70): the lower the number (around 20), the more robust and reproducible results. Excessive number of components might split networks into subsystems.

Group and intergroup analyses can also be conducted by ICA [30, 31]. For instance, temporal concatenation supposes common spatial maps associated with unique time-courses across subjects. The subject-specific maps and time-course can, then, be back-reconstructed on the basis of the group-level components. Finally, several algorithms enable intersession or intergroup comparisons such as dual regression and randomization. It is worth emphasizing that algorithmic results can vary in function of the population size and because of its probabilistic nature.

Finally, other methods can be applied in the "resting" brain such as:

- Detection of (fractional) "amplitude of low-frequency (0.01–0.1 Hz) fluctuations (ALFF)" calculating the intra-voxel power spectrum of the BOLD signal [32],
- "Regional homogeneity" (ReHo) approach measuring synchronization between time-course of BOLD signal of neighboring voxels using the Kendall's coefficient of concordance [33].

However, the resulting functional maps reflect local processing and not large-scale functional circuits.

Resting-State Network Identification The main step after having obtained the ICA components relies on the selection of components representing the true neural networks. Some criterion can be usefully applied and automated. First, only components whose frequency is comprised between 0.01 and 0.1 Hz or, at least, whose more than 50% of the spectrum remains within this interval, must be retained. Second, noisy components must be well-identified (structured noise, head and ocular motion, breathing, cardiac beating, scanner drift, etc.) and removed. Several typical aspects of noise must be kept in mind: *structural noise* like ring pattern of pixel around the encephalon due to head motion, diffuse spotty pattern over the brain, large or tighter bilateral frontal and orbitofrontal clusters with "spike" or "sawtooth" motif appearing in the corresponding time-course, respectively, clusters located in the ventricles, and *improper anatomical regions* like white matter, cisterns (especially around the brainstem), superior longitudinal sinus, or eyes [34]. This can be achieved by (operator-dependent) visual inspection or by automated classification algorithms. It is possible to take advantage of the extraction of noisy components in single-subject components, to denoise the raw data, what will improve further group analysis. Third, most of the RNs have been listed and can be easily recognized according to anatomical localization of the network nodes [35, 36] (Fig. 2.3):

1. The default mode network (DMN) (bilateral medial prefrontal, posterior cingulate, retrosplenial, precuneal, lateral inferior parietal, and parahippocampal cortices, cerebellar amygdalae, and cerebellar lobule VII) involved in consciousness, unconstrained cognition ("mind wandering"), episodic memory, emotional processing, and self-projection in future or past
2. The left executive control network (LECF) (left dorsolateral prefrontal and superior parietal cortices, contralateral cerebellar lobules HVI and HVII) involved in syntactic cognition and language
3. The right executive control network (RECN) (right dorsolateral prefrontal and superior parietal cortices, contralateral cerebellar lobules HVI and HVII) involved in holistic and visuospatial cognition
4. The dorsal attentional networks (DAN) (bilateral intraparietal sulcus, prefrontal, precentral cortices) for top-down focus of attention and goal-directed cognition
5. The salience network (SN) (prefrontal operculum, anterior insula, dorsal anterior cingulate, hypothalamus, and cerebellar lobules HVI and HVII) involved in interoception, emotion, and attentional reallocation based on stimuli salience
6. The sensorimotor network (bilateral S1-M1, supplementary motor area, cerebellar anterior lobe and lobule VIII)

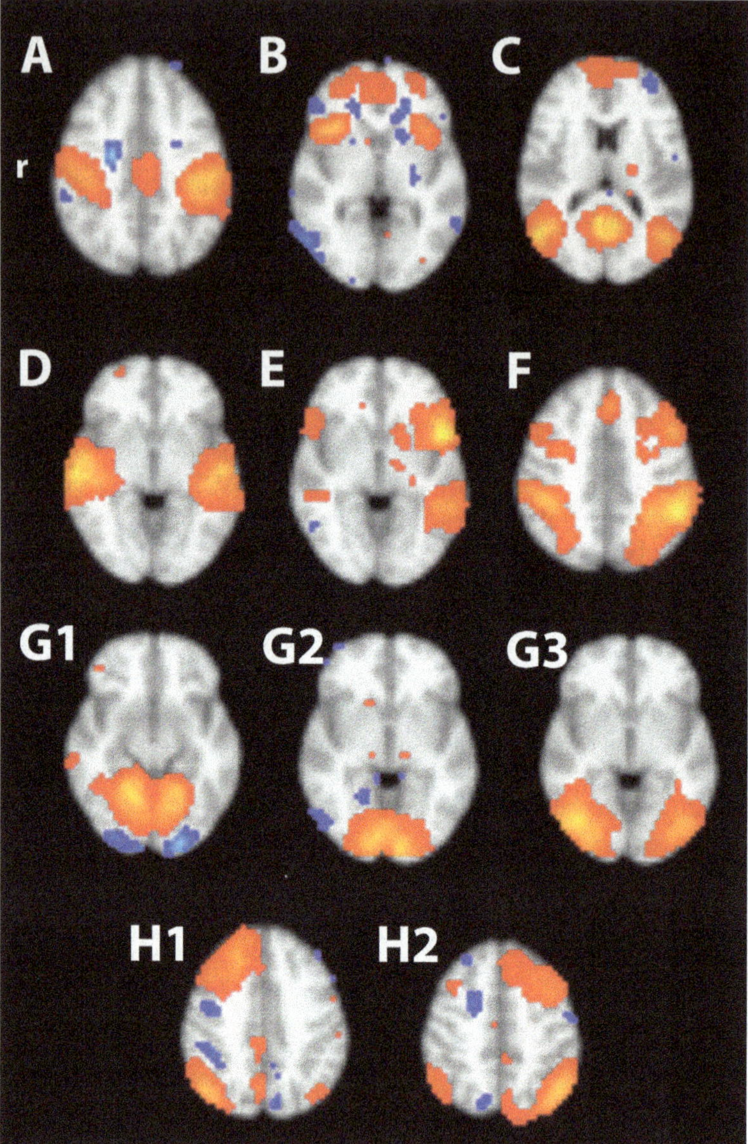

Fig. 2.3 Axial slices of the brain showing fMRI resting-state networks computed by independent component analysis. (**a**) Motor network. (**b**) Salience network. (**c**) Default mode network. (**d**) Auditory network. (**e**) Language network. (**f**) Dorsal attentional network. (**g**) Medial (G1), caudal (G2), and lateral (G3) visual networks. (**h**) Right (H1) and left (H2) executive networks. r: right

7. The medial primary visual cortex
8. The polar visual cortex
9. The lateral (extrastriate) visual cortex

10. The auditory network (superior temporal, insular, and postcentral cortices)
11. The linguistic network (superior and medial temporal, inferior frontal, and angular cortices)
12. The striatum
13. The cerebellum

SN, LECN, RECN, and DAN are grouped in a generic external attention system (GEAS) in opposition to the DMN implicated in self-referential thinking. GEAS and DMN appear to work most often in counter-phase. More precisely, DMN is anticorrelated to the DAN but can be flexibly correlated, for example, during goal-directed mental simulation or anticorrelated to ECN. The dorsolateral prefrontal region of the ECN can inhibit the medial prefrontal part of the DMN [37]. It is suggested that the SN may exert a role in switching between these two internally and externally oriented systems in response to behaviorally relevant salient external stimuli. Moreover, a seed-based analysis has characterized a ventral right-lateralized attention system encompassing the ventral frontal cortex and the temporoparietal junction [38]. It turns out that the switching mechanism would rely on antagonistic connectivity between the right insula and the dorsal posterior cingulate part of the DMN. At rest, only part of these networks such as DMN and SN are currently active. Finally, hierarchical clustering using, for instance, partial correlation matrix could contribute to distinguish and rank RN and their subnetworks.

Time-Dependent Properties of Resting-State Networks These networks evolve across time. Although RNs are robustly detected in fetus, children, asleep or anesthetized subjects, and animals (monkeys, rats, and mice), the functional connectivity is age-dependent and context-dependent as it varies in function of several factors such as open versus closed eyes, mood, drowsiness, mental content, and previous cognitive effort. First, age-dependent changes mainly concern intra- and inter-network reorganization of connectivity. In fetus, cortico (M1)-cerebellar, thalamocortical, interhemispherical coherences have been observed [39]. Moreover, a proto-DMN interconnecting medial prefrontal and precuneal cortices has also been identified after 35 weeks. In children, bilateral sensorimotor, auditory, primary visual anterior prefrontal, and medial and lateral parieto-cerebellar networks are the first RN to be discernible by age 1. A bit later, DMN, SN, and executive control networks can be visualized by age 2 but will undergo further gradual maturation processes including increasing within-network coherence, eliminating some distant interconnection, and developing long-range connectivity. Motor and perceptual RNs are mature earlier than higher cognitive circuits. In older people, DMN shows the same global connectivity than in young adults but with weaker connectivity between prefrontal and ventral precuneal cortices and stronger correlation in the dorsal precuneal circuit [40]. Second, functional connectivity is not only influenced by specific mental states, as mentioned above, but also by practice of motor, perceptual, and cognitive tasks. In particular, more or less transient synchronization may occur between performance-related brain areas and ECN. DMN exhibits task-induced deactivation and, more often, subsequent anticorrelation with networks subserving overt sensorimotor, attentional, or cognitive activity like stimulus-conscious perception: the weaker the anticorrelation, the poorer the performance likely due to interference between mind wandering and task-related attention. However, some nodes of the DMN, such as the right posterior cingulate cortex, can collaborate with ECN during memory recollection and improve rapidity of the task. Therefore, competition or cooperation between RN and strengthening interconnections may reflect transient "scaffolding" connectivity (attention to the task) and neuroplastic changes caused by practice (off-line consolidation, memory trace, and performance ability), respectively [41]. Third, classical studies provide a set of apparent stationary RN during the experiment. However, more refined post-processing using a sliding time window has demonstrated time-varying interactions of RN subnetworks and between RNs [42]. Moreover, a component related to drowsiness and light sleep has been characterized including breakdown of DMN connectivity, reduced thalamocortical coherence, and increased subcortical connectivity. In summary, the whole set of RNs is common to all subjects; their complex pattern of interactions shaped by experience-dependent plasticity may reflect individual idiosyncratic functional and structural connectivity and, consequently, personal cognitive profile.

Clinical Application Functional connectivity enables not only to explore normal brain functioning but also brain functional impairments in neurology (stroke, multiple sclerosis, pain, migraine, amnesia, spinocerebellar ataxia, Parkinson's disease, epilepsy, altered consciousness like vegetative state and coma), neuropsychology (sleep, aging), pharmacology (treatment evaluation, drug abuse), neuro-ophthalmology (glaucoma, visual deprivation), and psychiatry (attention-deficit/ hyperactivity disorder, obsessive and compulsive disorder, Tourette syndrome, post-traumatic stress disorder, depression, neglect, autism, schizophrenia, bipolar trouble, mild cognitive impairment, dementia), as well as potentially after neurostimulation (TMS, TDS) [43, 44]. It can also be used for presurgical planning and postsurgical follow-up helping to localize precisely (sensorimotor or language) functional areas. This method offers important clinical possibilities. It can be applied to sedated, unconscious, or uncooperative patient, especially patients unable to perform task-based fMRI. It permits to screen a wide range of functional networks either in an exploratory mode (ICA) or in focal seed-based mode and to determine qualitative (anatomical) and quantitative (inter-areal correlation values) RL alterations in comparison with matched normal subjects. These data can provide markers for diagnosis and prognosis and anatomo-clinical correlations.

Effective Connectivity

Effective connectivity can be defined as the influence that one neural system exerts over another, either at a synaptic level (synaptic efficacy) or a cortical level [45]. This hypothesis-driven approach emphasizes that determining effective connectivity requires a causal model of the interactions between the elements of the neural system of interest. Causal relationships are not inferred from the data but are assumed a priori. If fMRI and rsfMRI identify statistical covariation across functionally related neural nodes, they do not provide any information about direct causal influence between these nodes: "does activated node A directly cause or modulate activation of node B? or, alternatively, "does a third node D synchronize independent nodes A and B?". Effective connectivity tries to address these important points.

First, tractography may delineate which nodes are anatomically interconnected – however within the limits of spatial resolution and fiber crossing/convergence detection – since functional interconnection presupposes axonal interconnection. Second, several complementary methods inferring internodal causality based upon fMRI data have been developed such as dynamic causal modeling (DCM), structural equation modeling (SEM), time series analyses, and brain perturbation (with transcranial magnetic stimulation, for instance) [46].

DCM roots in the hypothesis that brain can be regarded as a deterministic nonlinear dynamic system transforming inputs into outputs and taking into account modulations [47]; DCM models the effects of experimental, external, and modulatory inputs on network dynamics. Briefly, the first step of this method consists in building a biologically plausible neuronal model of interacting brain regions detected by fMRI. Each of these neural regions is characterized by its state z, representing its amount of activation, and its coupling with its interconnected regions. The time evolution of z is thus described by a nonlinear differential equation: $\dot{z} = dz/dt = [A + \sum B^i.u_i + \sum D^j.z_j]\, z + C.u$ which depends on the current state z, the intrinsic internodal coupling (A matrix), the transient changes in intrinsic coupling due to jth internal input (B matrix), the direct action of external inputs u upon the region (C matrix), and the D matrix (nonlinear state equation) which encodes how the n regions gate connections in the system. However, only the hemodynamic response underlied by the neuronal activity is measured using fMRI. Therefore, this neuronal model must be complemented by a hemodynamic model biophysically plausible in terms of vasodilatory signal which induces increased blood flow. This results in local changes in blood volume and deoxyhemoglobin (dHb) from which the BOLD signal can be predicted. The aim of the DCM is to estimate neural (A, B, C, D) and hemodynamic parameters (flow, volume, etc.) such that the modeled and measured BOLD signals are maximally similar. Afterward, the parameters of the neuronal model can be estimated using Bayesian framework. The oversimplification of the anatomical/connectivity architecture of the neuronal networks constitutes one of the main limitations of this method.

SEM is a multivariate method which assigns specific strength (or *path coefficient*) to anatomical connections linking functionally related brain areas. Unlike DCM, SEM assumes instantaneous (non-dynamic) interactions and the inputs stochastic and unknown. So, SEM models can be considered to be "static" as they model instantaneous interactions between regions and ignore the influence of previous states on current responses. Path coefficients are computed by minimizing the covariance matrix of the modeled (theoretical) network and the covariance matrix of the observed data (BOLD response) – the amount of statistical dependency between the time series of the BOLD signal extracted from pairwise associated brain regions. Here again the relevance of SEM depends on the anatomical accuracy of the model and on the linear assumption of the change in path coefficients and estimates of path coefficients limited to the hemodynamic level.

Coherence analysis studies the correlation in the frequency domain, using fast Fourier transform, between BOLD responses of interconnected brain areas. This spectral method can also determine the causal directionality by calculating the sign of the slope of the phase spectrum (for low frequencies), and the inter-areal time lag by the absolute value of the slope. Alternatively, Granger causality method uses a linear autoregressive prediction claiming that current state of activation of region A, at t, is predicted by a linear combination of its previous states measured at (t-nTR). It further hypothesizes that if activation of region A causes activation of region B, then adding the autoregressive model of A to autoregressive model of B will improve the predictability of B activation in comparison to the autoregressive model of B alone. Then, the difference between observed data and the predictive model, estimated by the variance, can be used as a measure of adequacy of the model and, up to some algebraic transformation, of inter-areal causation.

Coherence analysis and Granger causality suffer from several common and specific weaknesses which prevent their utilization in fMRI [48]. The main problem relies on the difference between the duration of neuronal processing on the order of milliseconds, the rate sampling of BOLD data on the order of seconds, and the total duration of BOLD response (20–30 s). Therefore, the DCM method seems the most appropriate one to characterize causal directionality between brain regions.

Finally, one last complementary method to test potential causal influence between two brain areas consists in actively perturbing the functioning of a given brain area. Magnetic or electric transcranial stimulation can be used to activate or inhibit a specific brain region and to study the influence of this perturbation on the activation of interconnected areas.

Graph Theory

Graph theory has already proven to be applicable to a considerable diversity of complex systems, including markets, ecosystems, computer circuits, and gene-gene interactomes. Complex network theory is particularly appealing when applied to the study of clinical neuroscience, where many cognitive and emotional disorders have been characterized as disconnectivity syndromes, as indicated by abnormal phenotypic profiles of anatomical and/or functional connectivity between brain regions. For example, in schizophrenia, a profound disconnection between frontal and temporal cortices has been suggested to characterize the brain; in contrast, people with autism may display a complex pattern of hyperconnectivity within frontal cortices but hypoconnectivity between the frontal cortex and the rest of the brain [48].

Graph theoretical analysis is potentially applicable to any scale, modality, or volume of neuroscientific data (e.g., diffusion tensor imaging (DTI), diffusion spectrum imaging (DSI), EEG, MEG, cortical thickness, resting state, task-related fMRI) providing new measures of human brain organization in vivo. It is conceptually easier to link the brain graphs derived from these different data types to each other than it would be if each imaging dataset were described in terms of some modality-specific measure of association between regions, e.g., tractographic connection probabilities from DSI or correlations between regional fMRI time series. Facilitation between-modularity translation of results can be important for methodological cross-validation and, more fundamentally, for informing our understanding of how functional networks might interact with the substrate of a relatively static structural network. For example, fMRI and DTI brain graphs consistently demonstrate some common global topological properties allowing a high efficiency at different spatial and temporal scales with a very low wiring and energy cost [49].

We summarize below few of the topological measures that have been most extensively investigated in the neuroimaging literature to date [49, 50]. Any complex dynamical network will be mathematically described as graphs that represent a set of n nodes or vertices associated with k connections or edges/lines between them.

The "small-world" network, the first mathematical model originally described in social networks, combines high levels of local clustering among nodes of the network and short paths that globally link all nodes of the networks. Small-world organization (many short path lengths, few long path lengths) is intermediate between the random networks (all nodes are related randomly with same probability for short- and long-distance connections) and the regular networks or lattices where all nodes are only related to their nearest neighbors (many short path lengths, no long path length). Path length is the minimum number of edges that must be traversed to go from one node to another. Random and complex networks have short mean path lengths (high global efficiency of parallel information transfer), whereas regular lattices have long mean path lengths. Efficiency is inversely related to path length but is numerically easier to use to estimate topological distances between elements of disconnected graphs. *Small-world* and *efficiency* use two metrics: the *clustering coefficient* (C) which is a measurement of the efficiency of local connectivity and the *path length* (L) which is a simplified measurement of the global efficiency of information transfer on the network. These two metrics enable to define the small-world properties, in which the network exhibits a clustering coefficient greater than the clustering coefficient of a random network and a path length about the same as the path length of a random network. The small-world scalar dependent on the calculation of a path length can be troublesome for networks that contain one or more disconnected nodes. The path length of a disconnected node is infinity; it cannot transfer information to any other node on the network. In a complementary formalism, the *global efficiency* is introduced as an alternative metric of global integration that is inversely proportional to the characteristics path length of the network, thus allowing computation of a finite value for graphs with disconnected nodes.

Many complex networks consist of a number of modules [49]. There are various algorithms that estimate the *modularity* of a network, many of them based on hierarchical clustering. Each module contains several densely interconnected

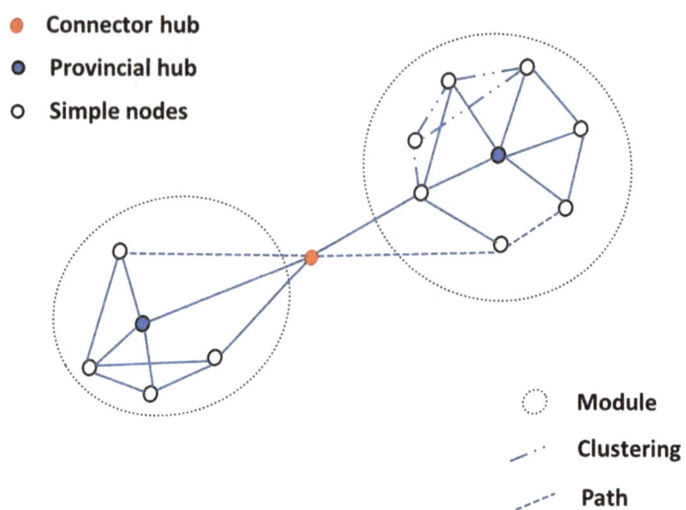

Fig. 2.4 Representation of network with modular organization. Provincial hubs are connected mainly to nodes in their own modules, whereas connector hubs are connected to nodes in other modules

nodes, and there are relatively few connections between nodes in different modules. A brain graph can be subdivided or partitioned into subsets or modules of nodes. In general, the aim is to find the partition that maximizes the ratio of intramodular to intermodular edges. The nodes in any module will be more densely connected to each other than to nodes in other modules. The *intramodular degree* is a measure of the number of connections a node makes with other nodes in the same module. The *participation coefficient* is a measure of the ratio of intramodular connectivity to intermodular connectivity for each node. Finally, these and related metrics can be used to define nodes as "connector hubs" with high intermodular connections or "provincial hubs" with high intramodular connections. Therefore, hubs can be described in terms of their roles in a community (modular) structure (cf. Fig. 2.4).

Degree and degree distribution indicate the likely presence of network hubs or highly connected nodes. The simplest topological measure is the *degree* of a node which is defined as the number of edges emanating from that node. Degree sometimes called degree centrality has been used to discriminate between nodes in the system that are well connected, i.e., so-called hubs, and nodes that are less well connected or non-hubs. Due to their relatively increased connectivity, high-degree nodes are likely to play an important role in the system's dynamics. The probability distribution for nodal degree is the *degree distribution* of the network. Brain graphs generally have heterogeneous or broad-scale degree distributions, meaning that the probability of a highly connected hub is higher than in a comparable random network. Most studies have found that an exponentially truncated power law is the best form of degree distribution to fit to networks based on functional and structural MRI data.

Connectivity degree is one of the most basic and important measures of network analysis. The degree K_i of a node i is defined as the number of connections to that node. Nodes with a high degree are interacting with many other nodes in the network. The degree K of a graph is the average of the degree of all N nodes in the graph G:

$$K = \frac{1}{N} \sum_{iG} K_i$$

The connection strength is a measure quantifying how closely network nodes are connected in terms of showing a relationship in their time-course of activation. The overall connection strength S_i is calculated as:

$$S_i = \sum_{i \neq jG} Kr_{i,j}$$

Measures of functional segregation quantify the presence of functionally related, densely interconnected groups of brain regions, known as clusters within the network. The local (nodal) clustering coefficient C_i is defined as the number of existing connections among the node's neighbors divided by all their possible connections:

$$C_i = \frac{E_i}{K_i(K_i - 1)/2}$$

where E_i is the number of existing connections among the node's neighbors.

The clustering coefficient of a network is the average of the clustering coefficient of all nodes:

$$C = \frac{1}{N} \sum_{iG} C_i$$

in which C quantifies the extent of local connectivity of the network.

Measures of functional integration characterize the ability to rapidly combine specialized information from distributed brain regions and are commonly based on the concept of a path, with shorter paths implying stronger potential for integration. The mean path length L_i of a node i is:

$$L_i = \frac{1}{N - 1} \sum_{i \neq jG} L_{i,j}$$

in which $L_{i,\,j}$ is the smallest number of edges that must be traversed to make a connection between node i and node j. The average inverse shortest path length is a related measure known as global efficiency of a network.

In terms of clinical application, Redcay et al. [51], to assess whole-brain network properties in adolescents with autism, collected resting-state functional connectivity MRI (rs-fcMRI) data from neurotypical adolescents (NT) and adolescents with autism spectrum disorder (ASD). Task-independent studies of the resting brain provide a window with which to examine intrinsic functional network organization. Recent findings suggest connectivity differences in autism with evidence for both hypo- and hyperconnectivity for short- and long-distance connections depending partly on age of the participants.

In this study, the authors used graph theory metrics on rs-fcMRI data with 34 regions of interest (i.e., nodes) that encompass four different functionally defined networks: cingulo-opercular, cerebellar, frontoparietal, and DMN (default mode network). These networks were selected because previous research with these same networks has demonstrated a developmental pattern of progressive increases in long-distance connectivity between nodes of the same network and concurrent decreases in connectivity between anatomically proximal nodes of distinct networks. In addition, functions associated with these networks have all been implicated in autism. Thus, examining these networks allows for a more rigorous test of the hypothesis of reduced long-distance and increased local connectivity in autism, across multiple networks that support varied functions.

As mentioned above, graph theory methods can examine the topological properties of each region within the context of all other regions of interest, including measures of the integration (global efficiency, path length), segregation (local efficiency, clustering coefficient), and centrality (or betweenness centrality) of networks. These metrics provide a more robust test of the theory of reduced long-distance and increased local connectivity by testing differences in measures of whole-brain network integration and segregation.

In the Redcay's work [51], data were analyzed using SPM8 (http://www.fil.ion. ucl.ac.uk/spm/) and CONN functional connectivity toolbox (https://www.nitrc.org/ projects/conn). The unweighted ROI-to-ROI correlation matrices were first thresholded at a cost value of $k = 0.15$. Cost is a measure of the proportion of connections for each ROI in relation to all connections in the network. When cost is equated across participants, direct comparisons across groups of network property differences can be made. Small-world properties are observed in the range of costs $0.05 < k < 0.34$, where global efficiency is greater than that of a lattice graph and local efficiency is greater than that of random graph. A cost threshold of 0.15 has been demonstrated to provide a high degree of reliability when comparing session-specific estimates of graph theoretical measures across repeated runs or sessions (e.g., global efficiency $r = 0.95$, local efficiency $r = 0.9$). The specific measures of interest were those of integration (global efficiency), segregation (local efficiency), and centrality (betweenness centrality). Between-group t-tests were used to compare networks measures between groups with a FDR correction of $p < 0.05$.

Global efficiency is calculated as the average of the inverse of the shortest path length between each ROI (or node) and all other ROIs. The shortest path length is defined as the fewest number of connections (or correlations) between two nodes. Thus, a network with high global efficiency would be one in which nodes are highly integrated, so the path between nodes is consistently short. This measure (with cost kept constant) can be thought of as reflecting global, long-distance connections within the brain. Local efficiency is calculated as the average inverse of the shortest path length between the neighbors of any given node (or ROI). In other words, local efficiency measures the extent to which nodes are part of a cluster of locally, interconnected nodes. Betweenness centrality (measure of centrality) measures the fraction of all shortest path lengths in a network that pass through a given node. Thus, if a node is directly connected to many other nodes in the network, it will have a shorter overall path length and function as a hub within and between networks.

Contrary to their hypotheses, Redcay et al. found no differences in measures of global or local efficiency. Only betweenness centrality, which indicates the degree to which a seed (or node) functions as a hub within and between networks, was significantly different between groups, and it was greater for participants with autism for the right lateral parietal (RLatP) seed of the DMN only. Using RLatP region as a seed region, authors demonstrate significantly functional connectivity in the ASD than NT group within anterior medial prefrontal cortex (aMPFC) using a few cluster correction of p < 0.05; NT group showed higher connectivity between the RLatP seed and cerebellar tonsils (a region previously associated with the DMN). The author suggests that the higher betweenness centrality in ASD may be due to greater long-distance connectivity within the DMN (right lateral parietal-anterior medial prefrontal cortex). They conclude that greater connectivity within right parietal cortex could indicate less functional specialization of this region in ASD.

Machine Learning in Neuroimaging

Machine learning (ML) algorithms developed in artificial intelligence and in statistics are increasingly applied to neuroimaging data for automatic and adaptive detection, classification, and prediction of complex, brain structural and functional patterns [52, 53]. From a neurological and psychiatric standpoint, ML can be potentially used for image processing (e.g., anatomical segmentation, image registration), lesion detection, radiomics, diagnosis, and prognostics, while such methods would allow to decode/encode brain states, for instance, in the field of fundamental neurosciences.

As structural and functional brain images contain a huge number of data (voxels), often noisy and spatially correlated, robust multivariate methods applicable to high dimensional data space are requested to uncover and to interpret complex or subtle latent, highly informative, structure. ML algorithms are, therefore, particularly well suited to fulfill this task and, contrary to univariate investigations, offer the additional possibility to draw inference at the single-subject level. Notwithstanding, ML necessitates large samples of training data, such as MRI scans, so as to decrease the risk of overfitting (inability to generalize from the data used for training to new data) [53].

ML Brief Overview
Put in a nutshell, ML is trained with data so that the algorithm progressively learns how to identify specific structured patterns "buried" within these same data and how

to predict the characteristics of new data. For example, ML can be utilized as a classifier segmenting and categorizing discrete data into groups and ascribing new data to their most likely belonging class. The ML algorithm can be trained in either a supervised or in a (semi-)unsupervised manner.

Supervised ML, such as a specific type of perceptron called support vector machine (SVM) or multilayered perceptron, infers from the training data including both the input data $\mathbf{x}$ and the desired output data $\mathbf{y}$, the function $\mathbf{f(x)} = \mathbf{y}$, which enables to predict the features of new data $\mathbf{z}$: $\mathbf{f(z)} = \mathbf{y}$. When the input data are continuous, this function is tantamount to a regression. Conversely, unsupervised ML strives to determine the probability distribution of the inputs $\mathbf{x}$, based only on the training dataset. One important example of unsupervised ML consists in recent deep-learning algorithms [54], which construct automatically hierarchical representations of the input intrinsic organization, from simpler to more complex and abstract concepts, in contrast to supervised ML which needs relevant feature extraction by an instructor. Unsupervised, data-driven ML training has consequently the benefit of (partially) getting rid of any operator dependency.

Perceptrons are particular implementation of artificial neural networks (ANN). ANNs are composed of a set of successive connected layers, each layer containing a given number of artificial neurons. Artificial neurons are formal automata which transform the sum of their weighted $\mathbf{w.x}$ inputs into an output y: $\mathbf{y} = \mathbf{f}(\sum \mathbf{w.x} - \mathbf{\theta})$, where $\mathbf{f}$ and $\mathbf{\theta}$ denote the activation function such as a logistic sigmoid function and an activation threshold (or bias), respectively (Fig. 2.5). Artificial neurons are grouped into "juxtaposed" layers: an input layer, one or several hidden layers, and an output layer. For instance, the output layer could indicate the probability of an input to fall into a given class. In a feed-forward ANN, each neuron is connected to all neurons belonging to the next layer, so that activation propagates feed-forwardly from the input layer up to the output layer after having undergone successive nonlinear transformations within the hidden layers (Fig. 2.6). The number of hidden layer determines the "depth" of the network. Of course, a lot of ANN architectures exist in relation to the number of artificial neurons, their activation function, the network "depth," and the connectivity pattern (feed-forward, recurrent, or full-connected networks). Perceptrons are trained by data inputs $\mathbf{x}$, and the corresponding computed output $\mathbf{y}$ is compared with the desired output $\mathbf{o}$. The difference between these two outputs $(\mathbf{o} - \mathbf{y})$, called error, serves to correct backward and step-by-step the weights of the connections between neurons from the output layer down to the

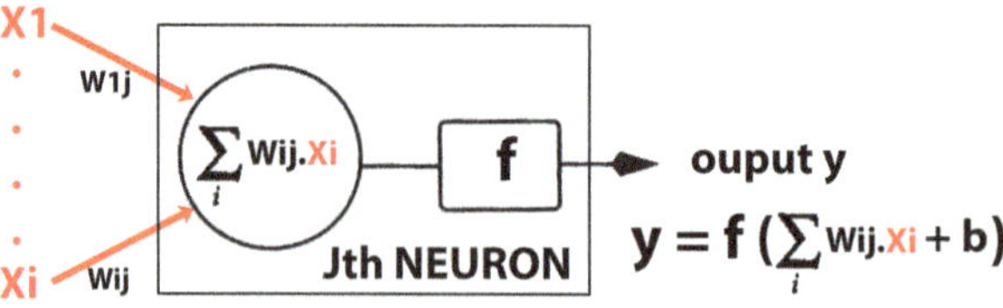

Fig. 2.5 Artificial neuron. The weighted sum of the n inputs i: $\sum w_{ij}.x_i$ is transformed by the nonlinear activation function into the jth neuron output: $y = f(\sum w_{ij}.x_i + b)$, which is sent to all next connected neuron. Note that b denotes a bias or an activation threshold

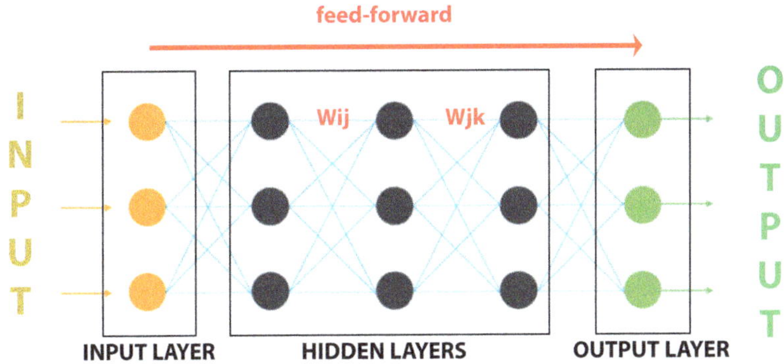

Fig. 2.6 Multilayer perceptron. It is a feed-forward artificial neural network composed of an input layer, several successive hidden layers, and an output layer. Each layer is composed of artificial neurons, the neurons of the hidden layers being connected with all the neurons of the previous and of the next layers

input layer. Stochastic gradient descent algorithm is used as the procedure of error minimization which determines how much to adjust iteratively the weights in order to lower the error. A bit more formally, a loss function measuring the error can be defined as: $\mathbf{E} = \frac{1}{2} \sum (\mathbf{o} - \mathbf{y})^2$. Then, minimizing E requires to find the steepest (ideally, the global) minima of the derivative of E with respect to the weights $\mathbf{w}$: **min** $[\partial\mathbf{E}/\partial\mathbf{w}]$ and to iteratively update these weights, using the gradient descent: $\mathbf{w}(\mathbf{t} + 1) = \mathbf{w}(\mathbf{t}) - \eta.\partial\mathbf{E}/\partial\mathbf{w}$, where η designates the learning rate. The main risk is to fall into a local minima. Stochastic gradient descent algorithms resort to a certain amount of noise injected in the descent gradient process in order to prevent this potential risk.

It is noteworthy that the first historical Rosenblatt's monolayered perceptron implemented a linear classifier, whereas later multilayered perceptrons functioned with nonlinearly separable data.

SVM can be regarded as a linear classifier partitioning a set of training examples $(\mathbf{x}, \mathbf{y})$, or vector, into several classes $\mathbf{y}$ well-delineated by a hyperplane [55, 56]. Examples $\mathbf{x}$ represent the most relevant features of the raw data retained by an instructor for the class attribution. If we restrict our example to two classes $\mathbf{y}$ denoted by -1 and $+1$, the equation of the boundary between these two classes, equivalent to a line, is given by: $\mathbf{f}(\mathbf{x}) = \mathbf{w}^{\mathbf{T}}\mathbf{x} + \mathbf{b} = \mathbf{0}$, where $\mathbf{w}$ is the discriminant vector normal to that line, and the function of decision assigning the data $\mathbf{x}$ to its corresponding class $\mathbf{y}$ is: $\mathbf{y} = \mathbf{sign}\,(\mathbf{w}^{\mathbf{T}}\mathbf{x} + \mathbf{b}) \in \{-1;\ +1\}$. Moreover, defining optimally the function: $\mathbf{y} = \mathbf{f}(\mathbf{x}) = \mathbf{w}^{\mathbf{T}}\mathbf{x} + \mathbf{b}$, which requires to minimize the classification error and the model complexity, supposes to accurately adjust the parameters $(\mathbf{w}^{\mathbf{T}}, \mathbf{b})$. This optimization step relies on the maximum-margin approach. The boundary separating the two classes is the line bisecting the margin demarcated by the closest points belonging to the opposite classes, called support vectors, characterized by: $\{\mathbf{x} \rceil\ \mathbf{f}(\mathbf{x}) = \mathbf{w}^{\mathbf{T}}\mathbf{x} + \mathbf{b} = -\mathbf{1}\}$ and $\{\mathbf{x} \rceil\ \mathbf{f}(\mathbf{x}) = \mathbf{w}^{\mathbf{T}}\mathbf{x} + \mathbf{b} = +\mathbf{1}\}$ (Fig. 2.7). Therefore, the optimizing procedure only takes into account the support

Fig. 2.7 Illustration of a support vector machine (SVM). SVM determines the decision boundary, the line B, which allows to partition the data into two distinct classes, using the support vectors

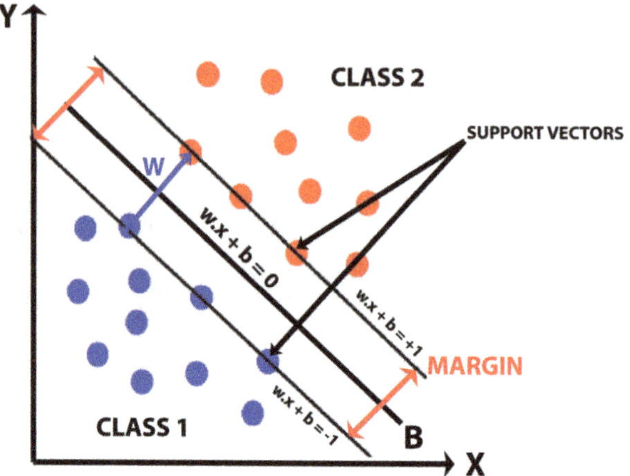

vectors and seeks maximization of the geometric margin, which is equivalent to minimizing $\|w\|^2$ under the following constraint: $\mathbf{y}.(\mathbf{w^T x + b}) \geq \mathbf{1}$. Finally, when data are not linearly separable, some mathematical manipulations, known as the "kernel trick," enable to convert easily a nonlinear boundary into a hyperplane by transforming nonlinearly the data space into a higher dimensional feature space, without explicitly calculating the nonlinear mapping. In conclusion, SVM can be used for classification of linearly and nonlinearly separable data. Since SVM computation is based on convex optimization, it is guaranteed to converge to global minimum. Thus, SVM appears particularly robust against overtraining when few samples of high dimensional data are processed. SVM was successfully tested for disease identification and cerebral state classification [57]. Davatzikos et al. [58] demonstrated the ability of a nonlinear SVM to distinguish cerebral activation patterns in relation with lying and truth-telling.

Unsupervised ML can be performed by artificial neural networks (ANN), such as stacked autoencoders (SAE), convolutional neural networks (CNN), deep belief networks (DBN), deep Boltzmann machine, or recurrent neural networks (RNN) [59]. As mentioned above, recent deep-learning algorithms build automatically from the raw data a set of internal and compositional representations of increasing abstraction. To achieve this goal, deep learning relies on multilayered ANNs encompassing an input layer, several hidden layers undergoing unsupervised training, a fully connected layer, and an output layer. The hidden layers carry out nonlinear transformations of their inputs and automatically extract relevant features in the outputs of their immediately previous layer. Upstream layers convey more and more complex and abstract characteristics of the empirical dataset. Endly, a supervised learning using backpropagating gradient is applied to the whole circuit to fine-tune the network parameters and to allow correct classification by comparing the computation of the last hidden layer with the desired output within a loss layer used during the training phase.

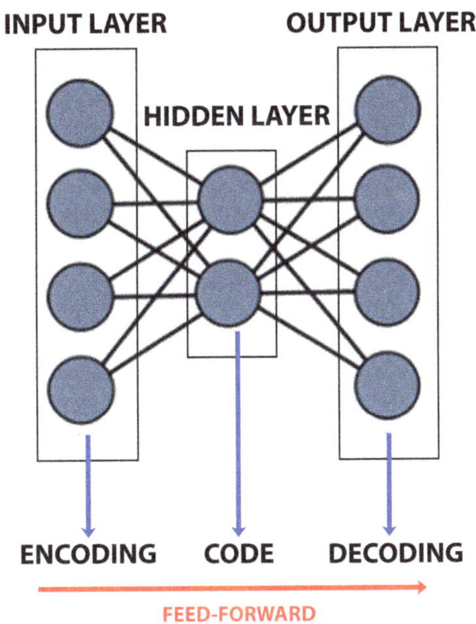

Fig. 2.8 Autoencoder (AE). A (shallow) AE is a three-layer feed-forward network encoding the input data in the hidden layer, of lower dimension, and decoding the corresponding latent representation in the output layer

A trainable autoencoder (AE) corresponds to a three-layer, feed-forward neural network mapping, deterministically or stochastically, an input to an output with a minimal amount of distortion. In other words, its hidden layer learns how to encode a compressed representation of the inputs transmitted by the (visible) input layer, by minimizing the quadratic error, and this latent representation, or code, allows the most reliable reconstruction of the inputs by the decoder layer (Fig. 2.8). Broadly speaking, AE accomplish (unperfect) copying tasks. Importantly, AE of lower dimension than the input (and output) layer permits the hidden layer to discriminate the most salient characteristics of the input data. AE can, therefore, be used for feature detection and dimension reduction, and sparse AE behave as classifiers. The computational ability of such ML can be augmented by hierarchically stacking AE so that the output of the previous AE serves as input of the next AE. In case of such SAE, the training procedure is based on a greedy-layer-wise unsupervised pretraining method that requires to train one layer at a time, the input of the next layer being the output of the previous one, using a backpropagating algorithm. A final supervised learning step can be added for improvement of the whole circuit performance.

CNN constitutes one of the most influential ML devoted to 2-D and 3-D image recognition, and more generally, to "grid-like topological" data. It forms a subclass of feed-forward multilayered neural network (perceptron). Its architecture includes two types of hidden layers. First, convolution layers act as simple feature detectors (first hidden layer) and more complex feature combination detectors (higher layers) applied to the output of the previous layer (Fig. 2.9). For instance, if the input is an image (voxels), three successive convolutional layers would detect, respectively:

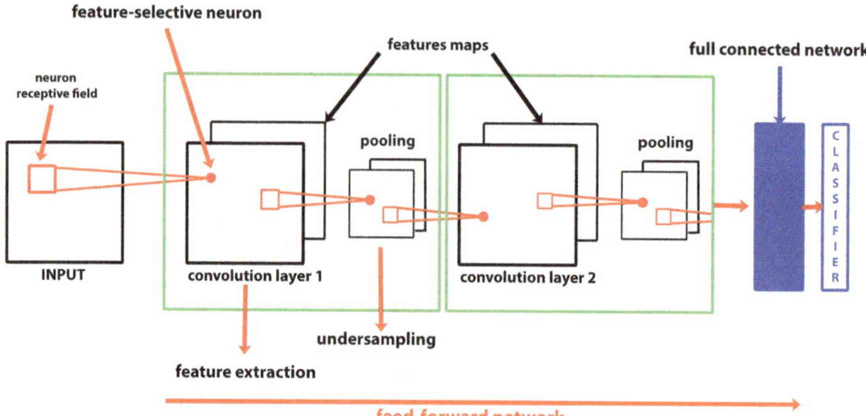

Fig. 2.9 Schematic representation of a convolutional neural network (CNN) implementing a representation learning algorithm. CNN is composed of the concatenation of several feed-forwardly connected blocks. Each block comprises a convolution layer and pooling layers, which perform feature extraction and undersampling (for dimensional reduction and more invariant representation), respectively. Each "deeper" block processes more and more compressed, complex and abstract conjunction of features. The last block of the chain is represented by a full-connected network and an output layer carrying out data classification. Note that a neuron in a convolution layer acts as a specific local feature detector within its receptive field. Mathematically speaking, these neurons perform first a convolution and afterward a nonlinear processing, such as linear rectifying, with respect to their inputs. All the neurons sharing the same feature (and the same weights) and whose receptive fields, all together, cover the whole previous input layer generate activation feature-specific maps. Therefore, each convolutional layer encompasses several feature maps. CNN are particularly well designed for image analysis and object and pattern recognition

(1) simple oriented edges; (2) apparent contours such as lines, curves, or corners; and (3) parts of object. Second, contrary to perceptron, a neuron of CNN is not connected to all the neurons of the previous layer. Each neuron of a convolutional layer analyzes a local spatial region of the image, so that a neuron is linked to a local subset of neurons, defining its receptor field. The neurons sensitive to the same feature share the same weight **w** and realize a spatial paving of the whole incoming data volume. They, thus, implement a feature-specific filtering (convolving) with a learnable kernel, sliding over the image data and generating feature-specific activation maps after nonlinear processing by an activation function, such as rectified linear unit. Second, pooling layers periodically interspaced and connected with convolution layers yield to subsampling, to "merge semantically similar features into one" [54] and to subsequent translational invariance. This processing contributes to reduce significantly the computational cost and to avoid overtraining. With such architecture, CNN is a powerful tool for object categorization in an image but needs a large sample of training data. In neuroimaging, Brosch et al. [60] successfully segmented lesions of multiple sclerosis using CNN. And other study pointed out the ability of CNN to distinguish resting-state fMRI between normal, mild

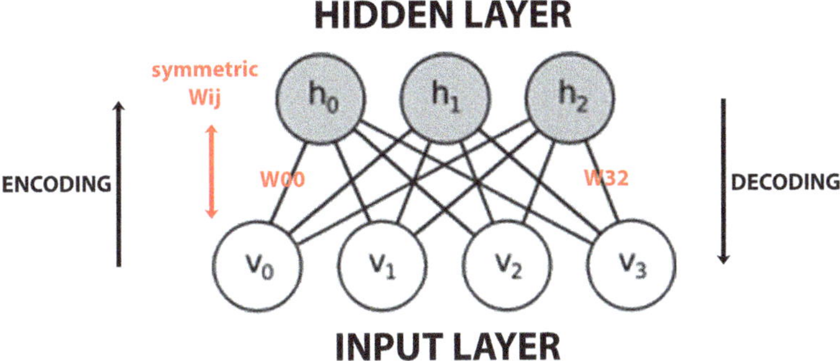

Fig. 2.10 Restricted Boltzmann machine (RBM). Shallow RBM is a two-layer neural network composed of stochastic neurons distributed in an input layer and a hidden layer and linked by weighted symmetric connections. However, there are no intralayer connections between neurons. A latent representation of the data is computed in the hidden layer, which finally is reconstructed, in a second time, in the input layer. RBM implements a probabilistic energy-based learning algorithm

cognitive impairment patients and Alzheimer disease patients [61], or normal and schizophrenia patients [62].

Other deep-learning algorithms are based on variant architectures such as DBN, a stochastic neural network composed of stacked restricted Boltzmann machines (RBM), using a contrastive divergence algorithm during the training, or such as RNN capable of processing sequential (temporal) data (language, speech, or writing). RBM is a two-layer network composed of a visible (input) layer and one hidden layer linked by symmetric connectivities, but neurons belonging to the same layer are not interconnected (Fig. 2.10). Thus, stochastic binary neurons of the hidden layer elaborate a latent probabilistic representation of the data, and, afterward, this later representation is fed backward in the visible layer to construct a generative model of the data. RBM behaves as a stochastic AE. The training algorithm aimed at reducing the difference between the probability distribution of the reconstructed and the original data vectors, by iteratively adjusting the connection weights. Recently, Plis et al. [63] demonstrated that (shallow) DBM and ICA equivalently identified resting-state intrinsic brain networks. Suk et al. [64] managed to discriminate between normal, mild cognitive impairment converter and nonconverter patients and Alzheimer patients, applying DBM processing to MRI, PET, and biochemical data. It could be also assumed that RNN could capture dynamical recruitment and synchronization of neural networks across time.

Finally, the choice of the ML depends on the specific characteristics of the data to be studied and on the results of the ML calculation: data categorization or statistical model. Therefore, after the training phase, it appears necessary to evaluate the performance of the algorithm in terms of accuracy, sensitivity, ability to properly generalize, reproducibility, robustness to noise, and adequate adjustment of hyperparameters (i.e., parameters of the network tuned by the instructor), for instance, and, among different tested algorithms, to select the algorithm generating the most realistic output. Cross-validation procedures can be exploited for performance estimation and algorithm selection [56].

ML in Neuroimaging

In a recent review of application of ML to structural and functional neuroimaging (MRI and PET data), Viera et al. [53] have concluded that ML provides a promising method to precisely classify some neurologic and psychiatric troubles, such as Alzheimer disease, attention-deficit/hyperactivity disorders, schizophrenia, spinocerebellar ataxia, and temporal epilepsy, as well as pain [65], and to predict disease state and disease evolution. They emphasized that deep learning seemed to exhibit better performance than SVM. However, further studies with larger cohorts are required, especially for deep learning. Finally, different ML algorithms can be combined to increase their performance and can be applied to multimodal data (images, biochemical data, clinical data, for instance). For fMRI and, especially, rsfMRI, ML proves to be efficient to localize and to identify activated networks, as well as to differentiate resting-state networks and subtle nonlinear relationships between specific brain areas. However, a clear methodology remains to be established for secure and efficient application of ML to fMRI [66].

Conclusion and Perspectives in Clinical Neuroimaging

In summary, clinical brain imaging benefits from advanced imaging and post-processing methods developed in fundamental neuroscience research. Voxel-based morphometry and measurement of cortical thickness enable to identify and to follow up cortical neuroplasticity and atrophy or disease progression, while tractography deals with white matter and, especially, anatomical tracts. FMRI and rsfMRI detect functional impairments or reconfiguration in specific task-recruited and canonical cerebral networks and can also track time-varying changes in functional brain connectivity using time-window analysis, such as windowed Fourier or wavelet transforms. Multiband acquisition allows for accelerating signal recording with echo time <1 s, improving time resolution of fMRI. Advanced mathematical post-processings complement these static or dynamical anatomo-functional data by objectivizing and quantifying disease-related modulatory, causal, and topological alterations. Machine learning and especially representation learning algorithms used in deep learning, applicable to structural MRI, fMRI, and rsfMRI, will also allow automatic and accurate identification of subtle normal or impaired cerebral patterns.

Moreover, utilization of ultrahigh field (>7 T) would not only increase significantly spatial resolution (μm) for morphological images but also would discriminate activation of cortical columns. This high-resolution fMRI would also permit to differentiate feed-forward thalamic activation of layer IV of cortical column from feed-back afferents impinging on supra- and infragranular layers [67]. This so-called laminar fMRI will open in vivo a new window, at the mesoscopic level, to computation of the functional unit of the neocortex and to the top-down and bottom-up influences upon this intracolumnar information processing.

References

1. Raichle ME (1998) Behind the scenes of functional brain imaging: a historical and physiological perspective. Proc Natl Acad Sci U S A 93:765–772
2. Hillman EMC (2014) Coupling mechanism and significance of the BOLD signal: a status report. Annu Rev Neurosci 37:161–181
3. Buxton RB (2013) The physics of functional magnetic resonance imaging (fMRI). Rep Prog Phys 76(9):096601
4. Drake CT, Iadecola C (2007) The role of neuronal signaling in controlling cerebral blood flow. Brain Lang 102:141–152
5. Tallini YN, Brekke JF, Shui B, Doran R, Hwang S-M, Nakai J, Salama G, Segal SS, Kotlikoff MI (2007) Propagated endothelial Ca^{2+} waves and arteriolar dilation in vivo. Circ Res 101:1300–1309
6. Marelli SP (2001) Mechanisms of endothelial $P2Y_1$- and $P2Y_2$-mediated vasodilation involve differential $[Ca^{2+}]i$ responses. Am J Physiol Heart Circ Physiol 28:H1759–H1766
7. Fergus A, Lee KS (1997) GABAergic regulation of cerebral microvascular tone in the rat. J Cereb Blood Flow Metab 17:992–1003
8. Li J, Iadecola C (1994) Nitric oxide and adenosine mediate vasodilatation during functional activation in cerebellar cortex. Neuropharmacology 33:1453–1461
9. Magistretti PJ, Pellerin L (1996) Cellular bases of brain metabolism and their relevance to functional brain imaging: evidence for a prominent role of astrocytes. Cereb Cortex 6:50–61
10. Magistretti PJ, Pellerin L (1999) Cellular bases of brain metabolism and their relevance to functional brain imaging: evidence for a prominent role of astrocytes. Philos Trans R Soc Lond Ser B Biol Sci 354:1155–1163
11. Ogawa S, Menon RS, Tank DW, Kim SG, Merkle H, Ellermann H, Ugurbil K (1993) Functional brain mapping by blood oxygenation level-dependent contrast magnetic resonance imaging. A comparison of signal characteristics with a biophysical model. Biophys J 64 (3):803–812
12. Buxton RB, Griffeth VE, Simon AB, Moradi F (2014) Variability of the coupling of the blood flow and oxygen metabolism responses in the brain: a problem for interpreting BOLD studies but potentially a new window on the underlying neural activity. Front Neurosci 8:139
13. Davis TL, Kwong KK, Weiskoff RM, Rosen BR (1998) Calibrated functional MRI: mapping the dynamics of oxidative metabolism. Proc Natl Acad Sci U S A 95(4):1834–1839
14. Shih Y-YI, Chen C-CV, Lin Z-J, Chiang Y-C, Jaw F-S, Chen Y-Y, Chang C (2009) A new scenario for negative functional magnetic resonance imaging signals: endogenous neurotransmission. J Neurosci 29(10):3036–3044
15. Logothetis NK, Pauls J, Augath M, Trinath T, Oeltermann A (2001) Neurophysiological investigation of the basis of the fMRI signal. Nature 412:150–157
16. Wang J, Aguirre GK, Kimberg DY, Roc AC, Li L, Detre JA (2003) Arterial spin labeling perfusion fMRI with very low task frequency. Magn Reson Med 49(5):796–802
17. Chen JJ, Jann K, Wang DJJ (2015) Characterizing resting-state brain function using arterial spin labeling. Brain Connect 5(9):527–542
18. Buxton RB (2016) Beyond BOLD correlations: a more quantitative approach for investigating brain networks. J Cereb Blood Flow Metab 36(3):461–462
19. Dai W, varma G, Scheidegger R, Alsop DC (2016) Quantifying fluctuations of resting state networks using arterial spin labeling perfusion MRI. J Cereb Blood Flow Metab 36(3):463–473

20. Biswal B, Yetkin FZ, Haughton VM, Hyde JS (1995) Functional connectivity in the motor cortex of resting human brain using echo-planar MRI. Magn Reson Med 34(4):537–541
21. Gusnard DA, Raichle MES (2011) Rearching for a baseline: functional imaging and the resting human brain. Nat Rev Neurosci 2:685–694
22. Raichle EM (2015) The restless brain: how intrinsic activity organizes brain function. Philos Trans B 370:1–11
23. Fox MD, Raichle M (2007) Spontaneous fluctuations in brain activity observed with functional magnetic resonance imaging. Nat Rev Neurosci 8:700–711
24. Mantini D, Perrucci MG, Del Gratta C, Romani GL, Corbetta M (2007) Electrophysiological signatures of resting state networks in the human brain. Proc Natl Acad Sci U S A 104:13170–13175
25. Brookes MJ, Woolrich M, Luckhoo H, Price D, Hale JR, Stephenson MC, Barnes GR, Smith SM, Moris PG (2011) Investigating the electrophysiological basis of resting state networks using magnetoencephalography. Proc Natl Acad Sci U S A 108(40):16783–16788
26. Shmuel A, Leopold DA (2008) Neuronal correlates of spontaneous fluctuations in fMRI signals in monkey visual cortex: implications for functional connectivity at rest. Hum Brain Mapp 29 (7):751–761
27. Leopold DA, Maier A (2012) Ongoing physiological processes in the cerebral cortex. NeuroImage 62:2190–2200
28. Greco G, Corbetta M (2011) The dynamical balance of the brain at rest. Neuroscientist 17:107–123
29. Margulies DS, Böttger J, Long X, Lv Y, Kelly C, Schäfer A, Goldhahn D, Abbushi A, Milham MP, Lohmann G, Villringer A (2010) Resting developments: a review of fMRI post-processing methodologies for spontaneous brain activity. MAGMA 23(5–6):289–307
30. Beckmann CF (2012) Modelling with independent components. NeuroImage 62:891–901
31. Calhoun VD, Liu J, Adali T (2009) A review of group ICA for fMRI data and ICA for joint inference of imaging, genetic and ERP data. NeuroImage 45:S163–S172
32. Zou Q-H, Zhu C-Z, Yang Y, Zuo X-N, Long X-Y, Cao Q-J, Wang Y-F, Zang Y-F (2008) An improved approach to detection of amplitudes of low-frequency fluctuation (ALFF) for resting-state fMRI: fractional ALFF. J Neurosci Methods 172(1):137–141
33. Zang Y, Jiang T, Lu Y, He Y, Tian L (2003) Regional homogeneity approach to fMRI data analysis. NeuroImage 22(1):394–400
34. Kelly RE, Alexopoulos GS, Wang Z, Gunning FM, Murphy CF, Morimoto SS et al (2010) Visual inspection of independent components: defining a procedure for artifact removal from fMRI data. J Neurosci Methods 189:233–245
35. Kalcher K, Huf W, Boubela RN, Filzmoser P, Pezawas L, Biswal B et al (2012) Fully exploratory network independent component analysis of the 1000 functional connectomes database. Front Hum Neurosci 6:1–11
36. Habas C, Kamdar N, Nguyen D, prater K, Beckmann CF, Menon V, Greicius MD (2009) Distinct cerebellar contribution to intrinsic connectivity networks. J Neurosci 29 (26):8586–8594
37. Chen AC, Oathes DJ, Chang C, Bradley T, Zhou Z-W, Williams LM et al (2013) Causal interactions between fronto-parietal central executive and default-mode networks in humans. Proc Natl Acad Sci U S A 110(49):19944–19949
38. Fox MF, Corbetta M, Snyder AZ, Vincent J, Raichle M (2006) Spontaneous neuronal activity distinguishes human dorsal and ventral attention systems. Proc Natl Acad Sci U S A 103:10046–10051
39. Hoff GEA-J, Van de Heuvel MP, Benders MJNL, Kersbergen KJ, de Vries LSD (2013) On the development of functional brain connectivity in the young brain. Front Hum Neurosci 7:650
40. Antonenko D, Flöel A (2014) Healthy aging by staying selectivity connected: a mini-review. Gerontology 60:3–9
41. Kelly C, Castallanos FX (2014) Strengthening connections: functional connectivity and brain plasticity. Neuropsychol Rev 24:63–76

42. Allen EA, Damaraju E, Plis SM, Erhardt EB, Eichele T, Calhoun VD (2014) Tracking whole-brain connectivity dynamics in the resting state. Cereb Cortex 24(3):663–676
43. Fox MD, Greicius M (2010) Clinical applications of resting state functional connectivity. Front Syst Neurosci 4:1–13
44. Rosazza C, Minati L (2011) Resting-state brain networks: literature review and clinical applications. Neurol Sci 32:773–785
45. Friston KJ (2011) Functional and effective connectivity: a review. Brain Connect 1(1):13–36
46. Penny WD, Stephan KE, Mechelli A, Friston KJ (2004) Modelling functional integration: a comparison of structural equation and dynamic causal models. NeuroImage 23(Suppl 1):S264–S274
47. Friston KJ, Harrison L, Penny W (2003) Dynamic causal modelling. NeuroImage 19(4):1273–1302
48. Ed B, Sporns O (2009) Complex brain networks: graph theoretical analysis of structural and functional systems. Nat Rev Neurosci 10:186–198
49. Rubinov M, Sporns O (2010) Complex network measures of brain connectivity: uses and interpretations. NeuroImage 52:1059–1069
50. Guye M, Bettus G, Bartolomei F, Cozzone PJ (2010) Graph theoretical analysis of structural and functional connectivity MRI in normal and pathological brain networks. MAGMA 23:409–421
51. Redcay E, Moran JM, Mavros PL, Tager-Flusberg H, Gabrieli JDE, Whitfield-Gabrieli S (2013) Intrinsic functional network organization in high-functioning adolescents with autism spectrum disorder. Front Hum Neurosci 7(573):1–11
52. Wang S, Summers RM (2012) Machine learning and radiology. Med Image Anal 16(5):933–951
53. Vieira S, Pinaya WHL, Mechelli A (2017) Using deep learning to investigate the neuroimaging correlates of psychiatric and neurological disorders: methods and applications. Neurosci Behavl Rev 74:58–75
54. LeCun Y, Bengio Y, Hinton G (2015) Deep learning. Nature 521:436–444
55. Wernick MN, Yang YY, Braaankov JG, Yourganov G, Strother SC (2010) Machine learning in medical imaging. IEEE Signal Process Mag 27(4):25–38
56. Lemm S, Blankertz B, Dickhaus T, Müller K-R (2011) Introduction to machine learning for brain imaging. NeuroImage 56:387–399
57. Cox DD, Savoy RL (2003) Functional magnetic resonance imaging (fMRI) "brain reading": detecting and classifying distributed patterns of fMRI activity in human visual cortex. NeuroImage 19(2):261–270
58. Davatzikos C, Ruparel K, Fan Y, Shen D, Acharyya M (2005) Classifying spatial patterns of brain activity with machine learning methods: application to lie detection. NeuroImage 28(8):663–668
59. Shen D, Wu G, Suk H-I (2017) Deep learning in medical images analysis. Annu Rev Biomed Eng 19:221–248
60. Brosch T, Tam R, Alzheimer's Disease Neuroimaging Initiative (2013) Manifold learning of brain MRIs by deep learning. In: International conference on medical image computing and computer-assisted intervention, Springer Berlin Heidelberg, pp 633–640
61. Suk H-I, Lee S-W, Shen D (2014) Hierarchical feature representation and multimodal fusion with deep learning for AD/MCI diagnosis. NeuroImage 101:569–582
62. Kim J, Calhoun VD, Shim E, Lee J-H (2016) Deep neural network with weight sparsity control and pre-training extracts hierarchical features and enhances classification performance: evidence from whole-brain resting-state functional connectivity patterns of schizophrenia. NeuroImage 124(Pt A):127–146
63. Plis SM, Hjelm DR, Salakhutdinov R, Allen EA, Bockholt H, Long JD, Johnson HJ, Paulsen JS, Turner JA, Calhoun VD (2014) Deep learning for neuroimaging: a validation study. Front Neurosci 8:229

64. Suk H-I, Lee S-H, Shen D (2015) Latent feature representation with stacked auto-encoder for AD/MCI diagnosis. Brain Struct Funct 220(2):841–859
65. Rosa MJ, Seymour B (2014) Decoding the matrix: benefits and limitations of applying machine learning algorithms to pain neuroimaging. Pain 155:864–867
66. Pereira F, Mitchell T, Botvinick M (2009) Machine learning classifiers and fMRI: a tutorial overview. NeuroImage 15(1 suppl):S199–S209
67. Self MW, van Kerkoerle T, Goebel R, Roelfsema PR (2017) Benchmarking laminar fMRI: neuronal spiking and synaptic activity during top-down and bottom-up processing in the different layers of cortex. Neuroimage. 2017 Jun 23. pii: S1053-8119(17)30517-7. doi: 10.1016/j.neuroimage.2017.06.045. [Epub ahead of print]

Chapter 3
Clinical Utility of Resting State Functional MRI

Mary Pat McAndrews and Alexander Barnett

Translational fMRI: Aims, Challenges, and Opportunities

The cognitive neuroscience of memory and language has been enhanced considerably by fMRI as a technique to probe and map these functions non-invasively. Such studies afford key insights into the neural architecture of these processes, with more recent studies focusing on networks of coherent activity that underlie cognitive operations. However, the translation of these insights into clinical applications has lagged behind, particularly when one considers evidence that goes beyond describing a difference between a particular patient group and the healthy brain controls. The clinical literature is typically hampered by small samples of heterogeneous patient groups, with little replication of findings across centres given adherence to different types of activation paradigms and variation in the clinical outcomes assessed. The emergence of a robust set of data using rsfMRI, where there are no task constraints, has begun to open new avenues for exploring more direct relationships between cognition and alterations in neural networks.

Translational work aims to understand which signal characteristics are meaningful in explaining variance in disease phenotype and severity, likely progression, treatment response, or reorganization capacity. In individuals with mTLE who experience recurrent seizures arising from one MTL, the concern is whether removal

M. P. McAndrews (✉)
Krembil Research Institute, University Health Network, Toronto, ON, Canada

Deparment of Psychology, University of Toronto, Toronto, ON, Canada

Neuropsychology Clinic, Toronto Western Hospital, Toronto, ON, Canada
e-mail: MaryPat.McAndrews@uhn.ca

A. Barnett
Krembil Research Institute, University Health Network, Toronto, ON, Canada

Deparment of Psychology, University of Toronto, Toronto, ON, Canada

© Springer International Publishing AG, part of Springer Nature 2018
C. Habas (ed.), *The Neuroimaging of Brain Diseases*, Contemporary Clinical Neuroscience, https://doi.org/10.1007/978-3-319-78926-2_3

of anterior temporal lobe structures, in attempting to eliminate seizures, will come at a significant cost in the form of postsurgical decline. In the domain of language, this is typically seen when the anterior temporal neocortex of the dominant (typically left) hemisphere is excised [1–3]. For memory, the functional adequacy of the epileptogenic MTL is key, in that better preoperative capacity is associated with greater decline [4, 5], typically with verbal memory associated with the left MTL and visuospatial memory with the right MTL [6–8]. In principle, fMRI is uniquely suited to investigate the functional anatomy of memory and language processes by providing a direct assay of adequacy of the epileptogenic tissue and the risk of potential post-operative morbidity, as well as the opportunity to characterize possible reorganization or compensation [9]. Another setting where fMRI can be an effective probe for functional adequacy of MTL is amnestic mild cognitive impairment (aMCI), a prodromal stage of Alzheimer's disease (AD) where memory is selectively impaired, in which the entorhinal cortex and hippocampus reveal earliest signs of structural pathology [10–13]. Indeed, the MTL has been the focus of much of the foundational work using fMRI as a biomarker of diagnosis and disease progression, as well as to predict the potential impact of treatments including pharmacotherapy and cognitive training [14, 15].

Much of the early translational application in these clinical contexts was driven by questions of task-related patterns of activation, with a particular focus on regions of interest. Relevant findings include reports showing less MTL activity during an encoding task for patients with mTLE [16–20] or MCI/AD [21, 22] compared to healthy controls. Similarly, the strength (magnitude, spatial extent) of activation in the left temporal lobe has been shown to predict the degree of decline in word retrieval following left temporal lobectomy in mTLE [23–25]. While this approach has been quite successful, there are also important limitations when one considers the long scanning times that may be required to produce reliable data, the concern about capacity for full task engagement in the patient population, the lack of consensus regarding sensitivity and specificity of particular metrics of activation, and the possibility of alternate or more diffuse patterns of organization within the disease group that might undermine the focus on one particular region of interest. Furthermore, there are examples in the literature showing that greater signal does not always signify better function, such as a study demonstrating strong bilateral hippocampal activation in a patient during an episode of transient global amnesia [26] and evidence that enhanced activity of the hippocampus may be a pathologic signal in aMCI [27, 28]. Moreover, we found that despite controls and aMCI patients showing similar task-related activation in the parahippocampal gyrus, activity in quite distinct networks was associated with good memory performance in the two groups [29]. While much effort has been aimed at identifying ideal task activation paradigms for clinical use [30–32], there is a growing scientific interest in the newly characterized 'resting state' networks that can be discerned when individuals are scanned doing nothing other than 'mind-wandering' in the scanner. We will briefly discuss the foundational work with healthy individuals before turning to the disease states.

Resting State Networks

The first described resting state network was termed the default mode (DMN) as it characterized a set of midline and lateral cortical regions that were co-activated during the 'rest' condition of a vast number of cognitive paradigms [33]. Subsequent work has shown that these regions show robust functional connectivity during introspective thought or 'mind-wandering' [34, 35]. Furthermore, the DMN can be fractionated into components that are collectively engaged during particular cognitive operations such as constructing a mental scene from memory and engaging in self-relevant decisions [36]. A number of other networks have been identified, characterized by intrinsic connectivity that can be captured from 6 to 10 min of scanning while participants are at rest, i.e. not engaged in any cognitive task beyond (typically) focusing on a fixation cross. The networks can be identified using seed-based techniques, which extract temporal correlations between a particular seed region and all other brain voxels, or using independent component analysis (ICA), which decomposes voxel-wise temporal correlations into spatially orthogonal components. The latter is the more common data-driven technique that has been used to identify a large number of robust cortical networks in the healthy brain, including those associated with attention (both dorsal and ventral), language, cognitive control, and more basic visual and sensorimotor functions [37–40]. By exploiting the massive high-quality rsfMRI data from the Human Connectome Project, it has been possible to further understand that large-scale networks can be decomposed into partially segregated subnetworks and how some regions in association cortex can participate in multiple networks to different degrees; such complexity undergirds the flexible behaviour that arises from those network interactions [41, 42].

rsfMRI and Memory in mTLE

There is a moderate record of success in the clinical application of paradigms eliciting focal activation in the MTL as applied to mTLE. These include verbal or visual memory paradigms in which the degree of activation in the epileptogenic MTL correlated with current memory performance or predicted the magnitude of post-operative memory decline [17–19, 43]. Several studies have shown that asymmetry of MTL activation during encoding is more strongly correlated with current memory function [20] or with decline in memory following surgery [44] than standard clinical parameters such as hippocampal volume. However, the success at predicting decline following right anterior temporal lobectomy (ATL) has been less successful than for left ATL, and at least one study demonstrated that activation associated with language lateralization was superior to memory-specific activation in predicting verbal memory decline following left ATL [31]. This may reflect relatively poor sensitivity or specificity to memory outcomes of the paradigms, the metrics, or both.

Resting state fMRI may be particularly well-suited to address questions of functional adequacy of the MTL because the hippocampus and parahippocampal gyrus are components of the DMN, which shows considerable overlap with the constellation of brain regions commonly engaged during episodic recollection [45, 46]. Furthermore, such functional connectivity (rsFC) measures may be superior to task-based measures as they eliminate variance or noise associated with different strategies and/or alternate networks being utilized during the activation task by patients in whom some degree of functional reorganization may have already taken place [47]. There have been a number of reports of abnormal connectivity between the MTL and other DMN regions in patients with mTLE [48–51], but only a few studies investigating the consequence of that disrupted connectivity to functional integrity as indexed by memory performance. In our initial study of this issue [52], we focused on two regions of interest in the DMN, the hippocampus and posterior cingulate cortex (PCC), which is considered one of the main hubs of the DMN and is commonly activated during recall and recognition. Relative to controls, we found reduced connectivity to the epileptogenic hippocampus and increased connectivity to the contralateral one in both left and right mTLE patients. Furthermore, stronger connectivity on the epileptogenic side was associated with better pre-surgical memory and greater postsurgical memory decline, whereas greater connectivity on the contralateral side was protective. We also found an increase in contralateral connectivity after removal of the epileptogenic side and that increase was also correlated with memory preservation suggesting compensatory plasticity. A subsequent study examined functional connectivity throughout 20 nodes of the DMN in relation to performance on clinical memory tests [53]. Although more complex, the pattern of connections was even more strongly correlated with memory measures than the simple two-node solution. Here, better memory was associated with increased posterior and interhemispheric connectivity (i.e. between MTL structures and medial and lateral parietal cortices), whereas poorer memory was associated with a pattern that emphasized stronger long-range posterior-to-anterior intrahemispheric connections. Of interest, a very similar pattern of stronger long-range connectivity was seen in a different cohort of left mTLE patients when they were actively attempting to retrieve personal autobiographical memories [54]. As this alternate pattern of connectivity was seen under different circumstances (resting state and directed memory retrieval) and with distinct analytic techniques (multivariate correlations and structural equation modelling), it may signify a robust change in memory networks in cases of insult to the MTL (Fig. 3.1).

Another group has investigated networks from both task and resting state fMRI in relation to memory processes. Their first study, using ICA on data collected during memory encoding, documented reduced functional connectivity that was associated with poorer memory performance in a network that included bilateral MTL and extended to bilateral occipital regions as well as left orbitofrontal cortex [50]. Of interest, focal activation in the orbitofrontal region, and not the ipsilateral hippocampus, was strongly correlated with task performance. In a more recent study using seed-based rsFC to investigate different parcellations of thalamo-cortical networks,

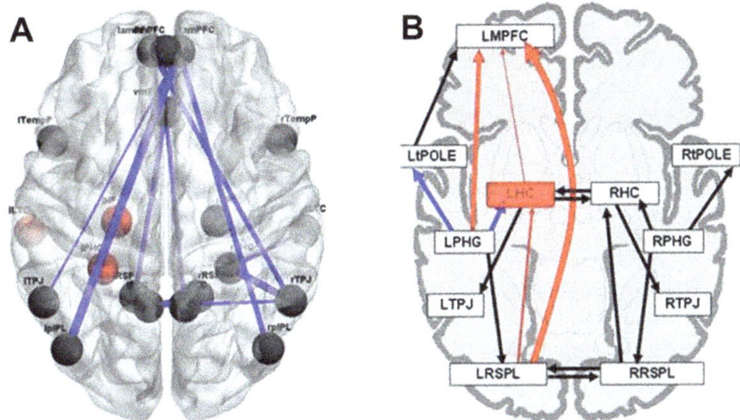

Fig. 3.1 Increased long-range intrahemispheric connectivity as a consequence of unilateral hippocampal damage in mTLE. (**a**): Blue lines indicate increased (relative to controls) long-range connectivity in rsfMRI from posterior medial and lateral parietal cortices to medial prefrontal regions in patients with left mTLE (Data from McCormick et al. [53]). (**b**): Altered effective connectivity (relative to controls) during autobiographical memory retrieval (Data from Addis et al. [54]). Major differences are increased connectivity (red lines) from retrosplenial to medial prefrontal cortex on the left and reduced connectivity (blue lines) involving left medial temporal regions. Red nodes in both figures indicate regions of reduced structural integrity of medial temporal cortex, and the width of lines indicates connectivity strength. (This figure is reproduced with permission from [115])

they reported short-term memory was related to thalamic connectivity with specific regions in frontal and parietal cortices in the contralateral hemisphere, whereas long-term memory was associated with the strength of rsFC between ipsilateral thalamus and entorhinal cortex [55]. A similar reduction in connectivity between ipsilateral temporal neocortex and the DMN was reported for left TLE based on ICA-derived networks in another study that did not directly examine memory performance [56]. Thus, connectivity amongst many cortical regions may need to be considered, as epileptic seizures may promote reorganization outside of canonical memory networks.

It is also important to know whether these functional connectivity measures go beyond other indicators of structural integrity or functional activation in explaining variation in memory performance; this speaks directly to their potential added value in clinical contexts. Of note, studies show wide variation in the correspondence between functional activation and structural measures of hippocampal integrity, ranging from no correlation between measures [43] to a moderately strong relationship [57]. In their study assessing thalamo-cortical connectivity and memory, Voets and colleagues noted that posterior hippocampal volume did add unique information to thalamic-frontal rsFC in explaining short-term memory performance, but this was not the case for the relationship between long-term memory and thalamic-medial temporal rsFC [55]. In both McCormick studies from our lab described above, performance on clinically relevant memory tests was more strongly predicted by rsFC than by structural damage to the MTL or other DMN nodes (assessed by hippocampal volume and grey matter density). Indeed, we have consistently found the influence of structural alterations on episodic memory capacity to be mediated by functional network capacity [20, 53]. These findings accord well with the concepts underlying the human brain connectome that structure undergirds but does not fully determine the brain's dynamic and flexible functional repertoire [58, 59]. Unfortunately there is no published work that directly compares the relative clinical utility of different fMRI measures such as focal activation, task-related connectivity, and resting state connectivity, but we can offer some unpublished observations from our own work. In the same group of ten patients with left mTLE, we compared focal left hippocampal activation during an autobiographical memory task that strongly engages the hippocampus and PCC [54, 60–62], rsFC between left hippocampus and PCC, and connectivity between those nodes during autobiographical recall [47]. When each measure was correlated with a standard clinical measure of verbal memory, we found the best correlate of behaviour to be rsFC ($r = 0.78$), followed by task-related FC ($r = 0.55$), followed by hippocampal activation ($r = 0.31$). We contend that similar comparisons in other clinical settings will be very important in advancing the predictive utility of fMRI.

Although these early results are promising, there are important complexities that must be addressed. For example, a few studies looking at connectivity and current memory capacity seeded from the hippocampus (rather than PCC as we had done) to explore widespread connectivity throughout the brain. One reported a pattern similar to ours, in that higher contralateral HC-PCC connectivity was associated with better episodic memory, and also found that connectivity between the ipsilateral or contralateral hippocampus and other MTL structures (entorhinal cortex and parahippocampal gyrus) was negatively associated with memory performance [63]. Another study reported that connectivity between the ipsilateral left hippocampus and left PCC/precuneus and inferior parietal lobule (IPL) was negatively associated with memory, whereas better performance was related to stronger connectivity between the epileptogenic hippocampus and contralateral PCC/precuneus and IPL [64]. Finally, in correspondence with Holmes but in contrast to McCormick papers, higher ipsilateral hippocampus to PCC in patients with left TLE was related to poorer verbal memory, whereas contralateral hippocampus connectivity with the medial frontal cortex was associated with better nonverbal memory in patients with right TLE in another study [65]. These findings indicate that not all alterations in connectivity are functionally efficient or compensatory, a lesson we have already learned from reports of increased focal activation, and also support the need for larger studies and especially those with prediction of change as the primary outcome.

rsfMRI and Memory in Other Patients with MTL Damage

Alterations in task-based activation in the medial temporal lobe have been reported in aMCI and AD, although the patterns are not invariably of a decline; for meta-analyses based on activation co-ordinates, see [66, 67]. Similar to the arguments for mTLE, however, assays of functional integrity in aMCI, AD, or other amnestic cases of focal MTL pathology that are based on extent or strength of activation in the MTL can yield incorrect inferences. For example, current data indicate that early aMCI is characterized by hyperactivation in the hippocampus, which appears to be pathological rather than compensatory [28, 68]. That is, there is an inflection point in the relationship between hippocampal activation and memory integrity as one proceeds from normal ageing, through very mild and more severe aMCI to AD [69]. Instead of hippocampal engagement, a more reliable indicator of encoding success and overall memory status in AD was reported to be the degree to which the medial parietal region (PCC, precuneus) was deactivated during memory encoding [70, 71]. These are two key nodes of the DMN, and there is compelling evidence of a global decrease in DMN integrity in aMCI and AD compared to age-matched controls; for reviews, see [72, 73].

Furthermore, a growing set of studies relate these alterations in rsFC to clinical memory impairment. Amnesic patients with bilateral MTL damage show decreased connectivity between PCC and MTL components of the DMN [74]. The degree of rsFC between these regions is correlated with memory performance and separates patients with amnestic MCI from non-amnestic MCI [75]. Other rsFC studies have described more widespread disruptions in connectivity between the hippocampus or PCC and other neocortical and subcortical regions that are correlated with the degree of memory impairments in MCI [76, 77]. As MCI and dementia will be discussed in more depth in another chapter in this volume, the foregoing is intended only to indicate that studies regarding fMRI biomarkers of memory integrity are being pursued in this disorder and that work from mTLE and aMCI may provide fruitful opportunities for cross-fertilization.

Hippocampal Parcellation and Connectivity

There is growing interest in examining connectivity patterns involving anterior and posterior segments of the hippocampus separately. Recent research has demonstrated that these segments have distinct multisynaptic patterns of connectivity with neocortical regions, which likely has considerable import for the type of memory processes they support [78–81]. Several different characterizations have been offered that distinguish these anterior and posterior networks. One focusses on the granularity of mnemonic contents with the more fine-grained representations involving posterior regions, enabling retrieval characterized by recollection of details rather than being limited to coarser 'gist' information [80]. Another model hypothesizes that encoding new information is the province of the anterior hippocampus-dorsal attention network connectivity, whereas retrieval of any sort depends more so on posterior hippocampus-DMN connectivity [82].

Although the histopathology of TLE is distributed along the longitudinal axis of the hippocampus [83], there is evidence that greater structural abnormalities on MRI are seen in the anterior half and there is better chance of seizure freedom in that setting [84–86]. Thus, it may be important to isolate contributions from the two segments in assessing functional capacity in this population. One recent study found similar general patterns of heightened connectivity amongst MTL regions for anterior and posterior segments, whereas only the latter region showed reduced connectivity to another cortical target (PCC) that was associated with defective memory in patients with left mTLE [63]. Of interest, our own work with healthy controls has also demonstrated different connectivity patterns, with the posterior hippocampus-to-neocortex networks showing a greater involvement with the type of relational memory processes that are particularly impaired in mTLE patients [87, 88]. There has not yet been much work on this anterior-posterior segmentation concept in the MCI/AD populations. In one study with no behavioural correlates, it was reported that reductions in spatial extent of these networks were found in aMCI patients compared to controls, with the most striking appearing in the left posterior hippocampus and right anterior MTL [89]. Clearly, this is an important avenue for future investigations in these populations, marrying rsfMRI to well-designed cognitive assays of clinically meaningful memory processes.

rsFC and Language in TLE

Resting state fMRI is also emerging as a tool for mapping the language network, composed of the inferior frontal gyrus, superior temporal gyrus, supramarginal gyrus, inferior parietal lobule, and premotor cortex. While these regions have been interrogated extensively using task-based fMRI [90], researchers have also found reliable rsFC amongst these regions [91, 92]. Here, we review the literature on rsfMRI and language network activity, discussing its findings and applications in healthy and disease populations. We will also discuss the similarities and differences between pre-surgical rsFC network characterization and traditional methods of pre-surgical language mapping, noting the potential advantages of each.

Language networks have been extracted from rsfMRI data using both seed-based [92] and ICA-based [91] approaches. Tomasi and Volkow [92] examined 970 healthy individuals collected from 22 research sites around the world. The authors used a standard seed-to-voxel approach with two regions of interest – Broca's and Wernicke's areas. Both seeds demonstrated robust connectivity to language network regions including inferior frontal gyrus (IFG, comprised of pars orbitalis, triangularis, and opercularis), superior temporal gyrus, inferior parietal cortex, middle frontal gyrus, inferior temporal cortex, superior frontal cortex, caudate, putamen, and cerebellum. This network connectivity was reliable across all sites and both seeds, although Broca's area was significantly more connected to anterior language regions, while Wernicke's area was significantly more connected to posterior regions. Furthermore, they found leftward lateralization of Broca's area and posterior Wernicke's area (angular gyrus) connectivity which supports lateralization of language to the left hemisphere as is expected in the general population. A subsequent study using the same procedures found these networks to be highly reliable within subjects across time intervals ranging from 45 min to 16 months [93]. This level of consistency in the network provides reassurance that low-frequency fluctuations at rest are related to intrinsic, stable network properties. In addition, there is evidence that connectivity in this network is associated with language performance. Reduced resting state connectivity was observed between posterior temporal regions related to reading (fusiform gyrus, inferior temporal gyrus, middle temporal gyrus, and superior temporal gyrus) and the left IFG in dyslexic readers compared to nonimpaired readers [94]. Furthermore, increased functional connectivity between the fusiform gyrus and left IFG was associated with better reading fluency.

The mounting evidence that rsfMRI can identify reliable networks and clinically relevant language functional capacity provides confidence that it may provide important information for pre-surgical language mapping for individuals with TLE. Prior to epilepsy surgery, it is imperative to establish hemispheric dominance and, occasionally, fine-grain localization of eloquent cortex. Traditionally this was established with invasive tests such as intracarotid injection of an anaesthetic agent such as sodium amobarbital (Wada procedure, [95]) or electrical stimulation mapping (ESM, [96]). More recently, fMRI using activation tasks has become a standard of practice in many epilepsy surgical centres, as it has been shown to have high concordance with Wada results [97–101]. Concordance with ESM has not been found as consistently [102], although using multiple language tasks does improve agreement [103]. Results from task-based fMRI are reliant on several factors such as the selection a threshold for activation [103, 104], calculation of laterality indices [105, 106], the type of task demands and control conditions [107–110], and the ability of the participant to perform the task, features that are not constraints for rsfMRI.

As with memory, there are only a few instances of this newer technique being applied in the pre-surgical mapping context. In one, language networks were extracted from resting state scans using a machine learning algorithm that was trained to classify voxels as being a part of one of seven canonical resting networks [111]. The patients also underwent ESM for language, and the electrode placement was co-registered to anatomical space from the MRI scans. They found strong agreement between ESM and rsFC language network classification. Similarly, moderate to strong concordance has also been found between resting state language network patterns and task-based fMRI activation patterns [112, 113]. Furthermore, greater left lateralization, determined by task activation, is associated with stronger connectivity between left IFG and other neocortical regions in the left hemisphere in controls and TLE patients [114].

While these findings are encouraging, more work is needed to validate that the information obtained from rsfMRI is informative for surgery if it is to replace invasive procedures or task-based fMRI. A large-scale comparison of rsfMRI laterality and Wada or other fMRI metrics has yet to be performed, and importantly, few studies have related rsFC to pre-surgical language performance or postsurgical language changes. We offer relevant observations from our own clinical series in which TLE patients had a resting state scan as well as a language activation task (involving naming to description compared to fixation control). Correlations with performance on a standard clinical measure (Boston Naming Test) were strongest for task-based functional connectivity between left inferior frontal and left middle temporal gyri ($r = 0.50$), weaker for rsFC between those regions ($r = 0.25$) and weakest for the task activation laterality index ($r = 0.16$); note the latter two values are previously unpublished data. However, using newer analytic techniques with larger language networks, we have begun to identify reasonably strong relationships between network integrity, as a function of similarity to controls, and post-operative changes in naming performance [115]. As with memory, there are compelling reasons to develop a strong clinical grounding for rsFC in language processes. Any patient may be incapable of performing our simple language tasks adequately or in the manner we designate, and thus our task-based patterns of activation may include regions that are engaged for the fMRI task but not truly indicative of language capacity.

Newer Metrics: Graph Theory and Dynamic Functional Connectivity

The number of interregional correlations that can be generated in a typical whole-brain rsfMRI study can yield a bewildering array of findings (e.g. 96 regions of interest will generate 4560 correlations). Graph theory is a mathematical technique that enables one to characterize complex patterns of pairwise correlations between objects (which could be brain regions, airport locations, genes, or people in a social network). Graphs are made up of vertices (nodes) which are connected by edges (connections), and vertices can interact through direct connections or indirectly across paths that are composed of multiple edges. There are a variety of metrics that have been applied to the study of brain organization, both at the level of structural and functional connectivity, aimed at characterizing typical features as well as those associated with development and pathology. A full accounting of the technique and specific metrics is well beyond the scope of this chapter, but the excellent work of Olaf Sporns, who carried out the foundational work and continues to explore its explanatory power in neurosciences, is recommended [116–119]. There are two principles or modes of brain connectivity that are key to cognition, and several metrics have been articulated to capture these. One is segregation, which refers to how neural elements form separate discrete clusters that can enable specialized local processing, and the second is integration, which refers to the capacity of the network as a whole to be interconnected to enable coordinated activity across neural populations. The manner in which individual nodes may contribute to the network's integrity and information flow is captured by influence measures. The most efficient interplay of segregation and integration is reflected in a 'small-world' organization based on functional connectivity patterns characteristic of the healthy primate brain [120, 121].

Graph metrics are now being used to characterize differences between the healthy adult brain and patients with MTL damage in mTLE [122–126] and MCI/AD [127–130]. To date, there have been few studies that have attempted to link alterations in graph characteristics to specific cognitive functions in these patient populations. Based on rsfMRI collected preoperatively, Doucet and colleagues examined characteristics of regions selected for their known involvement in specific cognitive domains (nine regions for verbal episodic memory, six for nonverbal episodic memory, ten for working memory/executive functions, and seven for expressive language) and their relationship to neuropsychological outcome in each domain following anterior temporal lobe resection [131]. They focused on properties of segregation (local efficiency), integration (distance), and influence or centrality (participation) for each of the nodes in an attempt to identify characteristics that might be selective to the functional domain and demonstrated strong utility of these measures as shown by significant prediction of outcome across most regression models. Considering language in left TLE, they found that integration amongst canonical subregions in the left inferior frontal cortex (pars orbitalis, pars triangularis, pars opercularis) was the most significant predictor of outcome but also that integration of the healthy (contralateral to side of seizure onset) hippocampus with other cortical regions was associated with better post-operative outcome. Surprisingly, despite the well-established involvement of the hippocampus in memory, they found that integration of the non-epileptogenic hippocampus with other neocortical regions was the most consistent predictor of cognitive outcome for all domains except for episodic memory. A somewhat complex pattern of findings characterized graph properties and verbal episodic memory, with increased integration and reduced centrality of the left inferior frontal cortex associated with better outcome for patients with left TLE. For patients with right TLE, the predictors of good verbal memory outcome were higher participation in the healthy left middle temporal cortex together with lower participation of the epileptogenic middle temporal cortex. The finding that no metrics involving either hippocampus was critical to predicting episodic memory outcome is somewhat at odds with our findings using simple correlation of DMN nodes [52]. These preliminary findings indicate that there is a substantial work ahead to determine the best methods for characterizing functionally relevant patterns of connectivity in mTLE and other clinical contexts.

Caveats and Limitations

While there is considerable promise to the clinical utility of rsfMRI, there are several limitations to the technique and concerns that need to be addressed. First, there are multiple analytic approaches that can be taken, and researchers and clinicians must decide which is most appropriate for their needs. A seed-based approach is easy to interpret but requires a priori selection of seed regions from the literature that will capture critical networks and may lead one astray if there has been substantial reorganization in the patient population of interest. For example, the selection of a damaged hippocampus as seed will likely not lead to identifying the effective memory network in mTLE. ICA approaches are data driven and will extract networks without a priori definition of seeds but can sometimes split networks and require selection of components of interest following analysis. A semiautomated approach has been proposed for identification of these networks based on training templates [91], and machine learning techniques can enable identification of atypical networks as long as there is sufficient homogeneity within the patient population of interest. Finally, issues of thresholding that are inherent to task-based fMRI can potentially be an issue in rsFC analyses. Arbitrary cut-offs such as $p < 0.05$ have the potential to eliminate meaningful voxels in single-subject analysis (e.g. fail to incorporate a region of meaningful connectivity into a language map), but without any form of thresholding, the signal will be noisy and uninterpretable. Examination of connectivity between two regions of interest in order to predict postsurgical change may avoid this but is again heavily reliant on proper ROI selection.

In addition to the techniques for characterizing the rsFC networks, one needs to be concerned about reliability of these measures, particularly at the individual subject level. Certainly, concerns have been raised about test-retest reliability in activation paradigms, particularly for complex tasks that can induce different strategies [132, 133]. There is evidence that rsFC is more reliable than activation magnitude or extent during a task [134, 135], but it is important to note that variables such as participant age, scan length, and other acquisition parameters can have modest influences [136, 137]. In addition, the extent to which the pathology in mTLE or MCI/AD or use of medications in these conditions may influence (either globally or locally) the neurovascular coupling that is crucial for observing the BOLD effect is just beginning to be investigated [138–141]. A recent review of limitations in clinical application can be found in [142]. A final crucial step in advancing the clinical utility of rsfMRI, particularly for the individual patient, is establishing sensitivity, specificity, and positive predictive value for these metrics. Fortunately the barrier to conducting resting state studies is considerably lower than with activation tasks in terms of amount of data required, the complexity of task influences, and the possibility of pooling data from multiple centres, making it more likely that we will make rapid progress on some of these issues.

Conclusions

Overall, there is very good evidence that rsfMRI is developing into a useful clinical tool for mapping language networks and characterizing functional integrity of memory networks in clinical populations and for providing predictions following surgical or other interventions. At present, however, considerable work still needs to be done in terms of comparisons to relevant 'gold standards' in clinical practice, determining ideal analytic strategies and decision algorithms, and relating network properties to the behaviours we are most interested in measuring and predicting.

References

1. Hermann BP, Wyler AR, Somes G, Clement L (1994) Dysnomia after left anterior temporal lobectomy without functional mapping: frequency and correlates. Neurosurgery 35(1):52–56. discussion 6–7. PubMed PMID: 7936152
2. Davies KG, Bell BD, Bush AJ, Hermann BP, Dohan FC Jr, Jaap AS (1998) Naming decline after left anterior temporal lobectomy correlates with pathological status of resected hippocampus. Epilepsia 39(4):407–419. PubMed PMID: 9578031
3. Schwarz M, Pauli E, Stefan H (2005) Model based prognosis of postoperative object naming in left temporal lobe epilepsy. Seizure 14(8):562–568. PubMed PMID: 16236531
4. Chelune GJ (1995) Hippocampal adequacy versus functional reserve: predicting memory functions following temporal lobectomy. Archives Clin Neuropsychol: Official Journal of the National Academy of Neuropsychologists 10(5):413–432. PubMed PMID: 14588901
5. Harvey DJ, Naugle RI, Magleby J, Chapin JS, Najm IM, Bingaman W et al (2008) Relationship between presurgical memory performance on the Wechsler memory scale-III and memory change following temporal resection for treatment of intractable epilepsy. Epilepsy Behav: E&B 13(2):372–375. PubMed PMID: 18556247
6. Milner B (1972) Disorders of learning and memory after temporal lobe lesions in man. Clin Neurosurg 19:421–446. PubMed PMID: 4637561
7. McAndrews MP, Cohn M (2012) Neuropsychology in temporal lobe epilepsy: influences from cognitive neuroscience and functional neuroimaging. Epilepsy Res Treatment 2012:925238. PubMed PMID: 22957249
8. Jones-Gotman M, Smith ML, Risse GL, Westerveld M, Swanson SJ, Giovagnoli AR et al (2010) The contribution of neuropsychology to diagnostic assessment in epilepsy. Epilepsy Behav: E&B. 18(1–2):3–12. PubMed PMID: 20471914
9. Alessio A, Pereira FR, Sercheli MS, Rondina JM, Ozelo HB, Bilevicius E et al (2013) Brain plasticity for verbal and visual memories in patients with mesial temporal lobe epilepsy and hippocampal sclerosis: an fMRI study. Hum Brain Mapp 34(1):186–199. PubMed PMID: 22038783
10. Braak H, Braak E (1996) Development of Alzheimer-related neurofibrillary changes in the neocortex inversely recapitulates cortical myelogenesis. Acta Neuropathol 92(2):197–201. PubMed PMID: 8841666
11. Whitwell JL, Jack CR Jr, Pankratz VS, Parisi JE, Knopman DS, Boeve BF et al (2008) Rates of brain atrophy over time in autopsy-proven frontotemporal dementia and Alzheimer disease. NeuroImage 39(3):1034–1040. PubMed PMID: 17988893
12. Jack CR Jr, Shiung MM, Gunter JL, O'Brien PC, Weigand SD, Knopman DS et al (2004) Comparison of different MRI brain atrophy rate measures with clinical disease progression in AD. Neurology 62(4):591–600. PubMed PMID: 14981176

13. Risacher SL, Saykin AJ, West JD, Shen L, Firpi HA, McDonald BC et al (2009) Baseline MRI predictors of conversion from MCI to probable AD in the ADNI cohort. Curr Alzheimer Res 6 (4):347–361. PubMed PMID: 19689234

14. Goekoop R, Rombouts SA, Jonker C, Hibbel A, Knol DL, Truyen L et al (2004) Challenging the cholinergic system in mild cognitive impairment: a pharmacological fMRI study. NeuroImage 23(4):1450–1459. PubMed PMID: 15589109

15. Rosen AC, Sugiura L, Kramer JH, Whitfield-Gabrieli S, Gabrieli JD (2011) Cognitive training changes hippocampal function in mild cognitive impairment: a pilot study. J Alzheimer's Dis: JAD 26(Suppl 3):349–357. PubMed PMID: 21971474

16. Detre JA, Maccotta L, King D, Alsop DC, Glosser G, D'Esposito M et al (1998) Functional MRI lateralization of memory in temporal lobe epilepsy. Neurology 50(4):926–932. PubMed PMID: 9566374

17. Vannest J, Szaflarski JP, Privitera MD, Schefft BK, Holland SK (2008) Medial temporal fMRI activation reflects memory lateralization and memory performance in patients with epilepsy. Epilepsy Behav: E&B 12(3):410–418. PubMed PMID: 18162441

18. Richardson MP, Strange BA, Thompson PJ, Baxendale SA, Duncan JS, Dolan RJ (2004) Pre-operative verbal memory fMRI predicts post-operative memory decline after left temporal lobe resection. Brain: A Journal of Neurology. 127(Pt 11):2419–2426. PubMed PMID: 15459025

19. Powell HW, Richardson MP, Symms MR, Boulby PA, Thompson PJ, Duncan JS et al (2008) Preoperative fMRI predicts memory decline following anterior temporal lobe resection. J Neurol Neurosurg Psychiatry 79(6):686–693. PubMed PMID: 17898035

20. Barnett AJ, Park MT, Pipitone J, Chakravarty MM, McAndrews MP (2015) Functional and structural correlates of memory in patients with mesial temporal lobe epilepsy. Front Neurol 6:103. PubMed PMID: 26029159

21. Mandzia JL, McAndrews MP, Grady CL, Graham SJ, Black SE (2009) Neural correlates of incidental memory in mild cognitive impairment: an fMRI study. Neurobiol Aging 30 (5):717–730. PubMed PMID: 17963998

22. Johnson SC, Schmitz TW, Moritz CH, Meyerand ME, Rowley HA, Alexander AL et al (2006) Activation of brain regions vulnerable to Alzheimer's disease: the effect of mild cognitive impairment. Neurobiol Aging 27(11):1604–1612. PubMed PMID: 16226349

23. Binder JR (2011) Functional MRI is a valid noninvasive alternative to Wada testing. Epilepsy Behav: E&B. 20(2):214–222. PubMed PMID: 20850386

24. Bonelli SB, Thompson PJ, Yogarajah M, Vollmar C, Powell RH, Symms MR et al (2012) Imaging language networks before and after anterior temporal lobe resection: results of a longitudinal fMRI study. Epilepsia 53(4):639–650. PubMed PMID: 22429073

25. Sabsevitz DS, Swanson SJ, Hammeke TA, Spanaki MV, Possing ET, Morris GL 3rd et al (2003) Use of preoperative functional neuroimaging to predict language deficits from epilepsy surgery. Neurology 60(11):1788–1792. PubMed PMID: 12796532

26. Westmacott R, Silver FL, McAndrews MP (2008) Understanding medial temporal activation in memory tasks: evidence from fMRI of encoding and recognition in a case of transient global amnesia. Hippocampus 18(3):317–325. PubMed PMID: 18064704

27. Dickerson BC, Sperling RA (2008) Functional abnormalities of the medial temporal lobe memory system in mild cognitive impairment and Alzheimer's disease: insights from functional MRI studies. Neuropsychologia 46(6):1624–1635. PubMed PMID: 18206188

28. Bakker A, Albert MS, Krauss G, Speck CL, Gallagher M (2015) Response of the medial temporal lobe network in amnestic mild cognitive impairment to therapeutic intervention assessed by fMRI and memory task performance. NeuroImage Clin 7:688–698. PubMed PMID: 25844322

29. Protzner AB, Mandzia JL, Black SE, McAndrews MP (2011) Network interactions explain effective encoding in the context of medial temporal damage in MCI. Hum Brain Mapp 32 (8):1277–1289. PubMed PMID: 20845396

30. Barnett AJ, Marty-Dugas J, McAndrews MP (2014) Advantages of sentence-level fMRI language tasks in presurgical language mapping for temporal lobe epilepsy. Epilepsy Behav: E&B. 32:114–120. PubMed PMID: 24534479

31. Binder JR, Swanson SJ, Sabsevitz DS, Hammeke TA, Raghavan M, Mueller WM (2010) A comparison of two fMRI methods for predicting verbal memory decline after left temporal lobectomy: language lateralization versus hippocampal activation asymmetry. Epilepsia 51(4):618–626. PubMed PMID: 19817807

32. Towgood K, Barker GJ, Caceres A, Crum WR, Elwes RD, Costafreda SG et al (2015) Bringing memory fMRI to the clinic: comparison of seven memory fMRI protocols in temporal lobe epilepsy. Hum Brain Mapp 36(4):1595–1608. PubMed PMID: 25727386

33. Raichle ME, Snyder AZ (2007) A default mode of brain function: a brief history of an evolving idea. NeuroImage 37(4):1083–1090. discussion 97-9. PubMed PMID: 17719799

34. Buckner RL, Andrews-Hanna JR, Schacter DL (2008) The brain's default network: anatomy, function, and relevance to disease. Ann N Y Acad Sci 1124:1–38. PubMed PMID: 18400922

35. Andrews-Hanna JR, Reidler JS, Huang C, Buckner RL (2010) Evidence for the default network's role in spontaneous cognition. J Neurophysiol 104(1):322–335. PubMed PMID: 20463201

36. Andrews-Hanna JR, Reidler JS, Sepulcre J, Poulin R, Buckner RL (2010) Functional-anatomic fractionation of the brain's default network. Neuron 65(4):550–562. PubMed PMID: 20188659

37. Smith SM, Miller KL, Moeller S, Xu J, Auerbach EJ, Woolrich MW et al (2012) Temporally-independent functional modes of spontaneous brain activity. Proc Natl Acad Sci U S A 109(8):3131–3136. PubMed PMID: 22323591

38. Yeo BT, Krienen FM, Sepulcre J, Sabuncu MR, Lashkari D, Hollinshead M et al (2011) The organization of the human cerebral cortex estimated by intrinsic functional connectivity. J Neurophysiol 106(3):1125–1165. PubMed PMID: 21653723

39. Bertolero MA, Yeo BT, D'Esposito M (2015) The modular and integrative functional architecture of the human brain. Proc Natl Acad Sci U S A 112(49):E6798–E6807. PubMed PMID: 26598686

40. Damoiseaux JS, Rombouts SA, Barkhof F, Scheltens P, Stam CJ, Smith SM et al (2006) Consistent resting-state networks across healthy subjects. Proc Natl Acad Sci U S A 103(37):13848–13853. PubMed PMID: 16945915

41. Yeo BT, Krienen FM, Chee MW, Buckner RL (2014) Estimates of segregation and overlap of functional connectivity networks in the human cerebral cortex. NeuroImage 88:212–227. PubMed PMID: 24185018

42. Yeo BT, Krienen FM, Eickhoff SB, Yaakub SN, Fox PT, Buckner RL et al (2015) Functional specialization and flexibility in human association cortex. Cereb Cortex 25(10):3654–3672. PubMed PMID: 25249407

43. Bonelli SB, Powell RH, Yogarajah M, Samson RS, Symms MR, Thompson PJ et al (2010) Imaging memory in temporal lobe epilepsy: predicting the effects of temporal lobe resection. Brain: A Journal of Neurology. 133(Pt 4):1186–1199. PubMed PMID: 20157009

44. Sidhu MK, Stretton J, Winston GP, Symms M, Thompson PJ, Koepp MJ et al (2015) Memory fMRI predicts verbal memory decline after anterior temporal lobe resection. Neurology 84(15):1512–1519. PubMed PMID: 25770199

45. Rugg MD, Vilberg KL (2013) Brain networks underlying episodic memory retrieval. Curr Opin Neurobiol 23(2):255–260. PubMed PMID: 23206590

46. McDermott KB, Szpunar KK, Christ SE (2009) Laboratory-based and autobiographical retrieval tasks differ substantially in their neural substrates. Neuropsychologia 47(11):2290–2298. PubMed PMID: 19159634

47. McAndrews MP (2014) Memory assessment in the clinical context using functional magnetic resonance imaging: a critical look at the state of the field. Neuroimaging Clin N Am 24(4):585–597. PubMed PMID: 25441502

48. Frings L, Schulze-Bonhage A, Spreer J, Wagner K (2009) Reduced interhemispheric hippocampal BOLD signal coupling related to early epilepsy onset. Seizure 18(2):153–157. PubMed PMID: 18675555

49. James GA, Tripathi SP, Ojemann JG, Gross RE, Drane DL (2013) Diminished default mode network recruitment of the hippocampus and parahippocampus in temporal lobe epilepsy. J Neurosurg 119(2):288–300. PubMed PMID: 23706058

50. Voets NL, Adcock JE, Stacey R, Hart Y, Carpenter K, Matthews PM et al (2009) Functional and structural changes in the memory network associated with left temporal lobe epilepsy. Hum Brain Mapp 30(12):4070–4081. PubMed PMID: 19517529

51. Zhang Z, Lu G, Zhong Y, Tan Q, Liao W, Wang Z et al (2010) Altered spontaneous neuronal activity of the default-mode network in mesial temporal lobe epilepsy. Brain Res 1323:152–160. PubMed PMID: 20132802. Epub 2010/02/06. eng

52. McCormick C, Quraan M, Cohn M, Valiante TA, McAndrews MP (2013) Default mode network connectivity indicates episodic memory capacity in mesial temporal lobe epilepsy. Epilepsia 54(5):809–818. PubMed PMID: 23360362

53. McCormick C, Protzner AB, Barnett AJ, Cohn M, Valiante TA, McAndrews MP (2014) Linking DMN connectivity to episodic memory capacity: what can we learn from patients with medial temporal lobe damage? NeuroImage Clinical 5:188–196. PubMed PMID: 25068108

54. Addis DR, Moscovitch M, McAndrews MP (2007) Consequences of hippocampal damage across the autobiographical memory network in left temporal lobe epilepsy. Brain : a journal of neurology. 130(Pt 9):2327–2342

55. Voets NL, Menke RA, Jbabdi S, Husain M, Stacey R, Carpenter K et al (2015) Thalamo-cortical disruption contributes to short-term memory deficits in patients with medial temporal lobe damage. Cereb Cortex 25(11):4584–4595. PubMed PMID: 26009613

56. Voets NL, Beckmann CF, Cole DM, Hong S, Bernasconi A, Bernasconi N (2012) Structural substrates for resting network disruption in temporal lobe epilepsy. Brain: A Journal of Neurology 135(Pt 8):2350–2357. PubMed PMID: 22669081

57. Bigras C, Shear PK, Vannest J, Allendorfer JB, Szaflarski JP (2013) The effects of temporal lobe epilepsy on scene encoding. Epilepsy Behav: E&B. 26(1):11–21. PubMed PMID: 23207513

58. Honey CJ, Kotter R, Breakspear M, Sporns O (2007) Network structure of cerebral cortex shapes functional connectivity on multiple time scales. Proc Natl Acad Sci U S A 104 (24):10240–10245. PubMed PMID: 17548818

59. Sporns O (2013) The human connectome: origins and challenges. NeuroImage 80:53–61. PubMed PMID: 23528922

60. Maguire EA (2001) Neuroimaging studies of autobiographical event memory. Philos Trans R Soc Lond Ser B Biol Sci 356(1413):1441–1451. PubMed PMID: 11571035

61. Svoboda E, McKinnon MC, Levine B (2006) The functional neuroanatomy of autobiographical memory: a meta-analysis. Neuropsychologia 44(12):2189–2208. PubMed PMID: 16806314

62. Denkova EJ, Manning L (2014) FMRI contributions to addressing autobiographical memory impairment in temporal lobe pathology. World J Rad 6(4):93–105. PubMed PMID: 24778771

63. Voets NL, Zamboni G, Stokes MG, Carpenter K, Stacey R, Adcock JE (2014) Aberrant functional connectivity in dissociable hippocampal networks is associated with deficits in memory. J Neurosci: Official Journal of the Society for Neuroscience. 34(14):4920–4928. PubMed PMID: 24695711

64. Holmes M, Folley BS, Sonmezturk HH, Gore JC, Kang H, Abou-Khalil B et al (2014) Resting state functional connectivity of the hippocampus associated with neurocognitive function in left temporal lobe epilepsy. Hum Brain Mapp 35(3):735–744. PubMed PMID: 23124719

65. Doucet G, Osipowicz K, Sharan A, Sperling MR, Tracy JI (2013) Extratemporal functional connectivity impairments at rest are related to memory performance in mesial temporal epilepsy. Hum Brain Mapp 34(9):2202–2216. PubMed PMID: 22505284

66. Nellessen N, Rottschy C, Eickhoff SB, Ketteler ST, Kuhn H, Shah NJ et al (2015) Specific and disease stage-dependent episodic memory-related brain activation patterns in Alzheimer's disease: a coordinate-based meta-analysis. Brain Struct Funct 220(3):1555–1571. PubMed PMID: 24633738
67. Schwindt GC, Black SE (2009) Functional imaging studies of episodic memory in Alzheimer's disease: a quantitative meta-analysis. NeuroImage 45(1):181–190. PubMed PMID: 19103293
68. Bakker A, Krauss GL, Albert MS, Speck CL, Jones LR, Stark CE et al (2012) Reduction of hippocampal hyperactivity improves cognition in amnestic mild cognitive impairment. Neuron 74(3):467–474. PubMed PMID: 22578498
69. Sperling R (2007) Functional MRI studies of associative encoding in normal aging, mild cognitive impairment, and Alzheimer's disease. Ann N Y Acad Sci 1097:146–155. PubMed PMID: 17413017
70. Miller SL, Celone K, DePeau K, Diamond E, Dickerson BC, Rentz D et al (2008) Age-related memory impairment associated with loss of parietal deactivation but preserved hippocampal activation. Proc Natl Acad Sci U S A 105(6):2181–2186. PubMed PMID: 18238903
71. Vannini P, O'Brien J, O'Keefe K, Pihlajamaki M, Laviolette P, Sperling RA (2011) What goes down must come up: role of the posteromedial cortices in encoding and retrieval. Cereb Cortex 21(1):22–34. PubMed PMID: 20363808
72. Chhatwal JP, Sperling RA (2012) Functional MRI of mnemonic networks across the spectrum of normal aging, mild cognitive impairment, and Alzheimer's disease. J Alzheimer's Dis: JAD 31(Suppl 3):S155–S167. PubMed PMID: 22890098
73. Krajcovicova L, Marecek R, Mikl M, Rektorova I (2014) Disruption of resting functional connectivity in Alzheimer's patients and at-risk subjects. Curr Neurol Neurosci Rep 14(10):491. PubMed PMID: 25120223
74. Hayes SM, Salat DH, Verfaellie M (2012) Default network connectivity in medial temporal lobe amnesia. J Neurosci: Official Journal of the Society for Neuroscience. 32(42):14622–14629. PubMed PMID: 23077048
75. Dunn CJ, Duffy SL, Hickie IB, Lagopoulos J, Lewis SJ, Naismith SL et al (2014) Deficits in episodic memory retrieval reveal impaired default mode network connectivity in amnestic mild cognitive impairment. NeuroImage Clin 4:473–480. PubMed PMID: 24634833
76. Han SD, Arfanakis K, Fleischman DA, Leurgans SE, Tuminello ER, Edmonds EC et al (2012) Functional connectivity variations in mild cognitive impairment: associations with cognitive function. J Int Neuropsychol Soc 18(1):39–48. PubMed PMID: 22005016
77. Wang Z, Liang P, Jia X, Qi Z, Yu L, Yang Y et al (2011) Baseline and longitudinal patterns of hippocampal connectivity in mild cognitive impairment: evidence from resting state fMRI. J Neurol Sci 309(1–2):79–85. PubMed PMID: 21821265
78. Libby LA, Ekstrom AD, Ragland JD, Ranganath C (2012) Differential connectivity of perirhinal and parahippocampal cortices within human hippocampal subregions revealed by high-resolution functional imaging. J Neurosci: Official Journal of the Society for Neuroscience 32(19):6550–6560. PubMed PMID: 22573677
79. Ritchey M, Libby LA, Ranganath C (2015) Cortico-hippocampal systems involved in memory and cognition: the PMAT framework. Prog Brain Res 219:45–64. PubMed PMID: 26072233
80. Poppenk J, Evensmoen HR, Moscovitch M, Nadel L (2013) Long-axis specialization of the human hippocampus. Trends Cogn Sci 17(5):230–240. PubMed PMID: 23597720
81. Kahn I, Andrews-Hanna JR, Vincent JL, Snyder AZ, Buckner RL (2008) Distinct cortical anatomy linked to subregions of the medial temporal lobe revealed by intrinsic functional connectivity. J Neurophysiol 100(1):129–139. PubMed PMID: 18385483
82. Kim H (2015) Encoding and retrieval along the long axis of the hippocampus and their relationships with dorsal attention and default mode networks: the HERNET model. Hippocampus 25(4):500–510. PubMed PMID: 25367784
83. Thom M, Sisodiya SM, Beckett A, Martinian L, Lin WR, Harkness W et al (2002) Cytoarchitectural abnormalities in hippocampal sclerosis. J Neuropathol Exp Neurol 61(6):510–519. PubMed PMID: 12071634

84. Babb TL, Brown WJ, Pretorius J, Davenport C, Lieb JP, Crandall PH (1984) Temporal lobe volumetric cell densities in temporal lobe epilepsy. Epilepsia 25(6):729–740. PubMed PMID: 6510381

85. Babb TL, Lieb JP, Brown WJ, Pretorius J, Crandall PH (1984) Distribution of pyramidal cell density and hyperexcitability in the epileptic human hippocampal formation. Epilepsia 25 (6):721–728. PubMed PMID: 6510380

86. Woermann FG, Barker GJ, Birnie KD, Meencke HJ, Duncan JS (1998) Regional changes in hippocampal T2 relaxation and volume: a quantitative magnetic resonance imaging study of hippocampal sclerosis. J Neurol Neurosurg Psychiatry 65(5):656–664. PubMed PMID: 9810933

87. Adnan A, Barnett A, Moayedi M, McCormick C, Cohn M, McAndrews MP (2016) Distinct hippocampal functional networks revealed by tractography-based parcellation. Brain Struct Funct 221(6):2999–3012. PubMed PMID: 26206251

88. McCormick C, St-Laurent M, Ty A, Valiante TA, McAndrews MP (2015) Functional and effective hippocampal-neocortical connectivity during construction and elaboration of autobiographical memory retrieval. Cereb Cortex 25(5):1297–1305. PubMed PMID: 24275829

89. Das SR, Pluta J, Mancuso L, Kliot D, Yushkevich PA, Wolk DA (2015) Anterior and posterior MTL networks in aging and MCI. Neurobiol Aging 36(Suppl 1):S141–S150. S50 e1. PubMed PMID: 25444600. Pubmed Central PMCID: 4342050

90. Price CJ (2010) The anatomy of language: a review of 100 fMRI studies published in 2009. Ann N Y Acad Sci 1191:62–88. PubMed PMID: 20392276

91. Tie Y, Rigolo L, Norton IH, Huang RY, Wu W, Orringer D et al (2014) Defining language networks from resting-state fMRI for surgical planning- a feasibility study. Hum Brain Mapp 35:1018–1030

92. Tomasi D, Volkow ND (2012) Resting functional connectivity of language networks: characterization and reproducibility. Mol Psychiatry 17:841–854. PubMed PMID: 22212597

93. Zhu L, Fan Y, Zou Q, Wang J, Gao JH, Niu Z (2014) Temporal reliability and lateralization of the resting-state language network. PLoS One 9:1–14. PubMed PMID: 24475058

94. Schurz M, Wimmer H, Richlan F, Ludersdorfer P, Klackl J, Kronbichler M (2015) Resting-state and task-based functional brain connectivity in developmental dyslexia. Cereb Cortex 25:3502–3514. PubMed PMID: 25169986

95. Wada JA (1997) Clinical experimental observations of carotid artery injections of sodium amytal. Brain Cogn 33(1):11–13. PubMed PMID: 9056272. Epub 1997/02/01. eng

96. Hamberger MJ (2007) Cortical language mapping in epilepsy: a critical review. Neuropsychol Rev 17(4):477–489. PubMed PMID: 18004662

97. Adcock JE, Wise RG, Oxbury JM, Oxbury SM, Matthews PM (2003) Quantitative fMRI assessment of the differences in lateralization of language-related brain activation in patients with temporal lobe epilepsy. NeuroImage 18:423–438. PubMed PMID: 12595196

98. Dym RJ, Burns J, Freeman K, Lipton ML (2011) Is functional MR imaging assessment of hemispheric language dominance as good as the Wada test?: a meta-analysis. Radiology 261:446–455. PubMed PMID: 21803921

99. Gutbrod K, Spring D, Degonda N, Heinemann D, Nirkko A, Hauf M et al (2011) Determination of language dominance: Wada test and fMRI compared using a novel sentence task. J Neuroimaging: Official Journal of the American Society of Neuroimaging:1–9. PubMed PMID: 21883628

100. Janecek JK, Swanson SJ, Sabsevitz DS, Ta H, Raghavan M, E Rozman M et al (2013) Language lateralization by fMRI and Wada testing in 229 patients with epilepsy: rates and predictors of discordance. Epilepsia 54:314–322. PubMed PMID: 23294162

101. Lehéricy S, Cohen L, Bazin B, Samson S, Giacomini E, Rougetet R et al (2000) Functional MR evaluation of temporal and frontal language dominance compared with the Wada test. Neurology 54:1625–1633. PubMed PMID: 10762504

102. Roux FE, Boulanouar K, Lotterie JA, Mejdoubi M, LeSage JP, Berry I et al (2003) Language functional magnetic resonance imaging in preoperative assessment of language areas:

correlation with direct cortical stimulation. Neurosurgery 52:1335–1347. PubMed PMID: 12762879

103. Rutten GJM, Ramsey NF, Van Rijen PC, Noordmans HJ, Van Veelen CWM (2002) Development of a functional magnetic resonance imaging protocol for intraoperative localization of critical temporoparietal language areas. Ann Neurol 51:350–360. PubMed PMID: 11891830

104. Benjamin CF, Walshaw PD, Hale K, Gaillard WD, Baxter LC, Berl MM et al (2017) Presurgical language fMRI: mapping of six critical regions. Hum Brain Mapp 38 (8):4239–4255

105. Branco DM, Suarez RO, Whalen S, O'Shea JP, Nelson AP, da Costa JC et al (2006) Functional MRI of memory in the hippocampus: laterality indices may be more meaningful if calculated from whole voxel distributions. NeuroImage 32(2):592–602. PubMed PMID: 16777435

106. Wilke M, Lidzba K (2007) LI-tool: a new toolbox to assess lateralization in functional MR-data. J Neurosci Methods 163(1):128–136. PubMed PMID: 17386945

107. Barnett A, Marty-Dugas J, McAndrews MP (2014) Advantages of sentence-level fMRI language tasks in presurgical language mapping for temporal lobe epilepsy. Epilepsy Behav: E&B 32:114–120. PubMed PMID: 24534479

108. Binder JR, Gross WL, Allendorfer JB, Bonilha L, Chapin J, Edwards JC et al (2011) Mapping anterior temporal lobe language areas with fMRI: a multicenter normative study. NeuroImage 54(2):1465–1475. PubMed PMID: 20884358

109. Binder JR, Swanson SJ, Hammeke TA, Sabsevitz DS (2008) A comparison of five fMRI protocols for mapping speech comprehension systems. Epilepsia 49:1980–1997. PubMed PMID: 18513352

110. Bradshaw AR, Thompson PA, Wilson AC, Bishop DVM, Woodhead ZVJ (2017) Measuring language lateralisation with different language tasks: a systematic review. Peer J 5:e3929. PubMed PMID: 29085748

111. Mitchell TJ, Hacker CD, Breshears JD, Szrama NP, Sharma M, Bundy DT et al (2013) A novel data-driven approach to preoperative mapping of functional cortex using resting-state functional magnetic resonance imaging. Neurosurgery 73:969–983. PubMed PMID: 24264234

112. Branco P, Seixas D, Deprez S, Kovacs S, Peeters R, Castro SL et al (2016) Resting-state functional magnetic resonance imaging for language preoperative planning. Front Hum Neurosci 10:11. PubMed PMID: 26869899

113. Sair HI, Yahyavi-Firouz-Abadi N, Calhoun VD, Airan RD, Agarwal S, Intrapiromkul J et al (2016) Presurgical brain mapping of the language network in patients with brain tumors using resting-state fMRI: comparison with task fMRI. Hum Brain Mapp 37:913–923. PubMed PMID: 24686109

114. Doucet GE, Pustina D, Skidmore C, Sharan A, Sperling MR, Tracy JI (2015) Resting-state functional connectivity predicts the strength of hemispheric lateralization for language processing in temporal lobe epilepsy and normals. Hum Brain Mapp 36:288–303. PubMed PMID: 25187327

115. Barnett A, Audrain S, McAndrews MP (2017) Applications of resting-state functional MR imaging to epilepsy. Neuroimaging Clin N Am 27(4):697–708. PubMed PMID: 28985938

116. Bullmore E, Sporns O (2009) Complex brain networks: graph theoretical analysis of structural and functional systems. Nat Rev Neurosci 10(3):186–198. PubMed PMID: 19190637

117. Sporns O (2013) Structure and function of complex brain networks. Dialogues Clin Neurosci 15(3):247–262. PubMed PMID: 24174898

118. Sporns O (2012) From simple graphs to the connectome: networks in neuroimaging. NeuroImage 62(2):881–886. PubMed PMID: 21964480

119. Sporns O, Betzel RF (2016) Modular brain networks. Annu Rev Psychol 67:613–640. PubMed PMID: 26393868

120. Stam CJ, Reijneveld JC (2007) Graph theoretical analysis of complex networks in the brain. Nonlinear Biomed Phy 1(1):3. PubMed PMID: 17908336

121. Achard S, Bullmore E (2007) Efficiency and cost of economical brain functional networks. PLoS Comput Biol 3(2):e17. PubMed PMID: 17274684

122. Wang J, Qiu S, Xu Y, Liu Z, Wen X, Hu X et al (2014) Graph theoretical analysis reveals disrupted topological properties of whole brain functional networks in temporal lobe epilepsy. Clin Neurophysiol 125(9):1744–1756. PubMed PMID: 24686109

123. Chiang S, Stern JM, Engel J Jr, Levin HS, Haneef Z (2014) Differences in graph theory functional connectivity in left and right temporal lobe epilepsy. Epilepsy Res 108 (10):1770–1781. PubMed PMID: 25445238

124. Song J, Nair VA, Gaggl W, Prabhakaran V (2015) Disrupted brain functional Organization in Epilepsy Revealed by graph theory analysis. Brain Connect 5(5):276–283. PubMed PMID: 25647011

125. Vlooswijk MC, Vaessen MJ, Jansen JF, de Krom MC, Majoie HJ, Hofman PA et al (2011) Loss of network efficiency associated with cognitive decline in chronic epilepsy. Neurology 77(10):938–944. PubMed PMID: 21832213

126. Doucet GE, Sharan A, Pustina D, Skidmore C, Sperling MR, Tracy JI (2015) Early and late age of seizure onset have a differential impact on brain resting-state organization in temporal lobe epilepsy. Brain Topogr 28(1):113–126. PubMed PMID: 24881003

127. Toussaint PJ, Maiz S, Coynel D, Doyon J, Messe A, de Souza LC et al (2014) Characteristics of the default mode functional connectivity in normal ageing and Alzheimer's disease using resting state fMRI with a combined approach of entropy-based and graph theoretical measurements. NeuroImage 101:778–786. PubMed PMID: 25111470

128. Sanz-Arigita EJ, Schoonheim MM, Damoiseaux JS, Rombouts SA, Maris E, Barkhof F et al (2010) Loss of 'small-world' networks in Alzheimer's disease: graph analysis of FMRI resting-state functional connectivity. PLoS One 5(11):e13788. PubMed PMID: 21072180

129. Khazaee A, Ebrahimzadeh A, Babajani-Feremi A (2016) Application of advanced machine learning methods on resting-state fMRI network for identification of mild cognitive impairment and Alzheimer's disease. Brain Imaging Behav 10(3):799–817. PubMed PMID: 26363784

130. Xie T, He Y (2011) Mapping the Alzheimer's brain with connectomics. Front Psych 2:77. PubMed PMID: 22291664

131. Doucet GE, Rider R, Taylor N, Skidmore C, Sharan A, Sperling M et al (2015) Presurgery resting-state local graph-theory measures predict neurocognitive outcomes after brain surgery in temporal lobe epilepsy. Epilepsia 56:517–526. PubMed PMID: 25708625

132. Brandt DJ, Sommer J, Krach S, Bedenbender J, Kircher T, Paulus FM et al (2013) Test-retest reliability of fMRI brain activity during memory encoding. Front Psych 4:163. PubMed PMID: 24367338

133. Clement F, Belleville S (2009) Test-retest reliability of fMRI verbal episodic memory paradigms in healthy older adults and in persons with mild cognitive impairment. Hum Brain Mapp 30(12):4033–4047. PubMed PMID: 19492301

134. Shehzad Z, Kelly AM, Reiss PT, Gee DG, Gotimer K, Uddin LQ et al (2009) The resting brain: unconstrained yet reliable. Cerebr Cort 19(10):2209–2229. PubMed PMID: 19221144

135. Zuo XN, Kelly C, Adelstein JS, Klein DF, Castellanos FX, Milham MP (2010) Reliable intrinsic connectivity networks: test-retest evaluation using ICA and dual regression approach. NeuroImage 49(3):2163–2177. PubMed PMID: 19896537

136. Song J, Desphande AS, Meier TB, Tudorascu DL, Vergun S, Nair VA et al (2012) Age-related differences in test-retest reliability in resting-state brain functional connectivity. PLoS One 7 (12):e49847. PubMed PMID: 23227153

137. Birn RM, Molloy EK, Patriat R, Parker T, Meier TB, Kirk GR et al (2013) The effect of scan length on the reliability of resting-state fMRI connectivity estimates. NeuroImage 83:550–558. PubMed PMID: 23747458

138. Babiloni C, Vecchio F, Altavilla R, Tibuzzi F, Lizio R, Altamura C et al (2014) Hypercapnia affects the functional coupling of resting state electroencephalographic rhythms and cerebral

haemodynamics in healthy elderly subjects and in patients with amnestic mild cognitive impairment. Clin Neurophysiol 125(4):685–693. PubMed PMID: 24238990

139. Gomez-Gonzalo M, Losi G, Brondi M, Uva L, Sato SS, de Curtis M et al (2011) Ictal but not interictal epileptic discharges activate astrocyte endfeet and elicit cerebral arteriole responses. Front Cell Neurosci 5:8. PubMed PMID: 21747758

140. Rosengarten B, Paulsen S, Burr O, Kaps M (2009) Neurovascular coupling in Alzheimer patients: effect of acetylcholine-esterase inhibitors. Neurobiol Aging 30(12):1918–1923. PubMed PMID: 18395940

141. Yasuda CL, Centeno M, Vollmar C, Stretton J, Symms M, Cendes F et al (2013) The effect of topiramate on cognitive fMRI. Epilepsy Res 105(1–2):250–255. PubMed PMID: 23333471

142. Agarwal S, Sair HI, Pillai JJ (2017) Limitations of resting-state functional MR imaging in the setting of focal brain lesions. Neuroimaging Clin N Am 27(4):645–661. PubMed PMID: 28985935

Chapter 4
The Neuroimaging of Stroke: Structural and Functional Advances

Sara Regina Meira Almeida, Gabriela Castellano, Jessica Vicentini, and Li Li Min

Introduction

Advanced magnetic resonance imaging (MRI) techniques such as diffusion tensor imaging (DTI), functional magnetic resonance imaging (fMRI), magnetic resonance perfusion, magnetic resonance spectroscopy, and volumetric imaging are improving our understanding of normal brain development up to pathophysiology of several adverse processes [1], including cerebrovascular diseases. The application of these techniques for research offers the opportunity to ask questions about where, when, and how the abnormal pattern of anatomical connectivity and plastic changes occurs in stroke. Indeed, neuroimaging techniques, especially the multimodality MRI, have significantly contributed to the understanding of the mechanisms of stroke recovery by characterizing brain structural and functional changes after stroke [2]. Stroke lesions trigger several brain-wide processes to accommodate for tissue loss. MRI has been extensively used to investigate brain activation changes during recovery and has provided important information on monitoring of therapeutic strategies that promote brain repair and functional reorganization after stroke [3].

This review focuses on the advances brought to the understanding of stroke mechanisms by MRI-based neuroimaging techniques, particularly those related to connectivity assessment among brain areas. Stroke is the major cause of long-term

S. R. M. Almeida (✉) · J. Vicentini · L. L. Min (✉)
Department of Neurology, School of Medical Sciences, University of Campinas (UNICAMP), Campinas, SP, Brazil

Brazilian Institute of Neuroscience and Neurotechnology (BRAINN), Campinas, SP, Brazil
e-mail: limin@fcm.unicamp.br

G. Castellano
Neurophysics Group, Institute of Physics Gleb Wataghin, University of Campinas (UNICAMP), Campinas, SP, Brazil

Brazilian Institute of Neuroscience and Neurotechnology (BRAINN), Campinas, SP, Brazil

© Springer International Publishing AG, part of Springer Nature 2018
C. Habas (ed.), *The Neuroimaging of Brain Diseases*, Contemporary Clinical Neuroscience, https://doi.org/10.1007/978-3-319-78926-2_4

"

disability throughout the world [4], leaving more than half of the patients dependent on daily assistance. Nonetheless, most patients exhibit a certain degree of recovery in the weeks, months, and sometimes even years following stroke, which may be directly related to structural and functional modifications in surviving brain tissue. Several animal and human stroke studies have reported vicarious function of ipsilesional and contralesional brain regions [5], which may contribute to restoration of functions, although the exact mechanisms that lead to functional recovery remain largely unclear. Elucidation of the critical pathways in post-stroke recovery would not only provide important fundamental insight in brain function and plasticity but could also lead the way toward development of new rehabilitation strategies for recovering stroke patients [2].

Brain Connectivity from MRI Data

Connectivity models are based on the concept that the brain is organized by segregation of specialized and anatomically distinct brain regions that are functionally integrated in networks mediating cognitive, sensory, or motor processing [6]. Structural connectivity describes how spatially separated brain regions are physically linked, for example, as demonstrated by invasive tracing of single axons or by noninvasively measuring diffusion along major fiber bundles as in DTI. In contrast, functional and effective connectivity describe how anatomically connected areas interact with each other and can be estimated from noninvasive techniques such as electroencephalography, magnetoencephalography, near-infrared spectroscopy, or fMRI. The last two connectivity approaches, however, fundamentally differ in the way of how these interactions are estimated. Functional connectivity is defined as temporal correlation between spatially remote neurophysiological events. In contrast to this nondirectional, correlative nature of functional connectivity, effective connectivity refers to the causal influences that brain areas exert over another under the assumptions of a given mechanistic model [7].

Structural connectivity is obtained from MRI data through the aforementioned DTI technique [8]. This is a noninvasive MRI method that measures the random motion of water molecules in brain tissue and enables examination of white matter microstructure in vivo. Since white matter tracts are composed of highly oriented fibers, which cause relatively high anisotropy of diffusing tissue water, DTI is very suitable to measure effects on white matter integrity. DTI data consists of mathematical entities, called tensors, that give information on the amount of diffusion happening in every direction within a given voxel of the image. From these data, simpler, scalar measures, such as fractional anisotropy (FA), can be obtained. FA quantifies the extent to which water diffusion is directionally restricted and is influenced by a number of factors including axonal myelination, diameter, density, and orientational coherence [9].

Functional and effective connectivity are obtained from MRI data by means of the fMRI technique. Functional MRI methods have been traditionally sensitized to changes in cerebral hemodynamics in response to task or stimulus-triggered neuronal activity. The most frequently used type of fMRI data is that obtained by means of the blood oxygenation level-dependent (BOLD) contrast, measured through a T2*-weighted gradient-echo MRI sequence [10]. Although effective connectivity has been, indeed, obtained from task-based fMRI data, functional connectivity has been mostly obtained through resting-state fMRI (rs-fMRI) data.

Rs-fMRI data is obtained without the need of a stimulation paradigm, during wakeful rest, but in the absence of an active task performance [11, 12]. Spontaneous fluctuations in baseline ("resting-state") neuronal signaling are reflected in low-frequency fluctuations (<0.1 Hz) of the BOLD signal and show temporal coherence between anatomically connected brain regions within a particular neuronal network, such as the sensorimotor system [11]. Throughout the gray matter, the extent of synchronization between these low-frequency BOLD fluctuations has been related to functional connectivity. Correlation of these signals with electroencephalographic brain activity has indicated that these slow hemodynamic fluctuations are associated with neuronal function [13].

There are many approaches to estimate functional connectivity from rs-fMRI data. One of the most used has been the seed-based approach, where a representative time series is extracted from a region of interest (the seed), and this series is correlated to all voxel time series of the brain [14] (Fig. 4.1). Another widely used approach has been independent component analysis (ICA) that aims at retrieving the brain networks that produce similarly varying time series [15]. And more recently, functional connectivity from rs-fMRI has been modeled with the aid of graphs, by parcellating the brain in several (anatomical or functional) regions and computing correlations among all the representative time series of these regions. This allows building a mathematical entity, the graph, composed of nodes (the regions) and edges (the connections), from which many topological properties may be extracted that in turn serve to characterize the underlying brain network [16].

Some of the advantages of the rs-fMRI technique are that many different cortical systems may be studied with a single acquisition (as opposed to task-based fMRI, where only the system associated to the given task is studied) and that it allows increasing the number of subjects for a given study or performing studies with task-impaired populations, who would be otherwise excluded from a task-based study [17]. Rs-fMRI has been increasingly applied as a tool to study alterations in the brain's intrinsic functional architecture as potential physiological correlates of neurological disorders [18].

Differently from functional connectivity, effective connectivity, as already mentioned, is estimated from task-based fMRI data. Also, effective connectivity requires an a priori model of the brain regions involved in the task being studied, as well as their interaction. It is, therefore, a model-driven technique, as opposed to the functional connectivity technique, which is data-driven [19]. From the proposed model, effective connectivity approaches infer the strength and direction of the connections, giving some idea of temporal relations among the regions (e.g., which region acted before the others).

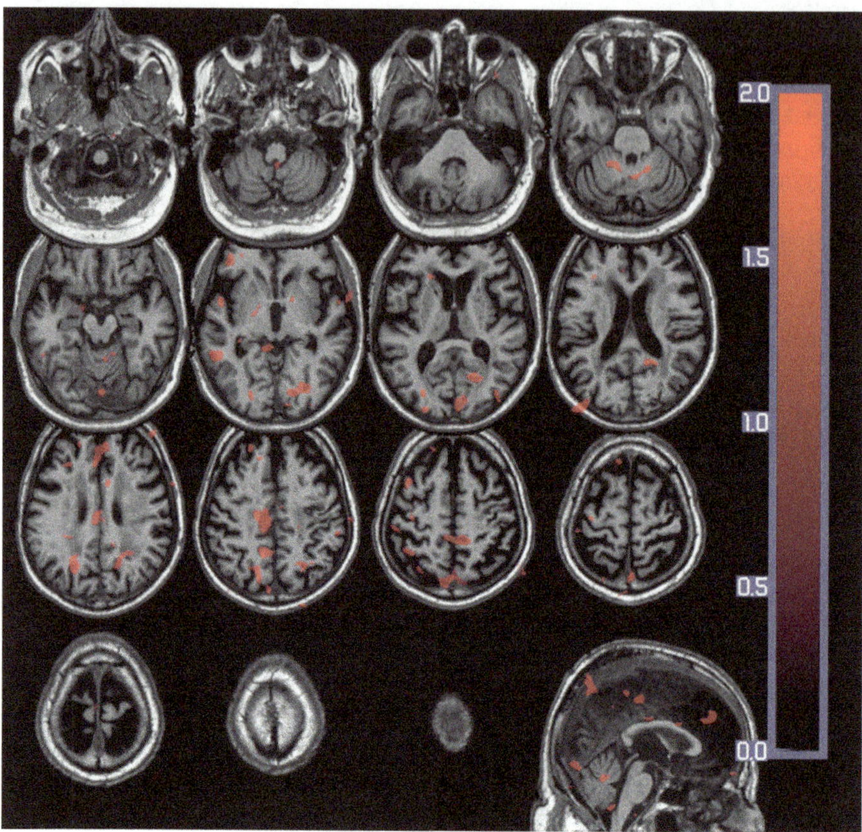

Fig. 4.1 Example of functional connectivity (rs-fMRI) data between cortical motor areas in stroke patients

Approaches to estimate effective connectivity include psychophysiological interactions [20], structural equation modeling (SEM) [21], Granger causality [22], and dynamic causal modeling (DCM) [23]. SEM is based on the translation of a network model with predefined regions linked by a set of directional paths into a linear regression model. The path coefficients are subsequently estimated using an iterative maximum likelihood algorithm to minimize the difference between observed and predicted covariance matrices [21]. DCM relies on a deterministic model that treats the brain as an input-output system of hidden neural dynamics [23]. This neural model describes changes in the system over time as a function of interactions between regional activity, known experimental inputs, and neuronal parameters [12].

All types of MRI data-derived connectivities (structural, functional, and effective) have been used to shed light on stroke mechanisms, and some of the main results are discussed below.

Insights into Stroke Mechanisms from Functional and Effective Connectivity Studies

Findings from functional connectivity obtained from rs-fMRI can be summarized by two major patterns of changes after stroke: reduced interhemispheric functional connectivity between cortical motor areas, which correlates with the severity of motor deficits, and reduced global network efficiency even in patients with good clinical recovery [12]. Studies in animals demonstrate that interhemispheric resting-state connectivity between ipsilesional primary sensorimotor cortex and its contralesional homologue significantly diminish in the first few days and subsequently increase while sensorimotor functions recover. However, interhemispheric connectivity remains reduced compared with assessments obtained prior to stroke [24].

Carter and colleagues [25] found correlations between rs-fMRI time series in a sensorimotor network consisting of M1, SMA, secondary somatosensory cortex, cerebellum, putamen, and thalamus in both hemispheres. The authors reported that particularly interhemispheric M1 connectivity positively correlates with motor performance at the subacute stage after stroke. In addition, stronger interhemispheric connectivity of ipsilesional M1 and contralesional areas such as thalamus, SMA, and middle prefrontal cortex within the first few days of onset predicts better motor recovery in the next 6 months post-stroke [26]. Also, patients with attention deficits had reduced interhemispheric connectivity between attention-related areas in parietal cortex or language areas in inferior frontal cortex [25].

In patients with sufficient integrity of ipsilesional sensorimotor cortex and corticospinal tract, motor recovery may occur rapidly after stroke and be mediated by reacquisition of normal dominance by ipsilesional sensorimotor cortex. However, in patients in whom the integrity of this sensorimotor cortex is insufficient to support good recovery, increased recruitment of contralesional sensorimotor cortices may be utilized to achieve motor recovery [27]. Nonetheless, the role of contralesional M1 for reorganization after stroke and mechanisms associated with shifts in interhemispheric functional connectivity motor cortices is still debatable.

Xu and colleagues [28] demonstrated decrease of the contralesional primary sensorimotor connectivity in the acute state; Rehme and colleagues [29] and Golestani and colleagues [30] showed the opposite. Greater activation of the contralesional hemisphere may occur to compensate for loss of connectivity in the ischemic region. Functional neuroimaging studies suggest that activity within the sensorimotor network, or ipsilesional motor cortex, is most abnormal early after hemiparetic stroke and motor recovery is related to normalization of its activity [27].

Animal models of stroke have provided evidence for a complex cascade of events enabling changes in structural connections and synaptic transmission [31]. These changes occur not only in the vicinity of the lesion but also in remote brain regions. For example, studies in rats demonstrated reduced interhemispheric resting-state connectivity with secondary myelin degeneration of transcallosal fibers within M1 [31]. In humans, it is possible to identify reduced integrity of transcallosal fiber tracts

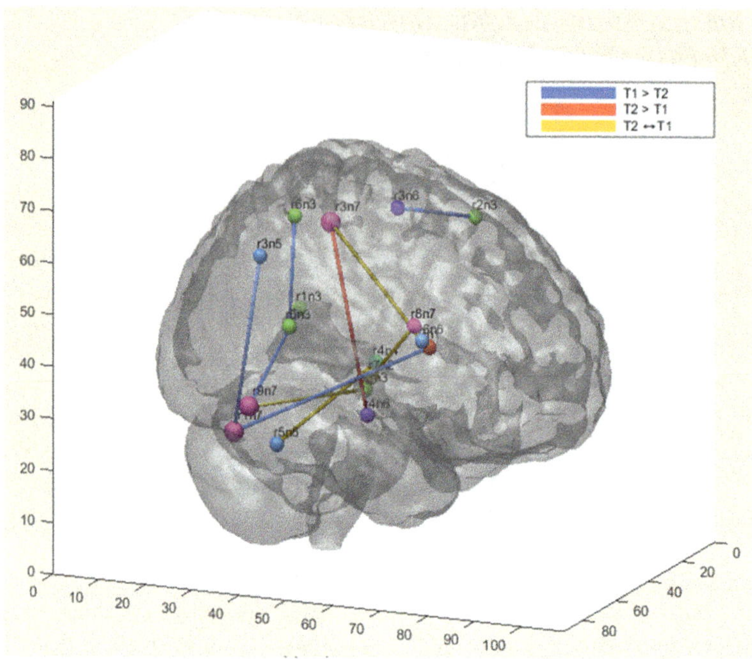

Fig. 4.2 Example of functional connectivity between the primary motor region and others networks in stroke patients, in longitudinal analysis – time 1 (T1) and time 2 (T2)

between motor areas that might result from a secondary degeneration of fibers connected to the lesion zone [32]. Lu and colleagues [33] reported reduced functional connectivity between M1 and the contralateral cerebellum in patients with pontine lesions.

We examined motor activation patterns intra- and interhemispheric in patients with stroke compared to healthy controls and the involvement of other functional networks besides the motor network. We observed that among patients with and without preserved function, functional connectivity between the primary motor region and the contralateral hemisphere was increased compared to controls. Nonetheless, only patients with decreased function exhibited decreased functional connectivity between executive control, sensorimotor, and visuospatial networks [34]. Possibly, functional recovery after stroke is associated with preserved functional connectivity of motor to nonmotor networks [34]. A recent study found that stroke patients with such greater capacity for global information integration achieved better performance in a sensorimotor skill training [35]. Thus, one hypothesis is that faster global information exchange may facilitate new functional networks' configurations [36] (Fig. 4.2).

Rehme and colleagues [12] believed that stroke lesions do not globally reduce connectivity in all functional systems of the brain but specifically alter connectivity of areas connected to that lesion. The alterations affect the communication efficiency in a given functional network which is closely related to behavioral deficits after stroke. Accordingly, the damage of hub regions has the strongest impact on local and global information transfer. In addition, more random network architectures with less local but high global efficiency seem to promote the relearning of sensorimotor skills but may also explain why performance is often less stable, even in well-recovered patients.

While functional resting-state correlations revealed a reduction of interhemispheric connections after stroke, the most consistent finding from effective connectivity analyses belongs to reduced intrahemispheric interactions in the ipsilesional hemisphere [12]. Grefkes and colleagues [37] showed that compared with healthy subjects, stroke patients with relatively poor motor performance exhibit an enhanced inhibitory influence from contralesional to ipsilesional M1 during movements of the paretic hand [37]. The hypothesis that this inhibition might contribute to the motor deficit of the patients is further substantiated by findings from intervention studies which demonstrated that reducing inhibitory influences from contralesional M1 via repetitive transcranial magnetic stimulation (rTMS) induces significant improvements in hand motor performance [38].

One factor that seems to determine the functional role of contralesional M1 for motor performance of the stroke-affected hand is the time that has elapsed since stroke onset. A longitudinal DCM study with recovering stroke patients showed that interhemispheric inhibitory influences from ipsilesional motor areas to contralesional M1 are significantly diminished in the first few days after onset [39]. After 2 weeks, this apparent disinhibition of contralesional M1 is accompanied by a promoting influence from contralesional to ipsilesional M1, particularly in patients with severe motor deficits. Hence, in the subacute phase, contralesional M1 seems to support activity of motor areas in the lesioned hemisphere. However, after 3–6 months, this supportive influence may turn into inhibition in those patients with incomplete motor recovery. Rehme and colleagues [12] suggest that motor deficits after stroke are not only caused by direct disruption of descending motor pathways but may also depend on a less effective communication between premotor areas and M1 in the lesioned hemisphere.

Another important aspect in stroke is the distributed networks for the control of behavior. It is entirely unknown what patterns of interaction within a network are most closely associated with behavioral deficits after injury [25]. Carter and colleagues [25] showed a critical behavioral significance of interhemispheric connectivity between homologous regions of a task-relevant network. In the dorsal attention network, the breakdown of interhemispheric functional connectivity correlated with difficulty in detecting targets in the contralesional visual field. Besides, connectivity scores in the dorsal attention network also correlated with measures of upper extremity and walking function. Physiological studies indicate that the dorsal attention network is important not only for stimulus selection but also for selection of limb responses [40], especially early after injury [41]. Therefore, the broader behavioral significance of the dorsal network can correspond to its involvement in a larger range of behavioral functions [25].

Vicentini and colleagues [42] found a relationship between increased default mode network (DMN) functional connectivity and depression and anxiety symptoms after stroke. DMN plays an important role in the emotional processing and is an anatomical-functional unit engaged in the processing of self-referential stimuli [43]. According to Vicentini and colleagues [42], patients with depression and anxiety symptoms show an increased connectivity in the left inferior parietal gyrus and left basal nuclei, when compared to patients without symptoms. Specific correlation between depression and anxiety scores and DMN functional connectivity indicates that depression symptoms are correlated with increased connectivity in the left inferior parietal gyrus, while anxiety symptoms are correlated with increased connectivity in the cerebellum, brainstem, and right middle frontal gyrus.

Insights into Stroke Mechanisms from Structural Connectivity Studies

The previous items deal with functional MRI methods that can be employed to identify altered patterns of brain activity after stroke. Changes in functional brain organization, however, are often closely associated with structural modification of neuronal elements in the brain. DTI offers a MRI-based means for the assessment of neuroanatomical changes associated with brain injury and repair [2]. Studies in patients and animals have reported loss of FA in ipsilesional white matter subacutely after stroke, which has been linked to demyelination or axonal loss [44, 45], and elevated FA in ipsilesional corticospinal tracts in chronic stroke patients, which could be associated with chronic reduction in edema or improved motor function [27, 44]. Other example was shown with recovery of FA values 3 years after stroke, which was observed in the internal capsule of patients with upper limb impairments subjected to rehabilitation [46]. Structural integrity of the corticospinal pathway appears critical for a favorable outcome in sensorimotor performance after stroke [2]. Decreases [47] and increases [44, 48] in FA have been observed, and though the exact mechanism of the increase remains unclear, it is possible that perpendicular diffusion is restricted hyperacutely [4].

Finally, other MRI-based techniques not related to brain connectivity assessment may also be used to understand the brain changes and repair mechanisms associated with stroke. For example, voxel-based morphometry (VBM) can be used to detect significant cortical gray matter volume changes in patients with stroke. Matsuoka and colleagues assessed the correlation between changes in cortical volumes and changes in neuropsychiatric symptoms during 6 months following a stroke. They found significant volume reductions in the anterior part of the posterior cingulate cortex and correlation between volume reductions and apathy scale. The delayed atrophy may reflect degeneration secondary to neuronal loss due to stroke. Such degeneration might have impaired control of goal-directed behavior, leading to the observed increase in apathy [49].

Conclusion

The field of experimental neuroimaging with MRI is rapidly expanding. Improvements in hardware and pulse sequences that decrease scan time while maintaining resolution will continue to impact the field. Post-processing strategies must evolve to encompass these increasingly complicated data sets. It also seems clear that multimodal imaging strategies are necessary to develop more detailed patient profiles that can be used to predict outcome [4]. Multicenter studies are increasingly needed to prove these technologies and their usability with stroke patients.

References

1. Van Horn JD, Pelphrey KA (2015) Neuroimaging of the developing brain. Brain Imaging Behav 9(1):1–4
2. Dijkhuizen RM, van der Marel K, Otte WM, Hoff EI, van der Zijden JP, van der Toorn A et al (2012) Functional MRI and diffusion tensor imaging of brain reorganization after experimental stroke. Transl Stroke Res 3(1):36–43
3. Corbetta M, Kincade MJ, Lewis C, Snyder AZ, Sapir A (2005) Neural basis and recovery of spatial attention deficits in spatial neglect. Nat Neurosci 8(11):1603–1610
4. Farr TD, Wegener S (2010) Use of magnetic resonance imaging to predict outcome after stroke: a review of experimental and clinical evidence. J Cereb Blood Flow Metab 30(4):703–717
5. Ward NS (2005) Neural plasticity and recovery of function. Prog Brain Res 150:527–535
6. Friston K (2002) Functional integration and inference in the brain. Prog Neurobiol 68 (2):113–143
7. Stephan KE, Harrison LM, Kiebel SJ, David O, Penny WD, Friston KJ (2007) Dynamic causal models of neural system dynamics:current state and future extensions. J Biosci 32(1):129–144
8. Le Bihan D (2003) Looking into the functional architecture of the brain with diffusion MRI. Nat Rev Neurosci 4(6):469–480
9. Basser PJ, Jones DK (2002) Diffusion-tensor MRI: theory, experimental design and data analysis – a technical review. NMR Biomed 15(7–8):456–467
10. Ogawa S, Lee TM, Kay AR, Tank DW (1990) Brain magnetic resonance imaging with contrast dependent on blood oxygenation. Proc Natl Acad Sci U S A 87(24):9868–9872
11. Biswal B, Yetkin FZ, Haughton VM, Hyde JS (1995) Functional connectivity in the motor cortex of resting human brain using echo-planar MRI. Magn Reson Med 34(4):537–541
12. Rehme AK, Grefkes C (2013) Cerebral network disorders after stroke: evidence from imaging-based connectivity analyses of active and resting brain states in humans. J Physiol 591(1):17–31
13. He BJ, Snyder AZ, Zempel JM, Smyth MD, Raichle ME (2008) Electrophysiological correlates of the brain's intrinsic large-scale functional architecture. Proc Natl Acad Sci U S A 105 (41):16039–16044
14. Fox MD, Raichle ME (2007) Spontaneous fluctuations in brain activity observed with functional magnetic resonance imaging. Nat Rev Neurosci 8(9):700–711
15. Beckmann CF (2012) Modelling with independent components. NeuroImage 62(2):891–901
16. Bullmore E, Sporns O (2009) Complex brain networks: graph theoretical analysis of structural and functional systems. Nat Rev Neurosci 10(3):186–198
17. Fox MD, Greicius M (2010) Clinical applications of resting state functional connectivity. Front Syst Neurosci:17(4)

18. Auer DP (2008) Spontaneous low-frequency blood oxygenation level-dependent fluctuations and functional connectivity analysis of the 'resting' brain. Magn Reson Imaging 26 (7):1055–1064
19. Friston KJ (2011) Functional and effective connectivity: a review. Brain Connect 1(1):13–36
20. Friston KJ, Buechel C, Fink GR, Morris J, Rolls E, Dolan RJ (1997) Psychophysiological and modulatory interactions in neuroimaging. NeuroImage 6(3):218–229
21. McLntosh AR, Gonzalez-Lima F (1994) Structural equation modelling and its application in network analysis in functional brain imaging. Hum Brain Mapp 2:2–22
22. Roebroeck A, Formisano E, Goebel R (2005) Mapping directed influence over the brain using Granger causality and fMRI. NeuroImage 25:230–242
23. Friston KJ, Harrison L, Penny W (2003) Dynamic causal modelling. NeuroImage 19 (4):1273–1302
24. van Meer MP, van der Marel K, Wang K, Otte WM, El Bouazati S, Roeling TA et al (2010) Recovery of sensorimotor function after experimental stroke correlates with restoration of resting-state interhemispheric functional connectivity. J Neurosci 30(11):3964–3972
25. Carter AR, Astafiev SV, Lang CE, Connor LT, Rengachary J, Strube MJ et al (2010) Resting interhemispheric functional magnetic resonance imaging connectivity predicts performance after stroke. Ann Neurol 67(3):365–375
26. Park CH, Chang WH, Ohn SH, Kim ST, Bang OY, Pascual-Leone A et al (2011) Longitudinal changes of resting-state functional connectivity during motor recovery after stroke. Stroke 42 (5):1357–1362
27. Schaechter JD (2004) Motor rehabilitation and brain plasticity after hemiparetic stroke. Prog Neurobiol 73(1):61–72
28. Xu HQW, Chen H, Jiang L, Li K, Yu C (2014) Contribution of the resting-state functional connectivity of the contralesional primary sensorimotor cortex to motor recovery after subcortical stroke. PLoS One 9:e84729
29. Rehme AK, Eickhoff SB, Rottschy C, Fink GR, Grefkes C (2012) Activation likelihood estimation meta-analysis of motor-related neural activity after stroke. NeuroImage 59 (3):2771–2782
30. Golestani AM, Tymchuk S, Demchuk A, Goodyear BG, Group V-S (2013) Longitudinal evaluation of resting-state FMRI after acute stroke with hemiparesis. Neurorehabil Neural Repair 27(2):153–163
31. van Meer MP, Otte WM, van der Marel K, Nijboer CH, Kavelaars A, van der Sprenkel JW et al (2012) Extent of bilateral neuronal network reorganization and functional recovery in relation to stroke severity. J Neurosci 32(13):4495–4507
32. Wang LE, Tittgemeyer M, Imperati D, Diekhoff S, Ameli M, Fink GR et al (2012) Degeneration of corpus callosum and recovery of motor function after stroke: a multimodal magnetic resonance imaging study. Hum Brain Mapp 33(12):2941–2956
33. Lu J, Liu H, Zhang M, Wang D, Cao Y, Ma Q et al (2011) Focal pontine lesions provide evidence that intrinsic functional connectivity reflects polysynaptic anatomical pathways. J Neurosci 31(42):15065–15071
34. Almeida SR, Vicentini J, Bonilha L, De Campos BM, Casseb RF, Min LL (2016) Brain connectivity and functional recovery in patients with ischemic stroke. J Neuroimaging 27(1):65–70
35. Buch ER, Modir Shanechi A, Fourkas AD, Weber C, Birbaumer N, Cohen LG (2012) Parietofrontal integrity determines neural modulation associated with grasping imagery after stroke. Brain 135(Pt 2):596–614
36. Beharelle AR, Kovačević N, McIntosh AR, Levine B (2012) Brain signal variability relates to stability of behavior after recovery from diffuse brain injury. NeuroImage 60:1528–1537
37. Grefkes C, Nowak DA, Eickhoff SB, Dafotakis M, Küst J, Karbe H et al (2008) Cortical connectivity after subcortical stroke assessed with functional magnetic resonance imaging. Ann Neurol 63(2):236–246

38. Grefkes C, Nowak DA, Wang LE, Dafotakis M, Eickhoff SB, Fink GR (2010) Modulating cortical connectivity in stroke patients by rTMS assessed with fMRI and dynamic causal modeling. NeuroImage 50(1):233–242
39. Rehme AK, Eickhoff SB, Wang LE, Fink GR, Grefkes C (2011) Dynamic causal modeling of cortical activity from the acute to the chronic stage after stroke. NeuroImage 55(3):1147–1158
40. Astafiev SV, Shulman GL, Stanley CM, Snyder AZ, Van Essen DC, Corbetta M (2003) Functional organization of human intraparietal and frontal cortex for attending, looking, and pointing. J Neurosci 23(11):4689–4699
41. Regnaux JP, David D, Daniel O, Smail DB, Combeaud M, Bussel B (2005) Evidence for cognitive processes involved in the control of steady state of walking in healthy subjects and after cerebral damage. Neurorehabil Neural Repair 19(2):125–132
42. Vicentini JE, Weiler M, Almeida SR, de Campos BM, Valler L, Li LM (2016) Depression and anxiety symptoms are associated to disruption of default mode network in subacute ischemic stroke. Brain Imaging Behav 11(6):1571–1580
43. Northoff G, Heinzel A, de Greck M, Bermpohl F, Dobrowolny H, Panksepp J (2006) Self-referential processing in our brain--a meta-analysis of imaging studies on the self. NeuroImage 31(1):440–457
44. Liu Y, D'Arceuil HE, Westmoreland S, He J, Duggan M, Gonzalez RG et al (2007) Serial diffusion tensor MRI after transient and permanent cerebral ischemia in nonhuman primates. Stroke 38(1):138–145
45. Assaf Y, Pasternak O (2008) Diffusion tensor imaging (DTI)-based white matter mapping in brain research: a review. J Mol Neurosci 34(1):51–61
46. Stinear CM, Barber PA, Smale PR, Coxon JP, Fleming MK, Byblow WD (2007) Functional potential in chronic stroke patients depends on corticospinal tract integrity. Brain 130 (Pt 1):170–180
47. Morita N, Harada M, Uno M, Furutani K, Nishitani H (2006) Change of diffusion anisotropy in patients with acute cerebral infarction using statistical parametric analysis. Radiat Med 24 (4):253–259
48. Schaechter JD, Fricker ZP, Perdue KL, Helmer KG, Vangel MG, Greve DN et al (2009) Microstructural status of ipsilesional and contralesional corticospinal tract correlates with motor skill in chronic stroke patients. Hum Brain Mapp 30(11):3461–3474
49. Matsuoka K, Yasuno F, Taguchi A, Yamamoto A, Kajimoto K, Kazui H et al (2015) Delayed atrophy in posterior cingulate cortex and apathy after stroke. Int J Geriatr Psychiatry 30 (6):566–572

Chapter 5
New Insights in Brain Tumor Magnetic Resonance Investigation

Remy Guillevin

During the last 25 years, conventional MRI had progressively replaced CT scanner for diagnostic and follow-up brain tumor imaging. Due to its intrinsic performance in brain tissue differentiation, this noninvasive technique is now available in every brain imaging centers. However, referring to traditional anatomic concepts, the so-called conventional MRI approach, using limited number of contrasts, has demonstrated its poor sensitivity and specificity. Today, the management of brain tumors in dedicated centers requires specific sequences for "basic" exploration as far as such lesions could not be explored anymore by conventional "anatomic" data. It appears mandatory to have minimum informations about the metabolic counterparts of growing lesions, whatever at their diagnostic and therapeutic stages. For these reasons, this chapter will focus on the recent and promising features given by both heuristic and technical advances.

Recent and huge increase of knowledge (and data!) in both genomic and metabolic fields of the brain and its tumoral counterparts has challenged the technical capabilities of MR investigation, leading to major developments in the metabolic exploration of the organs. Until recent past, these technics were reserved to experimental imaging on animals. Today, the availability of high-field (3 T) or ultra-high-field (7 T) strengths for clinical use opens wide heuristic possibilities using specific tools as multinuclear and multicontrast imaging, as well as local perfusion quantitation. Here arises the huge challenge for the neuroradiologist: to provide comprehensive analysis of the image content, e.g., the parameters beyond the images.

Last, the ultimate role for neuroimaging can be achieved by integration of these parameters into a transversal knowledge allowing hierarchical and logical understanding of the links between them. This requires to group into a same research team different competencies including genomic, neurochemistry/biology, MR physics,

R. Guillevin (✉)
Clinical Neuroimaging & Research team DACTIM-MIS/LMA CNRS 7348, CHU and University of Poitiers, 86020, Poitiers, France
e-mail: Remy.GUILLEVIN@chu-poitiers.fr; remy.guillevin@math.univ-poitiers.fr

© Springer International Publishing AG, part of Springer Nature 2018
C. Habas (ed.), *The Neuroimaging of Brain Diseases*, Contemporary Clinical Neuroscience, https://doi.org/10.1007/978-3-319-78926-2_5

neuroimaging, and mathematics, which appear to be the cross knowledge allowing (i) extraction of data sets arising from each type of sequence by specific tools and (ii) the only "global" provider for analysis of all the image content.

The BOLD Issue

Arising during the 1990s, this MR technique has been the most challenged one by neurosurgery issues. This technique indirectly assesses the changes of oxy- and deoxyhemoglobin present in the microvasculature. In patients with longstanding mass lesions, especially low-grade gliomas, physical distortion and compensatory reorganization may alter normal network and relationships [1–3]. Numerous publications have been provided upon the task-induced fMRI ability to identify preoperatively sensorimotor cortex. However, in the neuro-oncologic field, it must be kept in mind that this kind of sequences is "contrast" based on metabolic phenomenon and reflects the CMRO2 that may be altered by tumor vicinity from the cortical regions explored. Then, the neurovascular decoupling, due to tumor oxygen consumption, may contribute to BOLD level variations. Their results should be misinterpreted. In addition, several studies have suggested that signal distortion may arise from larger veins and thus induce a low spatial localization [4]. Last, head movements degrade image and data quality, as well as other susceptibility artifacts (air/liquid/bone). Moreover, the development of awaken surgery during the last 10 years in specialized centers allowed the availability of electrophysiology (direct cortical stimulation) and neuropsychology monitoring *during* the neurosurgery procedure.

A Conceptual Approach Using BOLD: Resting State fMRI

The major goal of preoperative planning remains to accurately predict the functional consequences of surgically removed lesions either immediately following surgery or in the long-term (i.e., accounting for plasticity-induced recovery).

The so-called virtual brain will incorporate localization and network-based approaches to neuroanatomy and, then, will model brain function in a holistic manner. Brain function is here considered as global and not limited to one region or network. The brain is then conceptualized as a collection of nodes that are connected via edges [5]. Connectome analysis described the brain as nodes that are circumscribed into brain regions and edges within the degree of synchronization of endogenous signals (e.g., as functional connectivity). The resulting organization balances local specialization with distributed connectivity and shortcuts between regions [6–8]. Connectome analysis of resting-state fMRI provides theoretical advantages over task-based fMRI in accounting for BOLD signal artifacts related to brain tumors. Whereas task-based fMRI is known to be vulnerable to tumor-related susceptibility artifacts, resting-state fMRI data allows removal of noise-related signal in a data-driven and physically principled manner using multi-echo-independent component analysis (ME-ICA).

This analysis allows neurosurgeons to obtain not only brain connectivity mapping but also intuitive modelling of lesions and plasticity, to gain an understanding of the effects at both a local and global level. Graph theory analysis of the connectome allows understanding of the importance of regions to network function and the consequences of their impairment or excision. The neurosurgeon can then explore mechanisms of putative plasticity using models such as connection rewiring, alternative routes for information flow, or reactivation of redundant pathways [9, 10].

New Standards for Brain Tumor Investigation

It is now well established that metabolic MR imaging required utilization of both spectroscopy and perfusion [11]. Increased resonance of both choline (^{1}H-MRS) and phosphoesters (^{31}P-MRS), as closely linked to membrane anabolism (whatever cellular or vascular), represents the metabolic counterpart of a tumoral process. The choline index is linearly correlated to Ki-67 index antigenic marker of tumoral proliferation [12]. NAA resonance decreases resulting from neuroaxonal replacement by tumoral tissue, thus providing a dynamic index of tumoral expansion, as choline to NAA index (CNI). While increased in low-grade gliomas, myoinositol resonance resulting from glial-astrocytic activation is not specific of an oncotype [12]. Free lipid resonance may occur in case of necrotic process (high-grade glioma metastasis) or in evolving low-grade gliomas in their pre-anaplastic form [12, 13]. This resonance is also specific in lymphoma referring to their usual aspect on basic images, e.g., without evidence of necrosis [14] (Fig. 5.1).

Dynamic aspects More than precise determination of the metabolite concentrations, temporal variation of metabolic profiles, as CNI and CCrI, may provide decisive arguments during therapeutic monitoring. Thus, initial evaluation of low-grade gliomas may allow the successive identification of increasing CNI, lactate resonance, free lipid resonance, increase of rCBV, and increasing pHi, all those parameters in their chronologic pattern highlighting pre-anaplastic evolution [15]. Moreover, lactate resonance arising is predictive of rCBV increase over 1.75 [16], which is predictive of overall survival dramatic decrease [17]. We can see here the input of "metabolic" MRI in transformative continuum of glioma (Fig. 5.2).

Therapeutic monitoring using ^{1}HMRS may be of interest by providing additional information to volumetric assessment predictive factors of LGG response under temozolomide treatment. Namely, the mean relative decrease of Cho/Cr ratio slope at 3 months after the beginning of chemotherapy by temozolomide is predictive of the tumor response over 14 months of follow-up [18].

During postradiation monitoring, normalized rCBV increase over 2.6 (significantly) in case of tumor progression [19]. A combination of thresholds for Cho/Cr

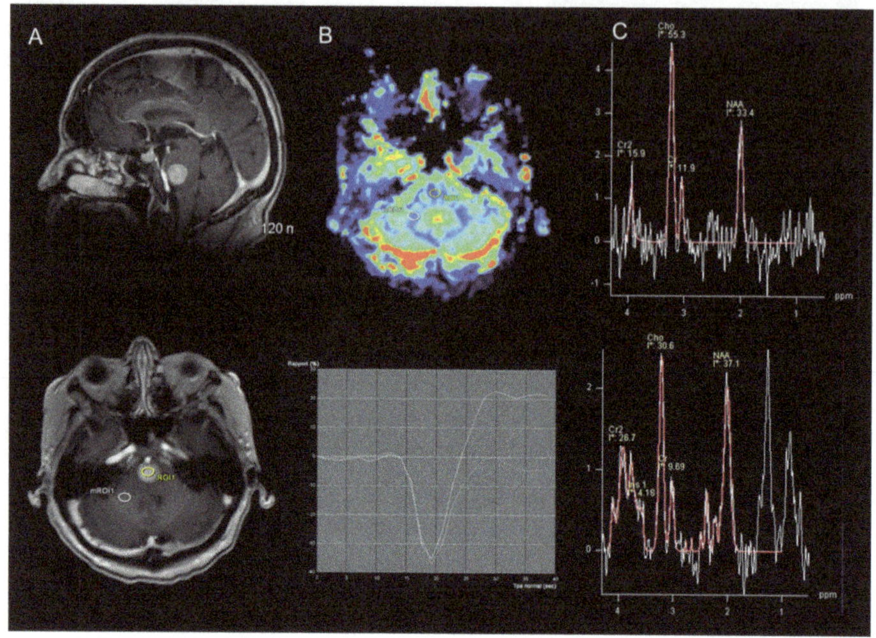

Fig. 5.1 Multiparametric MR examination of brainstem tumor including: (**a**) Intense and homogeneous without necrosis post-contrast enhancement of the tumor process. (**b**) No increase of rCBV with 150% PSR. (**c**) Strong resonance of free lipids and increase CCri at 2.1. *Diagnosis proposed PCNSL histologically proven*

and Cho/Naa may identify tumor progression with 83–84.5 specificity and 89–90% sensitivity [20, 21].

From Biometabolic Model to the Genomic profile Through the Metabolic Signal

Referring to the ability of proton MRS to detect and to quantify some of the metabolic counterparts of genomic mutations, it is possible for IDH1 mutation to detect and quantify 2-hydroxyglutarate resonance, a so-called oncometabolite, which accumulate [22]. Depending on magnetic field strength used, 2-hydroxyglutarate can be directly detected or using point-resolved spectroscopy within spectral difference editing. In addition, some authors have demonstrated the interest of longitudinal 2-hydroxyglutarate quantitation for monitoring treatment response in IDH1 mutant patients [23–25]. The authors developed a specific 3D sequence for oversampling the tumor and to avoid difficulties due to heterogeneity. It is potentially available in centers using routinely proton magnetic resonance spectroscopy.

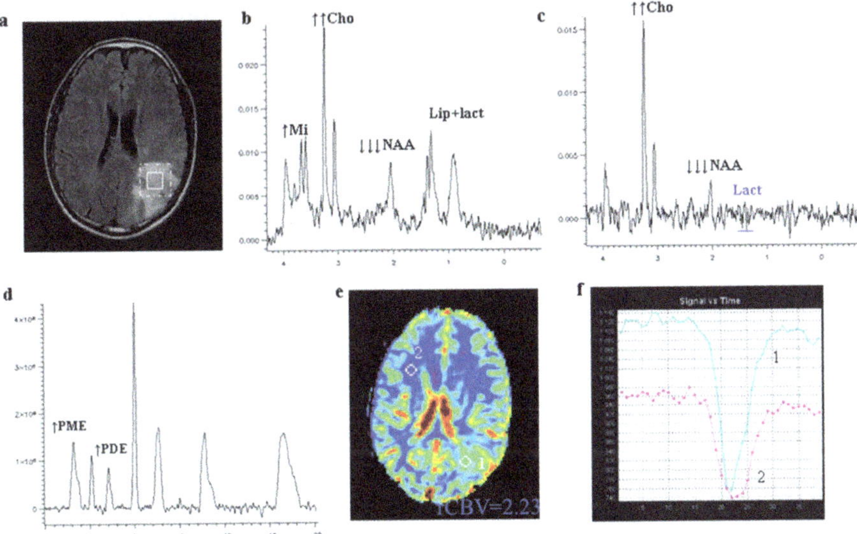

Fig. 5.2 (**a**) Location of proton voxel and phosphorus voxel on the left frontal lesion, T2 FLAIR sequence. (**b**) 1H MRS short TE showing an increase of choline, a dramatic decrease of NAA, and large presence of lipid/lactate. (**c**) 1H MRS long TE, presence of lactate. (**d**) 31P MRS, an increase of PMEs, deltaPi = 4.92, pHi = 7.08. (**e**) Perfusion MRI, a mapping of cerebral blood volume (rCBV). (**f**) Perfusion MRI, first pass curves: hyperperfusion with rCBV = 2.23

In a recent experimental work [26], the authors demonstrated that the IDH1 mutation may lead to a decrease of lactate production from pyruvate ("anti-Warburg effect"). Using multinuclear magnetic resonance spectroscopy, it is possible to fit this metabolic process: ^{31}P-MRS for steady-state for assessment of intracellular pH and PCr and ^{13}C-MRS with dynamic nuclear polarization for pyruvate-lactate fluxes calculation and *Warburg effect* assessment. Reduced flux of hyperpolarized [1-^{13}C] pyruvate to hyperpolarized [1-^{13}C] lactate was suggested to be due to reduced MCT1 and MCT4 expression in IDH1-mutated cells. Increased intracellular pH in both types of cells was not modified by the drop of intracellular lactate. The vascular counterpart of this "biochemical reprogramming" is also accessible by perfusion MRI, during the same time of examination. Kickingereder et al. [22] correctly predicted *IDH* mutation status in 88% of patients. It has been established that increasing levels of 2-hydroxyglutarate leads to indirectly decrease of hypoxia-inducible factor 1-alpha, thus limiting angiogenic growing [27].

Whereas some of those results are obtained in experimental conditions with limited extrapolation to human- in vivo conditions, we must noticed that they are partly consistent with theoretical conjectures expressed in previous papers [28, 29]. Yet, at this stage, before the recent knowledge assessing the links between genomic and metabolic modifications, alterations of MCT properties have been suggested by previously published models.

New "Metabolic" Contrasts: CEST Imaging or Molecular Imaging

CEST contrast is obtained after applying a saturation pulse at the specific resonance frequency of an exchanging proton site. The saturated spin is transferred to bulk water, and then specific molecular information can be obtained [30], within a so-called negative contrast [31]. In Amide Proton Transfer, chemical exchange occurs between protons of free tissue water (bulk water) and amide groups ($-NH$) of endogenous mobile proteins and peptides. Those exchangeable protons are more abundant in tumor tissues than in healthy tissues [32]. Other previous reports demonstrated that the APT signal increased by 3–4% in tumor compared with peritumoral brain tissue and human brain tumor at 3 T [33]. However, those chemical exchanges are dependent from other metabolic changes, as the reduction in intracellular pH after treatment with TMZ [27]. This would also result in a decrease in APT signal because $-NH$ proton exchange is known to be catalyzed by base [34], which is otherwise consistent with natural history of gliomas, as previously suggested in experimental and clinical studies [35, 36]. This point suggests that APT sequence may be used concomitantly with phosphorus spectroscopy for pHi monitoring.

The results of the different studies above-cited suggest that APT imaging may serve as a useful biomarker for monitoring treatment response during chemotherapy and follow-up after treatment, including for radionecrosis identification [30]. This can be implemented on a standard clinical scanner [33, 37].

Improving Data Analysis and Quantitation from Spectroscopy

Due to increasing importance of spectroscopic, even multinuclear, tools in brain tumor studies, the question arise of optimized quantitation with standardization of results coming from different imaging centers and then from different series and scanners, including different magnetic field strengths [38]. Moreover, the data post-processing is also heterogeneous as numerous software are available in many MR centers, whether commercial, freeware, or in-house. This may lead to important errors for differential diagnosis and follow-up characterization.

Therefore, automatic quantification, using software based on mathematical algorithms, is required to replace the interaction quantification for the sake of more efficiency and robustness. In their paper, Dou et al. [39] present a new automatic quantitative approach based on the convex envelope (AQoCE), compared with LCModel. This method leads to increase both specificity and sensitivity of MR spectroscopy, thus suggesting that automated processing of MR data may avoid intra-individual variability [40, 41]. Additionally, spectroscopic parameters are integrated in a global multiparametric analysis including contrast enhancement, related cerebral blood volume, and ADC values extracted from diffusion sequence. In a study on 120 patients, the authors found multivariate models with similar accuracy for predicting the grade II and the malignant transformation to grade III or IV. This "kinetic" approach is used in routine in some centers with expertise in the field, for the therapeutic follow-up and when the surgical resection is not possible [41]. Last, new proton high-resolution multivoxel sequences, so-called Laser or Mega Laser, allow 3D metabolic mapping, including 2-hydroxyglutarate and glutamate, thus marking the tumoral spatial heterogeneity [24]. Combined acquisitions allowing comparison of vessel compartment and permeability may provide simultaneously parametric maps of CBV, Vp, and K(trans), and the region(s) of highest value (hotspot) can be measured on each map and compared with histograms of rCBV, Vp, and K(trans), with global increased sensitivity of perfusion-weighted sequence [42].

Mathematic Tools: Statistical and Realistic Models

A New Generation of "Statistic" Approach: Machine Learning and Radiomics

While mostly reported in subjective and qualitative terms, clinical imaging includes a very large amount of data, which can be "mined" by machine learning tools and validated as quantitative imaging biomarkers. These data can be then used as predictive markers for diagnosis, prognosis, and therapeutic planning of adult brain tumors.

Imaging features in their functional aspect (image-based feature) are mandatory for radiomics model building. Anatomic information (location), physiologic data as cell density and diffusivity, blood flow, contrast enhancement, sodic compartmentation, and metabolic concentrations from spectroscopy allow discrimination of tumor regions with different environments, reflecting *tumor heterogeneity* [43]. Beyond these features the so-called computer vision allows extraction of quantitative data [44]. Computational image descriptors quantify different parameters at different scales from ROIs. The scale-invariant feature transform (SIFT) [45] is obtained from key point detection using a difference of Gaussian function and local image-gradient measurement, allowing quantitative assessment of tumor shape. Local binary patterns (LBP) [46] allow descriptor of pixel tested with its immediate neighborhood [47] sensitive to small gray-level differences not apparent to human observer.

Histogram of oriented gradients (HOG) 34 features have also proved to be efficient feature descriptors for quantifying image-gradient statistics that could be helpful in discriminating different tumor subtypes [48]. However the descriptors must (i) be able to identify specific patterns correlated with the clinical outcomes of patients, (ii) remain stable under various image acquisition parameters, and (iii) overcome the consequences of dynamic biological variations (test-retest and interobserver stability) [49]. These features may reflect tumor heterogeneity before treatment as well as after radiotherapy and then help to refine the outcome prognosis

[50, 51], providing helpful information regarding pseudoprogression and chemoradiation-induced necrosis [52], whereas conventional MRI information fail to discriminate. The so-called radiogenomics can provide quantitative noninvasive assessment of gene expression in glioblastomas [53], as well as their spatial distribution within the tumor [54]. During last years, different learning strategies have been developed in the brain tumor MRI field, as supervised, unsupervised, and semi-supervised forms, thus leading to extract specific informations. Depending on the availability of class labels or not, training samples are used for classifiers definition. Then, unsupervised clustering approach is used for automating brain tumor segmentation [55], which trend to become a challenging issue for brain tumor management. For this specific application, deep learning, by defining network architecture [56], is getting growing interest by introducing a so-called convolutional neural network-based approach [57]. Semi-supervised methods allow brain tumor prognosis even with partially staging labels missing [58]. However, it should be noted at this stage that large clinical data sets are required to make those different algorithms to be effective and pertinent, which emphasize the necessity of collecting them from multicenter cohorts for low-incident (sub) types of tumors (data sharing). For this reason, studies concerning those topics are mostly focused on glioblastomas.

The huge increase of both molecular data and radiomics features may then assess treatment.

Specific tools: texture and fractal analysis MR texture analysis. By quantification of histogram (with and without filtration) based on standard deviation (SD), which represents the width of the histogram or degree of variation from the mean pixel value, the distribution of gray levels within an image can be assessed to obtain texture features reflecting intralesional heterogeneity [59, 60]. The initial filtration step employs a Laplacian of Gaussian (LoG) band-pass filtration, thus extracting and highlighting image features at different sizes corresponding to spatial scale filter (SSF). The filtration step is mandatory for removing image heterogeneity due to noise and then magnifying biological important heterogeneity. The order (e.g., first or second) of histogram statistical parameters is determinant for the results. Some authors have selected different texture analysis parameters, such as entropy derived from ADC maps [27]; this point may be promising, as gliomas are known to exhibit thermodynamic and metabolic alterations with increased entropy rate. More specifically, as detailed in Vallée et al. [61, 62], upregulation of WNT/beta catenin pathway leads to cell proliferation via protein synthesis and angiogenesis. As MCT1 is also a target gene for beta catenin, accumulation of intracellular lactate, as a consequence of Warburg effect, will be also assessed by imaging, namely, proton spectroscopy [11]. Portrayed texture from necrosis, which may be important to detect during LGG transformation, may be thus extracted, whereas TA of T1 post-contrast may provide "true" quantitation of intralesional heterogeneity.

SWI-LIV Based on the following process after segmentation, intensity correction, and rescaling the intensity image, the SWI local image variance is calculated using the following formula: $LIV = G(X^2) - [G(X)]^2$ (X is the preprocessed image and G represents a Gaussian low-pass filtering). Grabner et al. established significant differences in SWI-LIV values dependent on the IDH1 mutational status and type of MRI contrast enhancement, thus leading to improve preoperative assessment of LGG. [63].

Fractal analysis The fractal dimension (FD) is a non-integer number that characterizes the morphometric variability of a complex and irregular shape [64]. After SW images automated computation, and extraction of both volume fraction of SWI signals within the tumors (signal ratio) and morphological self-similar features (fractal dimension [FD]), the results can be then correlated with each histopathological type of tumor and increase the accuracy of initial diagnosis (e.g., between WHO II and WHO III and malignant transformation of LGG).

Why Building Realistic Mathematical Models?

After various statistical methods, it appears important to challenge the dynamic aspect of biological interactions by realistic mathematical models. The linear correlation between the Lac/Cr ratio and increased rCBV is consistent with previous experimental results as well as a mathematical model of brain lactate kinetics based on the compartmentalization of brain energy metabolism between glia and neurons, published by Aubert et al. [65]. This model, integrating in vivo data measurements of WHO grade II gliomas, has suggested adaptation of their metabolism along with their evolution by modifying lactate transporters [66] and increasing anaerobic glycolysis to survive the hypoxic conditions [67, 68]. Furthermore, lactate clearance from brain tissue via the bloodstream plays only a minor role; these findings are also consistent with the Kuhr et al. study [69]. Increased lactate production is associated with increased aggressiveness, angiogenesis, and poor prognosis [70, 71].

As abovementioned in this chapter, conventional "radiologic" assessment of tumor changes may be difficult and requires repeated examinations within a long time to assess tumor growth, whatever the equivalent diameter for WHO grade II gliomas or larger tumor diameter on post-contrast T1-weighted images for upgrades. However, it must be noticed that (i) the delay of change in tumor grade is highly variable from one subject to another [72]; (ii) morphological changes are resulting from those of tumor bio-metabolism; and (iii) switching the metabolism of grade II glioma to a much closer metabolism to that of high-grade gliomas occurs very early during the natural history of gliomas [73]. Biomathematical model with simulations from computational metabolic data, including parameters of ^{1}H MRS, ^{31}P MRS, and the MRI perfusion, collected in vivo by magnetic resonance provided promising results (Fig. 5.3).

Considering previous definition of "metabolic" subclasses of WHO grade II gliomas [35], the arising question is the mechanism involved in the decrease in lactate. Regional blood flow, pH value, lactate transport which run via the MCT, and carriers lactate-H$^+$ appear to play a major role [74].

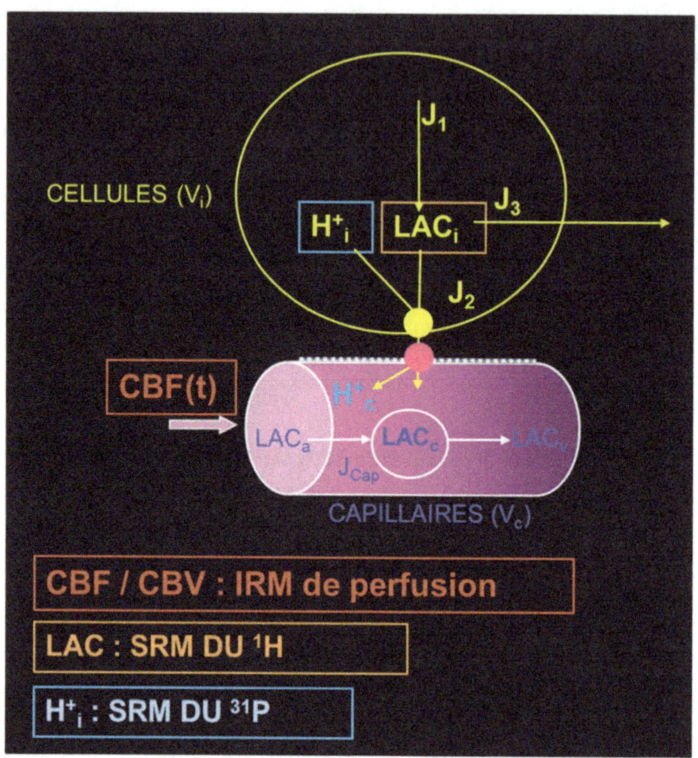

Fig. 5.3 Schematic representation of the model, with parameters input provided by MR (boxes)

Yet, lactate arising, increase of both CBF and pHi is a coherent result that is consistent with experimental study of Hubesh et al. [75]: despite a potential increase in glycolysis, the pHi of tumor cells is likely to increase due to alterations of membrane transporters proton/bicarbonate.

On the other hand, it seemed paradoxical that lactate, which the authors have shown that it would be a very early marker of changes of tumor metabolism [26], has very different values and sometimes a non-monotonic evolution. Taking into account intracellular pH and rCBF by the model of lactate allows to have a mathematic simulation of the high dispersion of results obtained from MRS and perfusion MRI without neither requiring the hypothesis of capillary recruiting highly unlikely in WHO grade II gliomas nor to the capillary proliferation excluded at this stage. However, these results strongly suggest changes in the transport of lactate across the blood-brain barrier and the membranes of tumor cells that are modifications of density and/or kinetic properties of MCT.

This hypothesis can be in relation with the resonance of free lipids observed in lesions showing no detectable lactate resonance. It is consistent with recent data from the literature which suggest also qualitative or quantitative changes of MCTs in glial tumors [76]. This prompted us to further study the properties of the model:

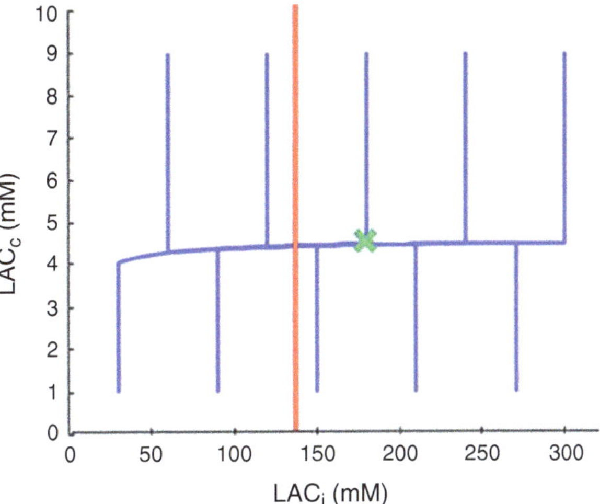

Fig. 5.4 Impact of a significantly decrease of CBF. The red line represents the maximum limit of the viability domain, with maximal acceptable value of LACi

whatever the parameter values, the model has a unique stationary point, which is asymptotically stable. Numerical computations using MATLAB software confirm this point [35]. In addition, this finding is consistent with a clinically observed fact that, within a short time scale from minutes to days, metabolites concentrations within the tumor appear nearly constant. Moreover, explicit and sufficient conditions are derived, ensuring that a stationary point is in a viability domain in the first quadrant (Fig. 5.4).

More generally, in a recent article, the authors [77, 78] studied the well-posedness and properties of a fast-slow system related with brain lactate kinetics. In particular, the existence and uniqueness of nonnegative solutions have been established and linear stability results obtained. Numerical simulations with different values of the small parameter were successfully compared with experimental data.

Conclusion

In conceptual approaches exposed in this chapter, the brain and its tumoral process are considered as a general multiparametrical, multidimensional, multiscaled and dynamic system, e.g., a mathematical object, moving in a viability domain, with specific ranges for values of each parameter. While numerous works are in progress even already published, this field of investigation seems to be at its beginning as the technical possibilities are increasing from year to year, especially in multinuclear magnetic resonance imaging, with the ability of studying different molecules according the nucleus studied. This method requires a transversal, multidisciplinary working team, including mathematicians, MR physicians, and radiologists, thus referring to dedicated centers for brain tumor management.

References

1. Iwasaki S, Nakagawa H, Fukusumi A, Kichikawa K, Kitamura K, Otsuji H et al (1991) Identification of pre- and postcentral gyri on CT and MR images on the basis of the medullary pattern of cerebral white matter. Radiology 179(1):207–213
2. Ojemann GA (1979) Individual variability in cortical localization of language. J Neurosurg 50 (2):164–169
3. Steinmetz H, Fürst G, Freund HJ (1990) Variation of perisylvian and calcarine anatomic landmarks within stereotaxic proportional coordinates. AJNR Am J Neuroradiol 11 (6):1123–1130
4. Tieleman A, Deblaere K, Van Roost D, Van Damme O, Achten E (2009) Preoperative fMRI in tumour surgery. Eur Radiol 19(10):2523–2534
5. Hart MG, Ypma RJF, Romero-Garcia R, Price SJ, Suckling J (2016) Graph theory analysis of complex brain networks: new concepts in brain mapping applied to neurosurgery. J Neurosurg 124(6):1665–1678
6. Bullmore E, Sporns O (2009) Complex brain networks: graph theoretical analysis of structural and functional systems. Nat Rev Neurosci 10(3):186–198
7. Achard S, Salvador R, Whitcher B, Suckling J, Bullmore E (2006) A resilient, low-frequency, small-world human brain functional network with highly connected association cortical hubs. J Neurosci 26(1):63–72
8. Watts DJ, Strogatz SH (1998) Collective dynamics of "small-world" networks. Nature 393 (6684):440–442
9. Fornito A, Zalesky A, Breakspear M (2015) The connectomics of brain disorders. Nat Rev Neurosci 16(3):159–172
10. Stam CJ (2014) Modern network science of neurological disorders. Nat Rev Neurosci 15 (10):683–695
11. Guillevin R, Menuel C, Abud L, Costalat R, Capelle L, Hoang-Xuan K et al (2012) Proton MR spectroscopy in predicting the increase of perfusion MR imaging for WHO grade II gliomas. J Magn Reson Imaging 35(3):543–550
12. Théberge J, Al-Semaan Y, Jensen JE, Williamson PC, Neufeld RWJ, Menon RS et al (2004) Comparative study of proton and phosphorus magnetic resonance spectroscopy in schizophrenia at 4 Tesla. Psychiatry Res 132(1):33–39
13. Hope PL, Costello AM, Cady EB, Delpy DT, Tofts PS, Chu A et al (1984) Cerebral energy metabolism studied with phosphorus NMR spectroscopy in normal and birth-asphyxiated infants. Lancet 2(8399):366–370
14. Taillibert S, Guillevin R, Menuel C, Sanson M, Hoang-Xuan K, Chiras J et al (2008) Brain lymphoma: usefulness of the magnetic resonance spectroscopy. J Neuro-Oncol 86(2):225–229
15. Azzopardi D, Wyatt JS, Cady EB, Delpy DT, Baudin J, Stewart AL et al (1989) Prognosis of newborn infants with hypoxic-ischemic brain injury assessed by phosphorus magnetic resonance spectroscopy. Pediatr Res 25(5):445–451
16. Graham GD, Blamire AM, Rothman DL, Brass LM, Fayad PB, Petroff OA et al (1993) Early temporal variation of cerebral metabolites after human stroke. A proton magnetic resonance spectroscopy study. Stroke 24(12):1891–1896
17. Kauv P, Ayache SS, Créange A, Chalah MA, Lefaucheur J-P, Hodel J et al (2017) Adenosine triphosphate metabolism measured by phosphorus magnetic resonance spectroscopy: a potential biomarker for multiple sclerosis severity. Eur Neurol 77(5–6):316–321
18. Guillevin R, Menuel C, Taillibert S, Capelle L, Costalat R, Abud L et al (2011) Predicting the outcome of grade II glioma treated with temozolomide using proton magnetic resonance spectroscopy. Br J Cancer 104(12):1854–1861
19. Zeng Q-S, Li C-F, Liu H, Zhen J-H, Feng D-C (2007) Distinction between recurrent glioma and radiation injury using magnetic resonance spectroscopy in combination with diffusion-weighted imaging. Int J Radiat Oncol Biol Phys 68(1):151–158

20. Matsusue E, Fink JR, Rockhill JK, Ogawa T, Maravilla KR (2010) Distinction between glioma progression and post-radiation change by combined physiologic MR imaging. Neuroradiology 52(4):297–306

21. Sugahara T, Korogi Y, Tomiguchi S, Shigematsu Y, Ikushima I, Kira T et al (2000) Posttherapeutic intraaxial brain tumor: the value of perfusion-sensitive contrast-enhanced MR imaging for differentiating tumor recurrence from nonneoplastic contrast-enhancing tissue. AJNR Am J Neuroradiol 21(5):901–909

22. Kickingereder P, Sahm F, Radbruch A, Wick W, Heiland S, Deimling A von, et al (2015) IDH mutation status is associated with a distinct hypoxia/angiogenesis transcriptome signature which is non-invasively predictable with rCBV imaging in human glioma. Sci Rep [Internet]. 2015 Nov 5 [cited 2016 Sep 1];5. Available from: http://www.ncbi.nlm.nih.gov/pmc/articles/PMC4633672/

23. de la Fuente MI, Young RJ, Rubel J, Rosenblum M, Tisnado J, Briggs S et al (2016) Integration of 2-hydroxyglutarate-proton magnetic resonance spectroscopy into clinical practice for disease monitoring in isocitrate dehydrogenase-mutant glioma. Neuro Oncol. 18(2):283–290

24. Andronesi OC, Loebel F, Bogner W, Marjańska M, Vander Heiden MG, Iafrate AJ et al (2016) Treatment response assessment in IDH-mutant glioma patients by noninvasive 3D functional spectroscopic mapping of 2-Hydroxyglutarate. Clin Cancer Res Off J Am Assoc Cancer Res. 22 (7):1632–1641

25. Jafari-Khouzani K, Loebel F, Bogner W, Rapalino O, Gonzalez GR, Gerstner E et al (2016) Volumetric relationship between 2-hydroxyglutarate and FLAIR hyperintensity has potential implications for radiotherapy planning of mutant IDH glioma patients. Neuro-Oncol. 18 (11):1569–1578

26. Viswanath P, Najac C, Izquierdo-Garcia JL, Pankov A, Hong C, Eriksson P et al (2016) Mutant IDH1 expression is associated with down-regulation of monocarboxylate transporters. Oncotarget 7(23):34942–34955

27. Ryu YJ, Choi SH, Park SJ, Yun TJ, Kim J-H, Sohn C-H (2014) Glioma: application of whole-tumor texture analysis of diffusion-weighted imaging for the evaluation of tumor heterogeneity. PLoS One 9(9):e108335

28. Costalat R, Francoise J-P, Menuel C, Lahutte M, Vallée J-N, de Marco G et al (2012) Mathematical modeling of metabolism and hemodynamics. Acta Biotheor 60(1–2):99–107

29. Duffau H (2013) Diffuse Low-Grade Gliomas in Adults [Internet]. Springer, London, [cited 2016 Sep 1]. Available from: http://link.springer.com/10.1007/978-1-4471-2213-5

30. Sagiyama K, Mashimo T, Togao O, Vemireddy V, Hatanpaa KJ, Maher EA et al (2014) In vivo chemical exchange saturation transfer imaging allows early detection of a therapeutic response in glioblastoma. Proc Natl Acad Sci U S A 111(12):4542–4547

31. Ren J, Trokowski R, Zhang S, Malloy CR, Sherry AD (2008) Imaging the tissue distribution of glucose in livers using a PARACEST sensor. Magn Reson Med 60(5):1047–1055

32. Zhou J, Lal B, Wilson DA, Laterra J, van Zijl PCM (2003) Amide proton transfer (APT) contrast for imaging of brain tumors. Magn Reson Med 50(6):1120–1126

33. Zhou J, Blakeley JO, Hua J, Kim M, Laterra J, Pomper MG et al (2008) Practical data acquisition method for human brain tumor amide proton transfer (APT) imaging. Magn Reson Med 60(4):842–849

34. Zhou J, Payen J-F, Wilson DA, Traystman RJ, van Zijl PCM (2003) Using the amide proton signals of intracellular proteins and peptides to detect pH effects in MRI. Nat Med 9 (8):1085–1090

35. Guillevin R, Menuel C, Vallée J-N, Françoise J-P, Capelle L, Habas C et al (2011) Mathematical modeling of energy metabolism and hemodynamics of WHO grade II gliomas using in vivo MR data. C R Biol 334(1):31–38

36. Colen CB, Shen Y, Ghoddoussi F, Yu P, Francis TB, Koch BJ et al (2011) Metabolic targeting of lactate efflux by malignant glioma inhibits invasiveness and induces necrosis: an in vivo study. Neoplasia (New York, N.Y.) 13(7):620–632

37. Wen Z, Hu S, Huang F, Wang X, Guo L, Quan X et al (2010) MR imaging of high grade brain tumors using endogenous protein and peptide-based contrast. NeuroImage 51(2):616–622

38. Mandal PK (2012) In vivo proton magnetic resonance spectroscopic signal processing for the absolute quantitation of brain metabolites. Eur J Radiol 81(4):e653–e664
39. Dou W, Zhang M, Zhang X, Li Y, Chen H, Li S, et al. (2015) Convex-envelope based automated quantitative approach to multi-voxel 1H-MRS applied to brain tumor analysis. PLoS ONE [Internet]. 2015 Sep 14 [cited 2016 Sep 1];10(9). Available from: http://www.ncbi.nlm.nih.gov/pmc/articles/PMC4569259/
40. Skogen K, Schulz A, Dormagen JB, Ganeshan B, Helseth E, Server A (2016) Diagnostic performance of texture analysis on MRI in grading cerebral gliomas. Eur J Radiol 85 (4):824–829
41. Jalbert LE, Neill E, Phillips JJ, Lupo JM, Olson MP, Molinaro AM et al (2016) Magnetic resonance analysis of malignant transformation in recurrent glioma. Neuro Oncol 18 (8):1169–1179
42. Santarosa C, Castellano A, Conte GM, Cadioli M, Iadanza A, Terreni MR et al (2016) Dynamic contrast-enhanced and dynamic susceptibility contrast perfusion MR imaging for glioma grading: preliminary comparison of vessel compartment and permeability parameters using hotspot and histogram analysis. Eur J Radiol 85(6):1147–1156
43. Gillies RJ, Kinahan PE, Hricak H (2016) Radiomics: images are more than pictures, they are data. Radiology 278(2):563–577
44. Lambin P, Rios-Velazquez E, Leijenaar R, Carvalho S, van Stiphout RGPM, Granton P, et al (2012) Radiomics: extracting more information from medical images using advanced feature analysis. Eur J Cancer Oxf Engl 1990 48(4):441–6.
45. Ambrosini RD, Wang P, O'Dell WG (2010) Computer-aided detection of metastatic brain tumors using automated three-dimensional template matching. J Magn Reson Imaging 31 (1):85–93
46. Ojala T, Pietikainen M, Maenpaa T (2002) Multiresolution gray-scale and rotation invariant texture classification with local binary patterns. IEEE Trans Pattern Anal Mach Intell 24 (7):971–987
47. Tuytelaars T, Mikolajczyk K (2008) Local invariant feature detectors: a survey. Found Trends Comput Graph Vis 3(3):177–280
48. Prasanna P, Tiwari P, Madabhushi A (2014) Co-occurrence of local anisotropic gradient orientations (CoLIAGe): distinguishing tumor confounders and molecular subtypes on MRI. Med Image Comput Comput-Assist Interv MICCAI 2014, vol. 17 (Pt 3): 73–80
49. Leijenaar RTH, Carvalho S, Velazquez ER, van Elmpt WJC, Parmar C, Hoekstra OS et al (2013) Stability of FDG-PET Radiomics features: an integrated analysis of test-retest and inter-observer variability. Acta Oncol Stockh Swed 52(7):1391–1397
50. Cao Y, Tsien CI, Nagesh V, Junck L, Ten Haken R, Ross BD et al (2006) Survival prediction in high-grade gliomas by MRI perfusion before and during early stage of RT [corrected]. Int J Radiat Oncol Biol Phys 64(3):876–885
51. Zhou M, Hall LO, Goldgof DB, Gillies RJ, Gatenby RA (2013) Survival time prediction of patients with glioblastoma multiforme tumors using spatial distance measurement. In: International society for optics and photonics; 2013 [cited 2018 Jan 29]. p. 867020. Available from: https://www.spiedigitallibrary.org/conference-proceedings-of-spie/8670/86702O/Survival-time-prediction-of-patients-with-glioblastoma-multiforme-tumors-using/10.1117/12.2007699.short
52. Siu A, Wind JJ, Iorgulescu JB, Chan TA, Yamada Y, Sherman JH (2012) Radiation necrosis following treatment of high grade glioma--a review of the literature and current understanding. Acta Neurochir 154(2):191–201. discussion 201
53. Zinn PO, Mahajan B, Majadan B, Sathyan P, Singh SK, Majumder S et al (2011) Radiogenomic mapping of edema/cellular invasion MRI-phenotypes in glioblastoma multiforme. PLoS One 6 (10):e25451
54. Cruz JA, Wishart DS (2007) Applications of machine learning in cancer prediction and prognosis. Cancer Informat 2:59–77

55. Clark MC, Hall LO, Goldgof DB, Velthuizen R, Murtagh FR, Silbiger MS (1998) Automatic tumor segmentation using knowledge-based techniques. IEEE Trans Med Imaging 17 (2):187–201
56. Szegedy C, Liu W, Jia Y, Sermanet P, Reed S, Anguelov D, et al (2015) Going deeper with convolutions. In: 2015 I.E. Conference on Computer Vision and Pattern Recognition (CVPR), pp 1–9.
57. Pereira S, Pinto A, Alves V, Silva CA (2016) Brain tumor segmentation using convolutional neural networks in MRI images. IEEE Trans Med Imaging 35(5):1240–1251
58. Cruz-Barbosa R, Vellido A (2011) Semi-supervised analysis of human brain tumours from partially labeled MRS information, using manifold learning models. Int J Neural Syst 21 (1):17–29
59. Davnall F, Yip CSP, Ljungqvist G, Selmi M, Ng F, Sanghera B et al (2012) Assessment of tumor heterogeneity: an emerging imaging tool for clinical practice? Insights Imaging 3 (6):573–589
60. Miles KA, Ganeshan B, Hayball MP (2013) CT texture analysis using the filtration-histogram method: what do the measurements mean? Cancer Imaging 13(3):400–406
61. Vallée A, Guillevin R, Vallée J-N (2018) Vasculogenesis and angiogenesis initiation under normoxic conditions through Wnt/β-catenin pathway in gliomas. Rev Neurosci 29(1):71–91
62. Vallée A, Lecarpentier Y, Guillevin R, Vallée J-N (2017) Thermodynamics in Gliomas: interactions between the Canonical WNT/Beta-Catenin Pathway and PPAR Gamma. Front Physiol 8:352
63. Grabner G, Kiesel B, Wöhrer A, Millesi M, Wurzer A, Göd S et al (2016) Local image variance of 7 Tesla SWI is a new technique for preoperative characterization of diffusely infiltrating gliomas: correlation with tumour grade and IDH1 mutational status. Eur Radiol 27 (4):1556–1567
64. Jiménez J, López AM, Cruz J, Esteban FJ, Navas J, Villoslada P et al (2014) A web platform for the interactive visualization and analysis of the 3D fractal dimension of MRI data. J Biomed Inform 51:176–190
65. Aubert A, Pellerin L, Magistretti PJ, Costalat R (2007) A coherent neurobiological framework for functional neuroimaging provided by a model integrating compartmentalized energy metabolism. Proc Natl Acad Sci U S A 104(10):4188–4193
66. Lamari F, La Schiazza R, Guillevin R, Hainque B, Foglietti M-J, Beaudeux J-L et al (2008) Biochemical exploration of energetic metabolism and oxidative stress in low grade gliomas: central and peripheral tumor tissue analysis. Ann Biol Clin (Paris) 66(2):143–150
67. Folkman J (1974) Tumor angiogenesis. Adv Cancer Res 19(0):331–358
68. Czernicki Z, Horsztyński D, Jankowski W, Grieb P, Walecki J (2000) Malignancy of brain tumors evaluated by proton magnetic resonance spectroscopy (1H-MRS) in vitro. Acta Neurochir Suppl 76:17–20
69. Kuhr WG, Korf J (1988) Extracellular lactic acid as an indicator of brain metabolism: continuous on-line measurement in conscious, freely moving rats with intrastriatal dialysis. J Cereb Blood Flow Metab Off J Int Soc Cereb Blood Flow Metab 8(1):130–137
70. Gruetter R (2003) Glycogen: the forgotten cerebral energy store. J Neurosci Res 74(2):179–183
71. Keenan MM, Chi J-T (2015) Alternative fuels for cancer cells. Cancer J (Sudbury Mass) 21 (2):49–55
72. Durmaz R, Vural M, Işildi E, Coşan E, Ozkara E, Bal C et al (2008) Efficacy of prognostic factors on survival in patients with low grade glioma. Turk Neurosurg 18(4):336–344
73. Hlaihel C, Guilloton L, Guyotat J, Streichenberger N, Honnorat J, Cotton F (2010) Predictive value of multimodality MRI using conventional, perfusion, and spectroscopy MR in anaplastic transformation of low-grade oligodendrogliomas. J Neuro-Oncol 97(1):73–80
74. Pellerin L (2008) Brain energetics (thought needs food). Curr Opin Clin Nutr Metab Care 11 (6):701–705
75. Hubesch B, Sappey-Marinier D, Roth K, Meyerhoff DJ, Matson GB, Weiner MW (1990) P-31 MR spectroscopy of normal human brain and brain tumors. Radiology 174(2):401–409

76. Lahutte-Auboin M, Guillevin R, Françoise J-P et al (2013) On a minimal model for hemodynamics and metabolism of lactate: application to low grade glioma and therapeutic strategies. Acta Biotheor 61:79–89. https://doi.org/10.1007/s10441-013-9174-8
77. Guillevin C, Guillevin R, Miranville A, Perrillat-Mercerot A (2018) Analysis of a mathematical model for brain lactate kinetics. Mathematical Biosciences and Engineering. IN PRESS;
78. Guillevin R, Miranville A, Perillat-Mercerot A (2017) Reaction-diffusion system associated with brain lactate kinetics. Electron J Differ Equ 23, JAN 18 2017

Chapter 6
Epilepsy Imaging

Charles Mellerio, Francine Chassoux, Laurence Legrand, Myriam Edjlali, Bertrand Devaux, Jean-François Meder, and Catherine Oppenheim

Introduction

Epilepsy is one of the most frequent chronic neurological disorders and is characterized by the recurrence of spontaneous seizures. Epilepsy imaging is considered in different situations:

- In patients without known epilepsy presenting with a first episode of acute seizure. In this case, the role of imaging is to detect and characterize a causal lesion and refer the patient for a specific treatment.
- In approximatively one third of patients with chronic focal epilepsy becoming drug-resistant and referred for presurgical evaluation. In this case, and especially if previous MRIs were negative, an optimized MR protocol and advanced imaging techniques are required in order to help in delineating the epileptogenic zone, one of the main conditions for an efficient surgery.
- In candidates for epilepsy surgery, neuroimaging is required before surgery to predict the functional outcome, including mapping of eloquent areas with fMRI; and, postoperatively, to evaluate the extent of the resection.

C. Mellerio (✉)
Department of Radiology, Centre Hospitalier Sainte-Anne, Université Paris Descartes Sorbonne, Paris, France

Department of imaging, Centre cardiologique du Nord, Saint-Denis, France
e-mail: C.MELLERIO@ch-sainte-anne.fr

F. Chassoux · B. Devaux
Department of Neurosurgery, Centre de Psychiatrie et Neurosciences, INSERM U894, Paris, France

L. Legrand · M. Edjlali · J.-F. Meder · C. Oppenheim
Department of Radiology, Centre Hospitalier Sainte-Anne, Université Paris Descartes Sorbonne, Paris, France

© Springer International Publishing AG, part of Springer Nature 2018
C. Habas (ed.), *The Neuroimaging of Brain Diseases*, Contemporary Clinical Neuroscience, https://doi.org/10.1007/978-3-319-78926-2_6

- In the research field, imaging helps in finding diagnostic and prognostic biomarkers in order to refine individually tailored treatment of epilepsy [1].

For all the above indications, MRI outperforms CT scanner, which may be limited to emergencies and will not be considered in this chapter. However, MRI is not needed in several typical syndromic epilepsies (such as absence epilepsy, juvenile myoclonic epilepsy, benign partial epilepsy of childhood with centrotemporal spikes, idiopathic generalized epilepsy) [2].

Despite advances in antiepileptic drugs treatment in the past 20 years, seizure freedom is not always reached and side effects can be difficult to tolerate. Intractability of epilepsy is defined as persistence of seizures in spite of two tolerated, appropriately chosen and used antiepileptic drugs, whether as monotherapies or in combination [3]. Uncontrolled epilepsy alters quality of life and cognitive function and increases the risks of sudden death. For these reasons, patients should be referred as early as possible for presurgical investigations. The role of neuroimaging is crucial in identifying the causal lesion, as its characterization may play a major role for referring the patients to surgery. In addition, identifying a lesional epilepsy is associated with a better prognosis after surgery [4, 5].

This chapter will review the role of neuroimaging and especially MRI in epilepsy, with a special focus on focal intractable epilepsies.

Which Protocol in MRI?

Choice of the Sequences

Epileptogenic lesions, especially malformations of the cortical development, may be small and subtle, hidden in the normal cortex. Therefore, a "standard" MR protocol early is ineffective for epilepsy [6]. By contrast, it is now clearly established that the seek for a causal lesion with brain MRI is improved when carried out by a neuroradiologist experienced in epilepsy imaging and when guided by clinical and electroclinical data [2, 7]. Moreover, an optimized protocol combining volumetric acquisition is required, with high spatial resolution allowing multiplanar reformat and gyral structures delineation and sequences with high contrast resolution in order to identify subtle cortical lesions [8]. This "minimal" optimized protocol is detailed in Table 6.1. Advanced sequences developed in the recent years can be added. Their selection is however subject to imaging duration and patient tolerance and should be chosen depending on the clinical data and the type of lesion expected. For example, when a focal cortical dysplasia (FCD) is suspected, 3D FLAIR [9] and/or 3D T2 double inversion recovery [10, 11] weighted sequences help in enhancing cortical contrast/noise ratio. For mesial temporal lobe epilepsy (TLE), other sequences are also important: thin coronal MRI slices, perpendicular to the long axis of the

Table 6.1 Imaging protocol for patients referred for epilepsy: basic sequences and additional/optional sequences

	Characteristic	Advantages
Volumetric FLAIR	Slice thickness <1.5 mm	Multiplanar reformat Excellent contrast resolution of subcortical anomalies Detect abnormal cells, inflammation, edema, gliosis (especially in hippocampal sclerosis), scar
Coronal spin echo T2	Slice thickness <3 mm If temporal lobe epilepsy: Oriented in the axis perpendicular to hippocampus In intercommissural axis otherwise	Excellent delineation resolution of the cortex and especially of hippocampus (digitations loss, dedifferentiation)
Axial T2 gradient echo (or susceptibility weighted imaging)		Hemosiderin/calcification sensitive sequences (trauma/scar, vascular malformations/cavernomas, tumor with calcifications)
Volumetric T1-weighted sequence	Slice thickness <1 mm isotropic (inversion recovery may optimize the gray-white matter contrast) and isotropic voxels <1 mm. Multiplanar reformats	Excellent spatial resolution Multiplanar/curvilinear/volumetric reformat May detect subtle sulco-gyral and cortical abnormalities Especially useful for malformation of cortical development
Additional/optional sequences		
Volumetric T2 with double inversion recovery	Suppression of the signal from both white matter and CSF Slice thickness <2 mm	May provide better contrast for cortical lesion detection such as focal cortical dysplasia
Arterial spin labeling	Post labeling delay depending on the age of the patient (1–1.5 s in children/>1.5 s in adults)	Quantitative maps of cortical perfusion Postictal hyperperfusion Interictal hypoperfusion of several cortical lesions
Volumetric T1-weighted sequence With contrast media injection		Rarely necessary, not systematic Helps characterization of epilepsy related tumors and vascular malformations
Functional MRI and diffusion tensor imaging	Paradigms depend on the location of the lesion	Surgery planning and prediction of the risk of corticectomy
T2 relaxometry		Quantitative MRI technique Detection of hippocampal sclerosis

hippocampus, T2-weighted, FLAIR, and T1-weighted with inversion recovery. Injection of a contrast media (Gadolinium) is in most cases not required except when tumor or vascular malformations are suspected, based on a high clinical hypothesis or on other sequences.

Choice of Magnetic Field and Head Coil

Higher field magnet is associated with higher signal-to-noise ratio and spatial resolution and could therefore benefit to brain MRIs in epilepsy, especially when a subtle cortical lesion is suspected. However, distortion and artifacts are more prominent. Studies comparing 3 T with 1.5 T MRI in patients with epilepsy suggested that even if the rate of lesion detected at 3 T is not significantly higher than at 1.5 T, increasing field is clearly associated with an improvement of lesion characterization an delineation [12–15]. Similarly, a few authors analyzed the benefit of 7 T magnets in epilepsy imaging, including a better delineation of anatomical details such as hippocampal subfields [16]. However, the clinical relevance of these findings is not established yet.

Increasing the number of channels in phased array coil increases the quality of images [17] but also increases the risk of inhomogeneity, with a low SNR in central deep cerebral region, distant to the peripheral head coils.

Overview of the Principle Causes of Refractory Epilepsy

Hippocampal Sclerosis (HS)

Hippocampal sclerosis (HS) is the most frequent lesion found in intractable temporal lobe epilepsy in adults. The MRI signature of HS relies on a simple pattern, combining volume loss of hippocampal formation better visualized on coronal T2w images and 3D T1 increased signal intensity on FLAIR sequences and loss of the internal hippocampal architecture better found on coronal T2 (Fig. 6.1). These features refer to the pathology, displaying, respectively, neuronal loss and gliosis, mainly located in the hilus of the dentate gyrus and in the CA1 and CA3 pyramidal cell layers.

In doubtful or unclear lesions, a loss of hippocampal head digitation, best identified in thin T2w slices perpendicular to the axis of the hippocampus

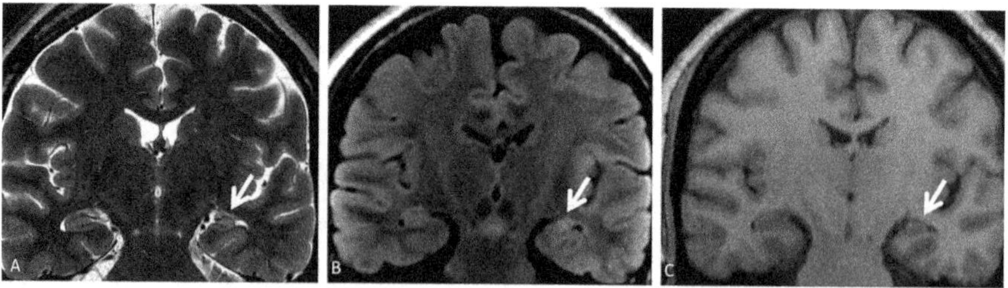

Fig. 6.1 Patient with left temporal lobe epilepsy. Coronal view perpendicular to the hippocampus long axis in T2 (**a**), FLAIR (**b**), and T1 (**c**) shows a left hippocampal atrophy, loss of the global architecture, and increased FLAIR signal in comparison to the right side, typical of HS

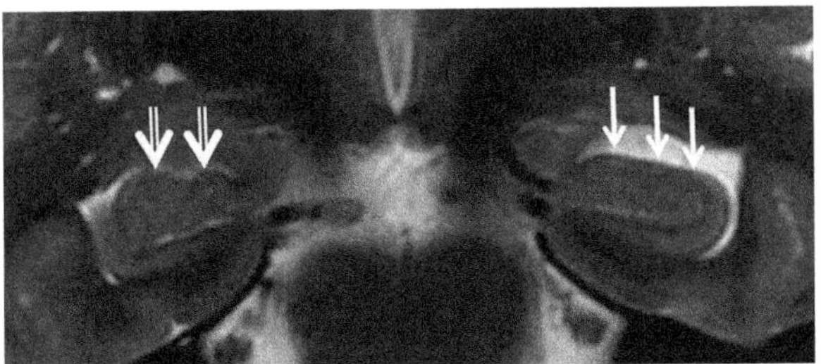

Fig. 6.2 Loss of the digitations of the left hippocampal head (arrows) on a T2w coronal view perpendicular to the hippocampal long axis in a case of left HS, in comparison to normal hippocampal digitations on the right side (double arrows)

(Fig. 6.2), can also support the diagnosis [18]. Associated with these anomalies limited to the hippocampus, modification of the architecture and MR signal can spread to other temporal and limbic region. Thus, FLAIR hyperintensity can also affect the ipsilateral lobe associated with a blurring of the temporal cortex. The atrophy can also affect the ipsilateral fornix, mammillary body, amygdala, cingulate gyrus, pulvinar, and contralateral cerebellum. The comparison between both hippocampi is an important tool to detect asymmetries, although one must keep in mind that about 10–20% of patients may have visually bilateral HS (Fig. 6.3).

When an extra-hippocampal cause of epileptogenic lesion is found, careful screening of the hippocampi is important. Indeed, a "dual pathology" [19, 20] can be found in up to 15% in association of HS and another temporal or extratemporal epileptogenic lesion (such as cortical dysplasia, developmental tumors, perinatal injury, etc.). Explanations for the coexistence of both lesions are controversial but may include the secondary development of HS from the symptomatic and prolonged seizures originating from another neocortical lesion.

Advanced Imaging and Temporal Lobe Epilepsy

When conventional MRI is considered normal, but clinical and electroclinical evaluations suggest a temporal lobe epilepsy, advanced MR techniques of imaging can be useful for lateralization, localization, and characterization of the epileptogenic zone.

Among the proposed advanced MR tools, quantitative analysis offers an objective and automated evaluation of the atrophy (volumetry) and of the altered signal (relaxometry) within the hippocampus formation and temporal lobe. These two techniques were recently enriched by the new robust techniques of automated segmentation of temporo-mesial structures. Thus, volumetry and morphometry, based on statistical parametric mapping to compare voxels against corresponding voxels derived from a normative database, help to detect volume loss and shape changes in patients with MR negative TLE [21]. T2 relaxometry provides a

Fig. 6.3 Coronal view perpendicular to the hippocampus long axis in T2 (**a**), FLAIR (**b**), and T1 (**c**) shows a bilateral hippocampal atrophy and increased FLAIR signal. Even if pronounced, this HS case might be difficult to diagnose because of the absence of asymmetry

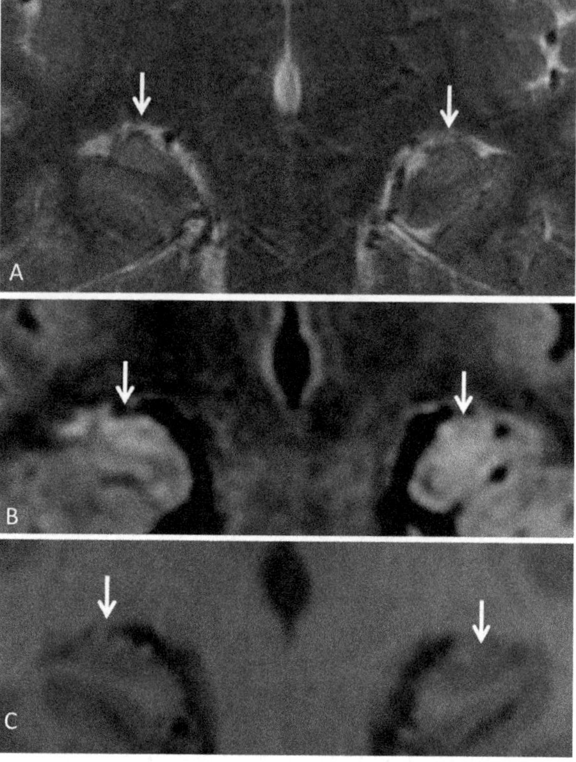

quantitative assessment of T2 relaxation time using a multi-echo sequence which can be performed on single subjects with TLE and may be a clinically useful tool for the detection of the seizure focus [22]. It provides quantitative data within extra-hippocampal adjacent structures in the temporal lobe and amygdala that are difficult to assess in conventional visualization only.

Arterial spin labeling (ASL) is now routinely used in MR protocols. This sequence uses magnetically labeled arterial blood water molecules as an endogenous contrast agent and provides, noninvasively and without any contrast media injection, a cartography of cerebral blood flow, related to cortical perfusion. In epilepsy protocols, ASL is now moving toward a "conventional" and is no longer an "advanced" MR sequence. Thus, it can also help localizing and defining the epileptic zone by showing, without any contrast media, an interictal mesial temporal hypoperfusion, with a good correlation, when present, to the hypometabolism found on interictal PET [23–25]. If the ASL sequence is acquired during or just after seizure (rarely observed in clinical practice), the injured hippocampus will be highly hyper-perfused (Fig. 6.4).

Proton MR spectroscopy, another quantitative analysis of cerebral structures, can be used to measure NAA/Cr in temporal lobes, with a poor diagnosis value related to a non-specific reduced NAA secondary to neural loss. However, a low prognosis

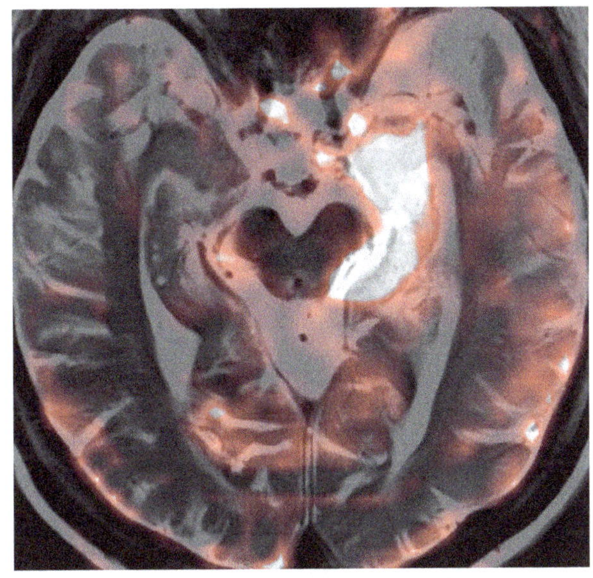

Fig. 6.4 Axial T2-weighted image in the hippocampal axis co-registered with ASL cerebral blood flow map (red-white scale) in a patient with left HS, 15 min after a temporal seizure, shows an increase of the blood flow within the injured hippocampus

value can be related to the widespread of abnormal spectrum in extra-hippocampal region and in contralateral temporal lobe [26].

Focal Cortical Dysplasia

Focal cortical dysplasia (FCD) is one of the main causes of extratemporal, drug-resistant partial epilepsy that is surgically curable in adults and the first cause in children. It designates a spectrum of histological abnormalities in the structure of the laminar cortex associated with the presence of abnormal cells, such as dysmorphic neurons and/or balloon cells. The latest classification subdivides FCD in three categories: Type I (abnormal lamination without abnormal cells), Type II (major cortical disorganization, presence of dysmorphic neurons with or without balloon cells), and Type III (Type I associated with another epileptic lesion) [27]. The major predictor of a favorable surgical outcome is complete removal of the dysplastic cortex, especially for FCD Type II. Therefore, accurate MR assessment of the lesion location and extent is critical for the outcome.

Typical MR features (Fig. 6.5) include abnormalities of the cortex and of the underlying white matter, for which isotropic volumetric with a high contrast resolution is essential.

- Cortical thickening and blurring of the gray-white matter interface, best seen on T1 wi, correspond to the presence of dysmorphic neurons and balloon cells in the cortex and the subcortical junction, ectopic neurons, or axonal loss in underlying white matter. This sign can be very subtle and needs to be observed in at least two

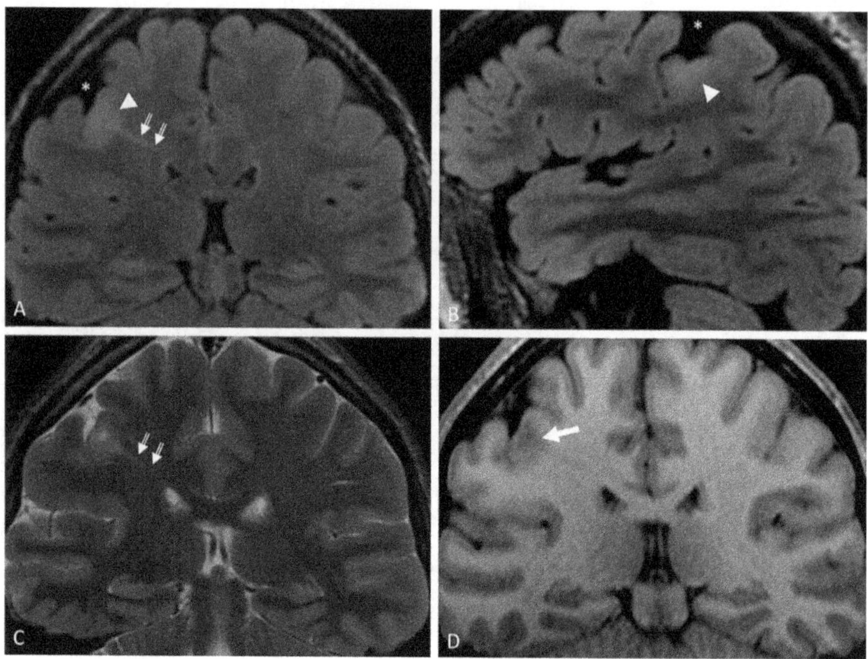

Fig. 6.5 Coronal and axial reformats of a 3D FLAIR (**a, b**), coronal T2 (**c**), and T1 (**d**) in a patient with right frontal lobe epilepsy show the typical pattern of a FCD: Cortical thickening and blurring (arrow), cortical and subcortical hypersignal (arrow head) tapering to the ventricle wall ("transmantle sign" – double arrow), and sulcal abnormality with enlarged sulcus (asterisk)

adjacent sections and in at least two different axes (Fig. 6.6) in order to avoid being confounded with pseudo-thickening [28].

- T2/FLAIR signal increase of cortical and more frequently of subcortical under-lying white matter [29] and related to a high density of balloon cells in the cortex [30]. Sequences with an increased gray-white matter contrast, such as T2 double inversion recovery (DIR) imaging, that suppresses signal from both CSF and white matter, can be of interest for the detection of these cortical abnormalities [11].
- Transmantle sign, best visible in FLAIR, is a linear extent of the subcortical signal increase to the ventricle surface, reproducing the path of migrating neuroblasts. This typical feature, a MR "signature" of FCD, is associated with a good postoperative prognostic [31]. Being better visible at 3 T [12], this sign does not systematically taper an orthogonal plane and highly benefits from 3D FLAIR multiplanar reformats (Fig. 6.7) [9].
- Sulco-gyral abnormalities might be the more subtle, difficult to assess and thus are underestimated in FCD. Yet, these subtle features might be the only detectable abnormality, all other signs being absent, in the so-called negative MRI FCD [29, 32] and have thus to be carefully searched. They are defined as unusual depth, angulation, or shape of a sulcus (Fig. 6.7). In central region, which is

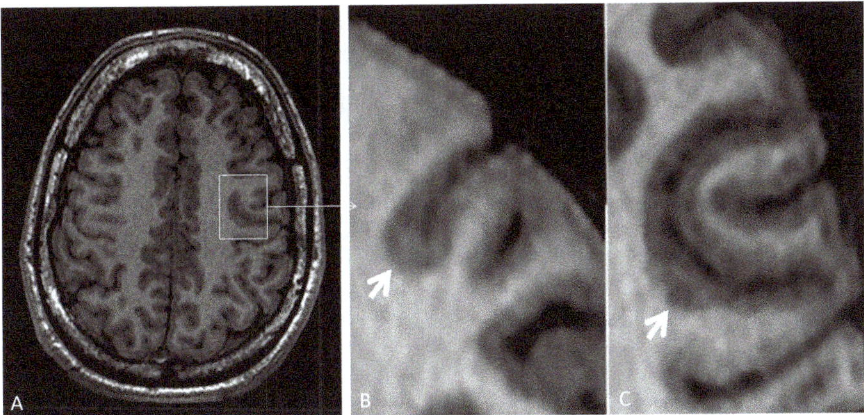

Fig. 6.6 Left precentral FCD with subtle cortical thickening and blurring if the gray-white matter interface in axial T1 (**a**). Reformat in two other axis (**b**, **c**) confirms that it does not correspond to a partial volume effect ("pseudo-thickening"). Of note, the unusual depth of the sulcus also helps to define the location of the FCD

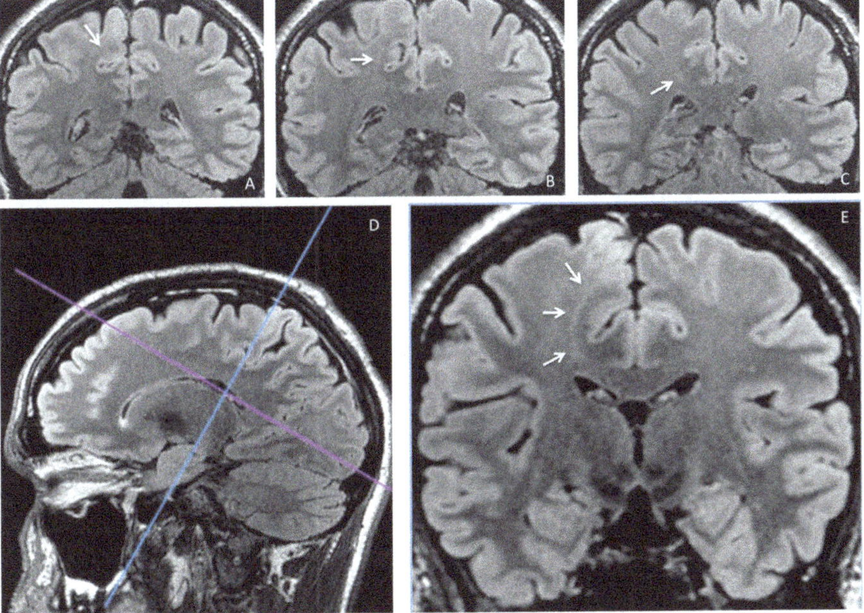

Fig. 6.7 Orthogonal coronal view (**a–c**) of a 3D FLAIR in a patient with a FCD Type II of the paracentral lobule: the transmantle sign is hardly visible, limited to a thin linear hyper signal. When reformatted in the axis of the dysplastic sulcus (**d**, **e**), the transmantle sign is much more obvious, tapering from the cortex to the ventricle wall

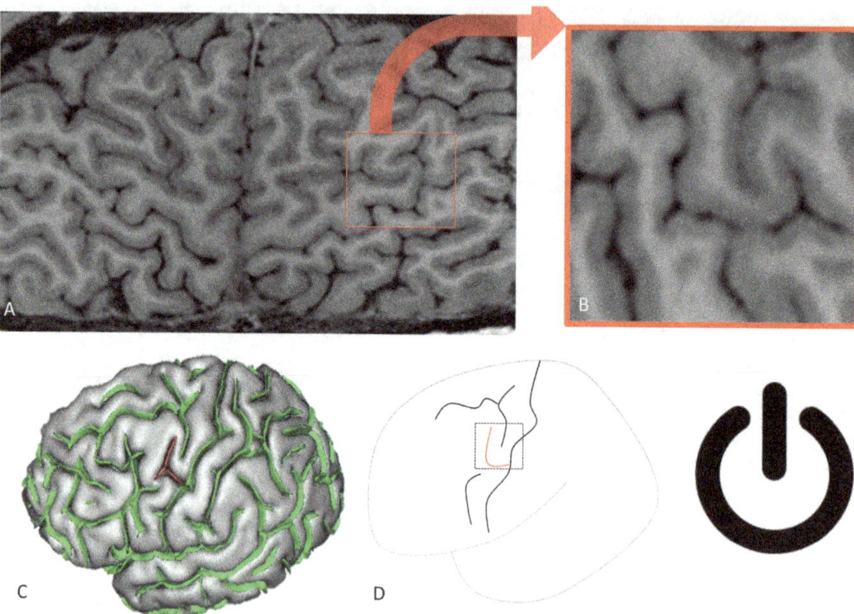

Fig. 6.8 Curvilinear reformat of the left and right central sulci (**a**) with magnification and rotation of the left central sulcus (**b**) corresponding to the "power button sign." 3D view of the left hemisphere surface (**c**) and schematic representation (**d**) of this pattern typical of the presence of a FCD in the homolateral central region

known to display a low interindividual variability, FCD might be associated in up to 62% of the case with a specific particular sulcal pattern reproducing the "power button" symbol (Fig. 6.8), even when all other previously described signs are absent [33].

Recent studies reveal that 20–40% of patients with proven FCD have negative MRI [29, 30, 34, 35], while other studies report that up to 60% of patients referred for epilepsy and negative MRI may present FCD [36, 37].

A negative MRI is known to be associated with a late referral to surgery and a reduced prognosis [5], suggesting that conventional MRI is not sufficient for evaluation of patients with a clinically suspected FCD. Thereby, multimodal advanced techniques present a special interest in this epileptogenic lesion.

A promising advanced processing is the morphometric analysis. Diffusion tensor imaging in the subcortical white matter helps in visualizing white matter alterations that are associated with FCD [38]. A reduction in the fractional anisotropy near the seizure focus as compared with the contralateral side can thus increase sensitivity for the detection of FCD. This feature is, however, not specific of any epileptogenic lesion.

Arterial spin labeling provides a cartography of the cerebral blood flow that is physiologically comparable to cortical metabolism obtained with PET, even if less sensitive. This technique can be easily added to the routine MRI evaluation, and, in

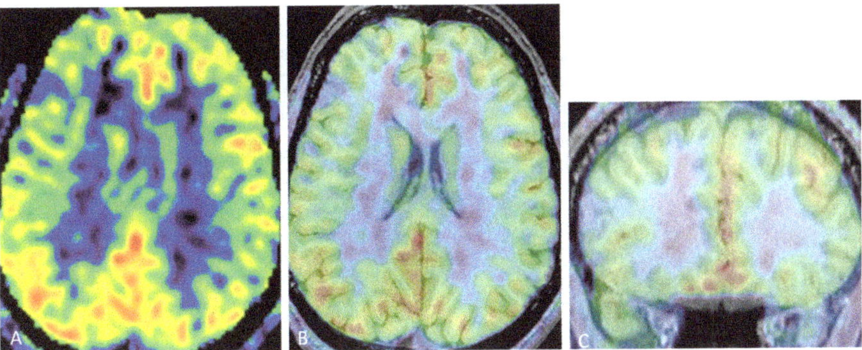

Fig. 6.9 Patient with a right frontal DCF. Interictal ASL with cerebral blood flow cartography in axial view (**a**). This CBF map is registered to T1 in axial (**b**) and coronal view (**c**) in order to better identify a cortical hypoperfusion in the right frontal lobe, matching with an underlying FCD

order to provide efficient data on cortical perfusion, this cartography needs to be overlapped with the anatomical 3D T1 or 3D FLAIR, acquired during the same MR session. As in HS, if the ASL sequence is acquired during or just after seizure, the epileptogenic focus will be highly hyper-perfused [39]. This latter condition is however rarely observed in clinical practice. Thus, the relevance of ASL relies during the interictal phase with the observation of a focal zone of cortex on structural MRI that is hypoperfused compared to the adjacent cortex (Fig. 6.9). This hypoperfusion, co-localized with FCD, can also give an additional clue on the location of subtle lesion [24]. However, as for DTI, this hypoperfusion can also be observed in other epileptogenic lesions [40].

Voxel-based morphometry (VBM) can also decrease the amount of negative MRI in FCD [41]. Morphometric analysis procedure methods produce a junction map to detect gray-white matter abnormal features and sulcal increased depth and direct the attention to suspicious regions of interest. Associated with conventional reading, these techniques may improve detection of subtle lesions [42]. However, these techniques have not entered to routine, yet. As for temporal lobe epilepsy, T2 relaxometry maps can also increase sensitivity for detection of FCD [43].

Gyral and sulcal abnormal patterns can also benefit from surface-based morphometry techniques that allow features such as cortical thickness to be measured [32] but also abnormal sulcal patterns to be recognized [44].

Epileptogenic Tumors

All different histological types of tumor can provide seizures as they involve the cortex. However, two developmental neoplasms that are part of the wide range of malformations of the cortical development, gangliogliomas and dysembryoplastic neuroepithelial tumors, are highly associated with intractable seizures in children

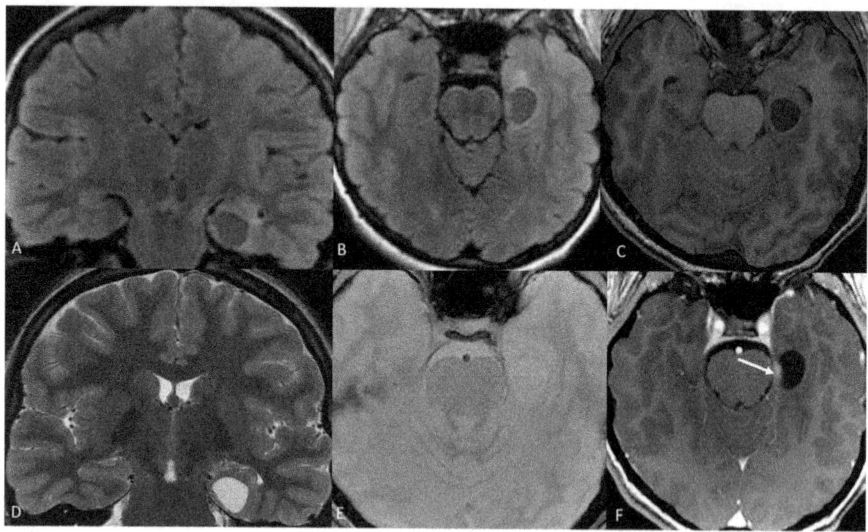

Fig 6.10 A 22-year-old male patient with intractable left temporal epilepsy. Coronal and axial FLAIR (**a**, **b**), axial T1 (**c**), coronal T2 (**d**), T2* (**e**), and axial T1 with contrast media (**f**) show a mixed cystic and solid cortical mass. The nodular portion presents with mild enhancement (arrow)

and young adults. These low-grade tumors are composed of neural and glial elements. They are usually slow growing, and their main clinical feature is epilepsy.

Surgery is the treatment of choice for glioneuronal tumors in most of the cases, providing good results even in eloquent areas [45]. Complete tumor resection is crucial for long-term favorable outcome.

Gangliogliomas

Gangliogliomas are glioneuronal tumors defined by the presence of large binuclear neurons within a glial component comprising inflammatory infiltrates. The association with other dysplastic cortical abnormalities is frequent.

Gangliogliomas are mainly supratentorial with a clear temporal predominance, although they can be found in all the lobes, always located within the cortex. The classical aspect (Fig. 6.10) consists in a mixed cystic and solid cortical mass, either mainly nodular with a poorly defined limit or mainly cystic with a mural nodule. The nodular portion presents with a moderate hyper T2 and FLAIR and is generally slightly and nonhomogeneously enhanced. The cystic portion, on the other hand, is homogeneous in highly hypo T1 and hyper T2 without enhancement [46]. Small calcifications can be found in up to 40% of the cases. Of note, surrounding edema and mass effect are rare due to the developmental origin of these tumors. Hemorrhagic forms are extremely rare. Malignant transformation may rarely occur and lead to anaplastic ganglioglioma. For this reason, a prolonged and systematic MR follow-up is required, especially in cases of surgical abstention or incomplete resection.

In diffusion, the ADC within gangliogliomas is high, reflecting a low tumor cellularity, contrary to purely glial tumors such as astrocytoma [47]. In spectroscopy, low NAA and variable choline are nondiscriminating [46]. In gradient echo perfusion, a small increase in relative cerebral blood volume can be observed without impairment of the permeability [48].

Dysembryoplastic Neuroepithelial Tumors

Dysembryoplastic neuroepithelial tumors (DNETs) are benign tumors, also classified in glioneuronal tumors, predominantly located in the temporal lobe and mostly revealed by partial seizures in children and young adults. DNETs are characterized by a specific glioneuronal component which is isolated (simple forms) or associated with multinodular glial proliferation as well as cortical disorganization (complex forms). A third subtype (non-specific) is composed by various types of glial proliferation: oligodendroglial, astrocytic, or mixed. As previously mentioned, DNET can be associated with other epileptogenic lesions such as hippocampal sclerosis ("dual pathology") or FCDs (Type III).

A recent study proposed an MRI-based scheme with three main features, associated with the optimal surgical resection strategy and well correlated with each histological subtype [49]:

- DNET type 1 ("cystic or polycystic-like") consists of a well-delineated mass with a liquid signal similar to that of the CSF, giving it a cystic (Fig. 6.11) or polycystic "bubbly" appearance (Fig. 6.12) within the cortex [50]. This subtype is very characteristic and easy to diagnose in most cases. When close to the skull inner table, a smooth bone remodeling can be observed.
- DNET type 2 ("nodular-like") is characterized by a heterogeneous and variable signal within a partially delineated nodular or multinodular appearance in the cortex. Calcifications are present in 60% of the cases (Fig. 6.13).
- DNET type 3 ("dysplastic-like") presents with a poorly delineated homogeneous infiltration with a very slight hypersignal on T2 FLAIR, sometimes limited to a poor gray-white matter demarcation (Fig. 6.14), and is thus more difficult to distinguish from other glial neoplasms. One of its characteristic patterns is the very strong predominance in mesial areas, especially in amygdalo-hippocampal complex.

Contrast enhancement is rare (10–20%), often partial and nodular, and equally found in each MRI subtype.

Type 1 are always associated with simple and complex subtypes; the cystic appearance is correlated to the high-water content of the specific glioneuronal component. However, types 2 and 3 correspond to non-specific forms. This classification may help to determine the extent of the resection as a strong relationship between the MRI feature, the pathological structure, and the extent of the epileptogenic zone has been demonstrated [51].

Tumor recurrence and malignant transformation are uncommon.

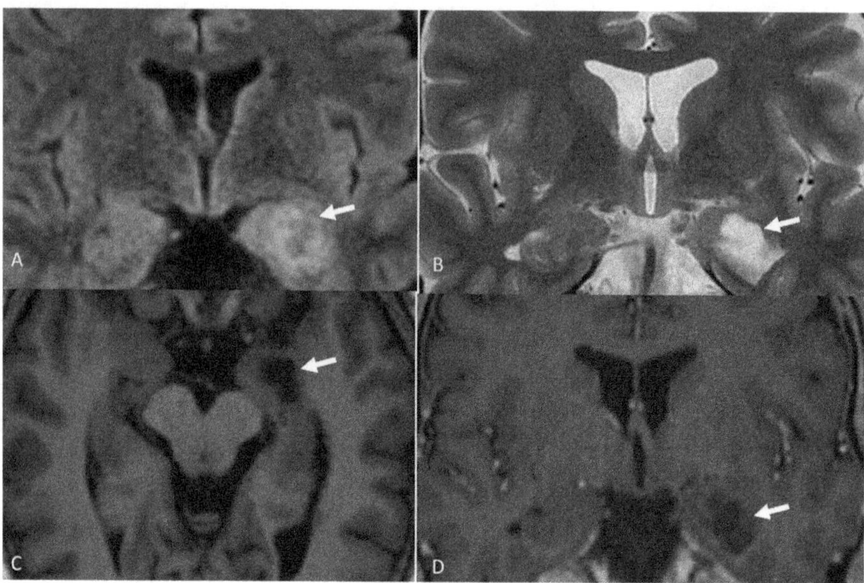

Fig. 6.11 Coronal view in FLAIR (**a**) and in T2 (**b**), axial T1 (**c**), and coronal T1 with contrast media (**d**) in a patient with left temporal drug-resistant epilepsy. Pseudocystic lesion of the left amygdalo-hippocampal complex without any mass effect. The signal in all sequence is similar to the CSF. Note the thin rim in hyper FLAIR. No enhancement is observed after gadolinium. This aspect is typical of DNET type 1

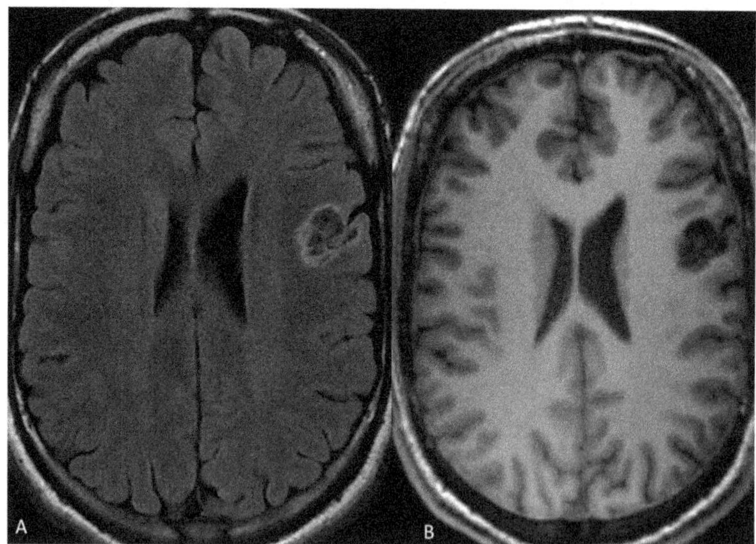

Fig. 6.12 Pseudo-polycystic "bubbly" appearance of a left frontal type 1 DNET in axial FLAIR (**a**) and T1 (**b**)

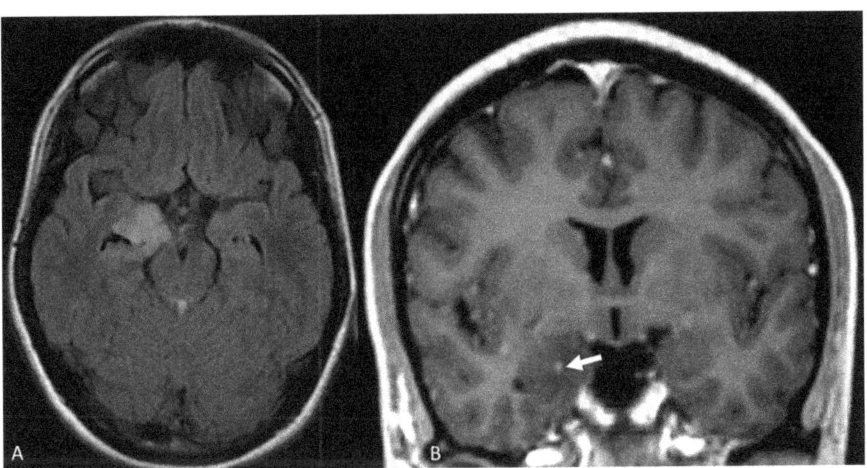

Fig. 6.13 Partially delineated nodular appearance of a type 2 DNT with signal increase in axial FLAIR (**a**) and hyposignal in T1 with a small punctate contrast enhancement after gadolinium infusion (**b** arrow)

Advanced imaging can help characterize DNETs, but as for conventional visualization [52], the more specific patterns are observed in type 1 forms rather than in non-specific subtypes. In diffusion, the ADC is very high due to the importance of water content in the extracellular space and superior to that observed in gliomas. In spectroscopy, a low increase or normal choline allows the differential diagnosis with a glioma. In perfusion (gradient echo or ASL), the decrease of local perfusion also distinguishes from an ordinary glioma (Fig. 6.15).

Other Malformations of Cortical Development

Classifications for malformations of cortical development (MCD) are numerous and complex. Several classifications are based on chronological elements (referring to the supposed stages of each developmental disorder) and/or morphological considerations. These classifications have the advantage to provide an approach of the pathogenic mechanisms affecting the normal development of the cerebral cortex at each stage from neuronal proliferation (focal cortical dysplasia, DNET and gangliogliomas, tuberous sclerosis...), to migration (heterotopia, lissencephaly...) and subsequent cortical organization (polymicrogyria, schizencephalia...) [53].

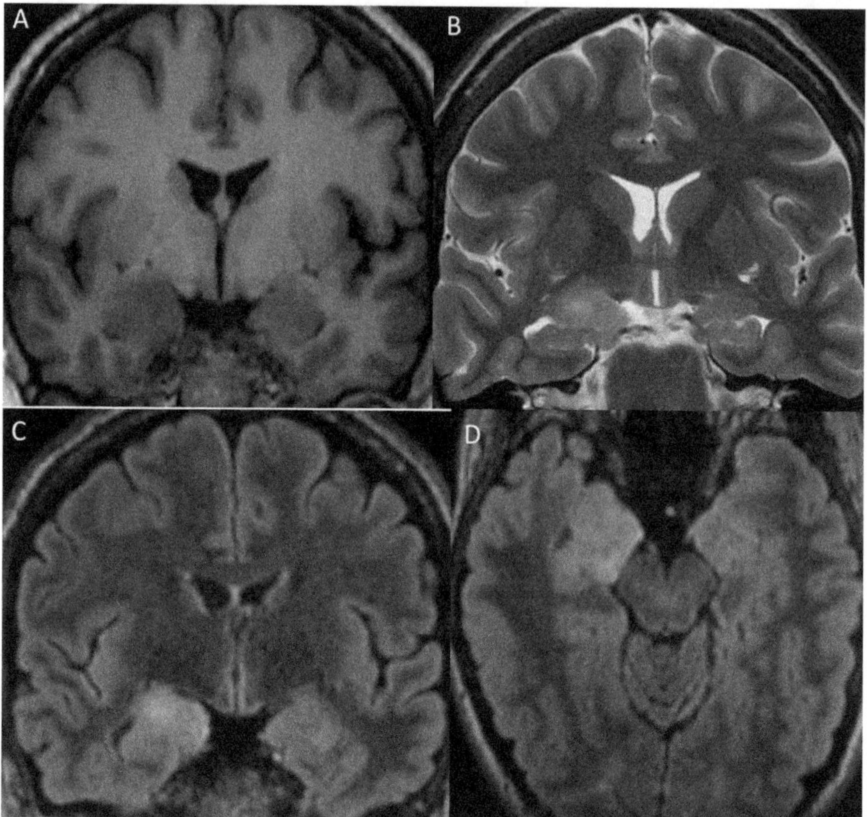

Fig 6.14 "Dysplastic" form of a right amygdalo-hippocampal DNET, hardly visible on coronal T1 (**a**), with a simple gray-white matter smoothing, in slight hyper signal on coronal T2 (**b**), coronal (**c**), and axial (**d**) FLAIR

Polymicrogyria

Polymicrogyria (PMG) is a common malformation of cortical development, characterized by numerous excessive small convolutions separated by shallow sulci, leading to a scalloped aspect of the cortex. This lesion can be multifocal or limited to a cortical region. The perisylvian location is classical (Fig. 6.16), often bilateral. It may be associated with other developmental lesions such as schizencephaly, callosum agenesis, or nodular heterotopia. Other neurological disorders may be associated to epilepsy such as cognitive deficiency and focal impairment. The severity of these neurological disorders depends on the extent and topography of the PMG as well as the associated abnormalities.

As polymicrogyria is the result of an abnormal cortical organization of "normal" neurons, there is no signal anomaly on conventional sequences. Three-dimensional and curvilinear reformatting from a volumetric T1 sequence helps characterize PMG

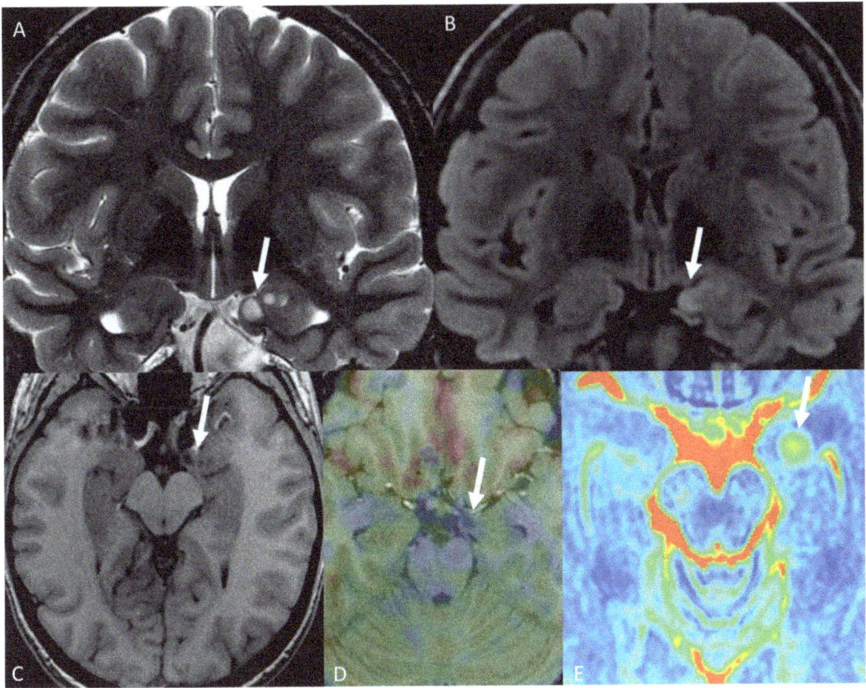

Fig. 6.15 Advanced imaging in a case of pseudocystic type 1 DNET of the left cerebral amygdala (arrow) on coronal T2 (**a**) and FLAIR (**b**) slices, axial T1 (**c**), ASL with co-registration of cerebral blood flow map on a T1 slice (**d**) showing a decrease of the local cerebral blood flow of the DNET in comparison to the right amygdala. Finally, the apparent diffusion coefficient map (**e**) shows an increased ADC of the DNET in comparison to adjacent cortex

MR features (irregular, scalloped, and discretely thickened cortex) and their extent, as well as associated anomalies of sulcal depth and orientation. In perisylvian polymicrogyria, for example, the lateral fissure may be in continuity with the central sulcus (Fig. 6.17).

The diffusion/spectral/perfusion parameters of the abnormal cortex are identical to those of the adjacent normal cortex during interictal period.

Gray Matter Heterotopias

Gray matter heterotopias correspond to unique or multiple clusters of normal neurons in an inappropriate position. They are the consequence of the disruption of neuronal migration and often associated with genetic anomalies. There are three types of gray matter heterotopia corresponding to different positions on the neuronal migration pathway during the fetal period [53] from the germinal (subependymal) zones to the cortex: periventricular nodular heterotopia, the most frequent, laminar

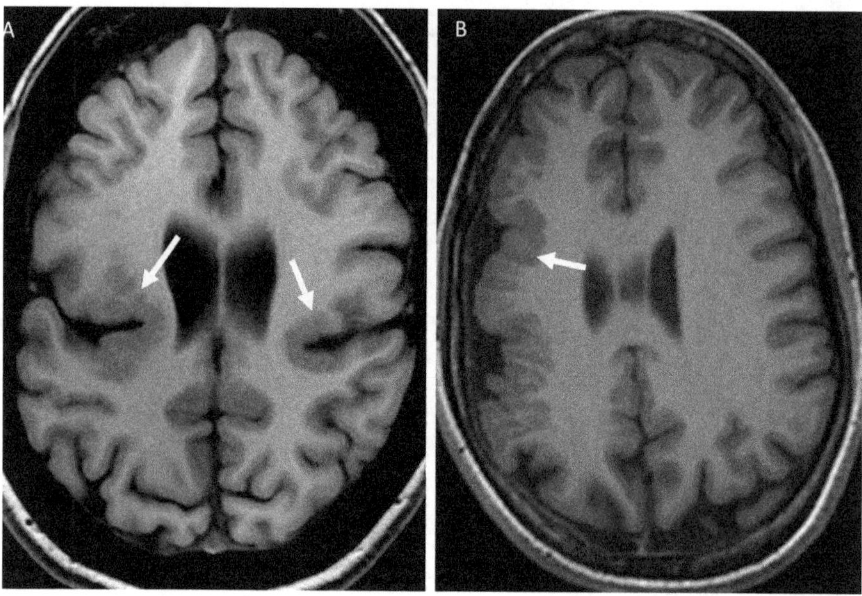

Fig. 6.16 Bilateral perisylvian polymicrogyria (**a**) and frontoparietal polymicrogyria in another patient (**b**) with a scalloped aspect of the cortex composed by numerous excessive small convolutions

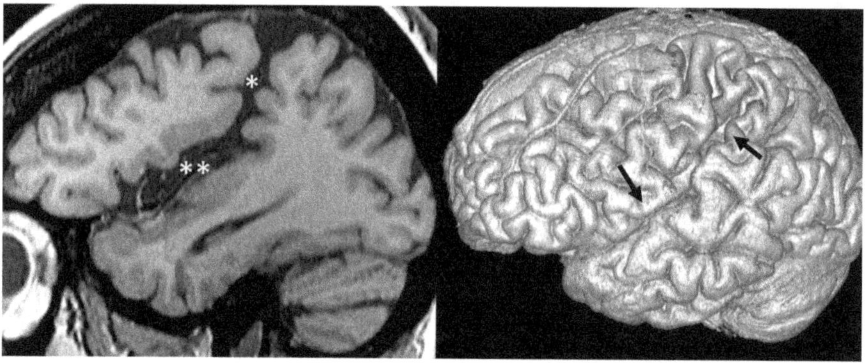

Fig. 6.17 Abnormal sulcal pattern in a patient with left perisylvian polymicrogyria. The lateral fissure (double asterisk) is in continuity with the central sulcus (asterisk)

heterotopia, and nodular subcortical heterotopia. There is a continuum between laminar heterotopia and lissencephaly (absence or decrease of cortical convolutions).

Periventricular nodular heterotopias can be found on the entire surface of the lateral ventricles. They are isolated or multiple, sometimes bilateral, and can be confluent with a scalloped appearance of the ventricle wall. The subcortical nodular heterotopia can be directly in contact with the cortex, with a pseudo-thickening of the gyrus.

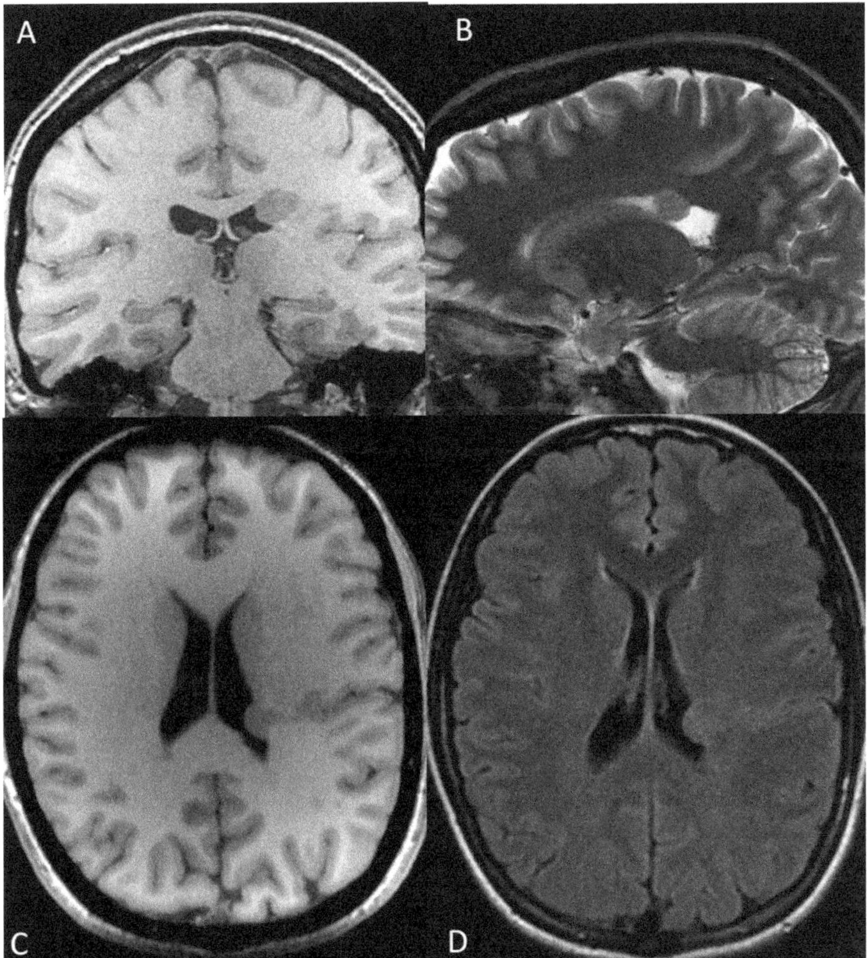

Fig. 6.18 Subependymal nodular gray matter heterotopia with a signal strictly identical to that of the cortex on T1 (**a**, **c**), T2 (**b**), and FLAIR (**d**) sequences

The gray matter heterotopia consists of normal neurons. Its signal is thus strictly identical to that of the cortex on all conventional sequences, including FLAIR (Fig. 6.18). There is no enhancement after contrast media injection. Nevertheless, a vascular structure can be observed through heterotopia in a juxtacortical position. The cortex is usually normal, but some heterotopia may be associated with other malformations of cortical development (e.g., polymicrogyria).

Functional imaging essentially allows for the differential diagnosis with brain tumors even if conventional sequences are often sufficient. In perfusion (ASL or gradient echo), the relative blood flow within heterotopia is identical to that of the cortex. In spectroscopy, unlike a tumor, the spectrum is that of the normal parenchyma. In fMRI, focal activity consistent with activation of the normal cortex during a task is rarely observed and corresponds to the possible presence of functional neurons within heterotopia.

Tuberous Sclerosis

Tuberous sclerosis (or Bourneville tuberous sclerosis) is a genetic disease classified among phacomatosis, involving multiple organs (including the brain, skin, kidney, heart, and lung). Brain lesions are characterized by disruptions of both proliferation and migration and organization during cortical development, leading to abnormalities located from the periventricular space to the cerebral cortex. MRI features are characterized by the combination of multiple cortical tubers that have similar presentation as FCD (including transmantle sign), subependymal calcified nodules and subependymal giant cell astrocytoma. The final diagnosis relies on the presence of major criteria and/or minor criteria. Each MRI feature can be present or absent (all MRI features are considered as "major" criteria).

Mapping Cortical Brain Functions with fMRI in Patients with Epilepsy

Identifying brain functions is crucial when planning a surgical treatment. This is helpful for determining the surgical strategy, including the need of intracranial recordings and for assessment of functional risks. Due to brain plasticity, long-term epilepsy is associated with inter- or intra-hemispheric and white matter connection reorganization, especially when the epileptic focus is located in functional areas. In this context of potential reorganization, MRI may provide information about the eloquent areas within the cortex with functional MRI (fMRI) and of the underlying subcortical neuronal bundles with diffusion tensor imaging (DTI). The anatomical location of the presumed epileptogenic zone will define which paradigms are required for the presurgical workup. For example, fMRI for fronto-central epilepsy will focus on sensorimotor cortex, while mesial temporal epilepsy will require language and memory tasks, and posterior epilepsies will investigate visual tasks.

Sensorimotor Cortex

Sensory-motor function in fMRI corresponds to one of the most robust network fMRI, highlighted with easily feasible paradigms such as finger tapping, foot flexion/extension, or lip movement tasks. It can be used to identify the primary motor cortex. Owing to its typical location in the frontal lobe and especially in the central region, FCD represents a major indication of sensory-motor task and corticospinal tract evaluation. However, other malformations of the cortical development may also affect the pre- and postcentral gyrus. In epilepsy presurgical imaging more than in other fMRI indications, each side has to be acquired separately in order to distinguish a reorganization of motor cortex near an epileptogenic lesion and to compare each response without being confound by the direct corticospinal tract contribution. Thus, the motor task for each limb has to be accomplished with the same frequency and strength.

It should be noted that a recent seizure may affect the motor cortical network with reduced responses on the side of the focus in comparison to the opposite side, in patients with extratemporal epilepsy [54]. This seizure-related alteration of the cortex function must be taken into account when interpreting motor fMRI in patients with frequent seizures.

Long-lasting epilepsy may also affect permanently the cortical organization of motor and sensory network, especially in patients with MCDs. The degree of reorganization depends on the period of the alteration during cortical development stages: early stage injuries (such as cortical dysplasias) will provide more substantial cortical reorganization than later lesions (such as gray matter heterotopia or polymicrogyria) [55, 56]. This reorganization can be present through a partial or total reduction of the response in the affected motor and sensory areas contralateral to the explored limb, in comparison to the normal hemisphere. A redistribution of the activation may also be observed with a permanent migration of the activated clusters to a different area, resulting from an adaptation process [57]. This relocation can be observed in the same hemisphere but in a different location, often in the vicinity of the expected anatomic area [58], or more rarely in the contralateral normal sensory-motor area. Consistent with this reorganization, activation within the dysplastic tissue itself is rarely observed, even if the lesion is directly located inside the central region (Fig. 6.19).

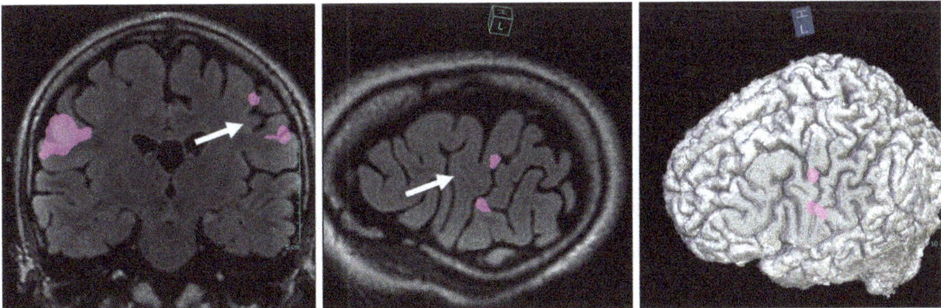

Fig. 6.19 Patient with a left precentral focal cortical dysplasia within the lateral primary motor cortex (arrow). The patient is asked to perform lip movements during a functional MRI acquisition. In comparison to the normal right side, the response near the lesion is less significant and fragmented on both side of the lesion

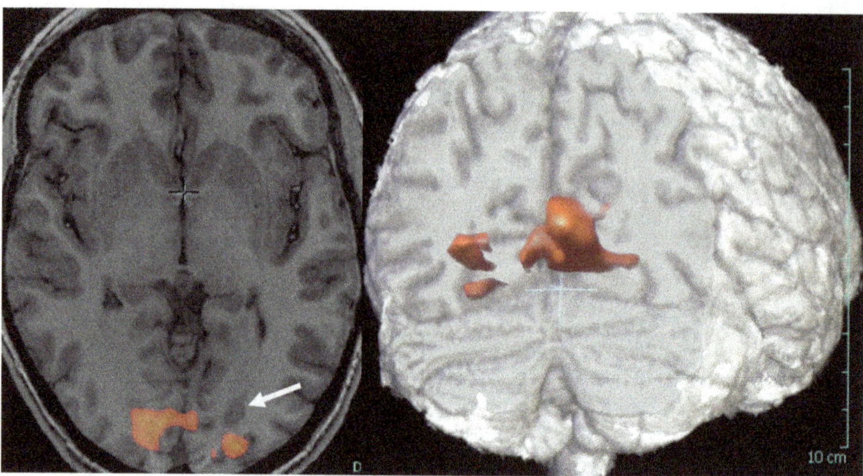

Fig. 6.20 Patient with a left occipital periventricular heterotopia. Using a blinking checkerboard as a paradigm, cortical response at the posterior part of the calcarine sulcus is more robust on the right side than on the side of the lesion

In MCDs of later stages such as heterotopia or polymicrogyria, the normal landmarks of the sensory-motor cortex may be disrupted and/or disorganized, especially in perisylvian polymicrogyria. fMRI is thus crucial before surgery or invasive procedure. Contrary to FCD however, functional activation may be preserved in polymicrogyric areas in which gyral and sulcal patterns are preserved, even if reduced in extent [59]. In patients with heterotopia, an activation within ectopic gray matter can be observed, usually coactivated with the functional areas directly overlying the heterotopia [60].

In epilepsy surgery, correlations between fMRI results and peroperative data during surgery are usually more reliable than in surgery for gliomas. However direct electrocortical stimulations are mandatory to prevent the risk of causing a lasting deficit.

Visual Cortex

In the same way as for sensory-motor cortex, primary and secondary visual cortex may reorganize in patient with occipital lesional epilepsy, leading to abnormal pattern of activation or absence of activation around calcarine region in the vicinity of the epileptogenic zone. This reorganization occurs in the presence of a MCD (Fig. 6.20) such as polymicrogyria or FCD [55, 61].

Language

There is now very good evidence that fMRI is able to determine hemispheric dominance for language in frontal (Broca's area) and temporal (Wernicke's area) regions, in line with results from the intracarotid amobarbital "Wada" test [62–65]. This agreement reaches up to 85% when using a combination of at least three language paradigms and is greater in right TLE with left language dominance, than in left TLE with left language dominance [66, 67]. This lateralization can be obtained using a visual appreciation of the numbers of activated clusters in each perisylvian region or by using a quantitative method such as the lateralization index [68]. Atypical lateralization is more frequent in patients with left hemisphere epilepsy [69]. As for sensory-motor reorganization, the atypical activations can widespread in the same hemisphere but in other areas than typical perisylvian regions or in the opposite hemisphere (Fig. 6.21), leading to the atypical dominance [70, 71]. This language "shift" is more likely observed in left-handed patients, in left TLE, and in long-lasting epilepsy with early onset [69].

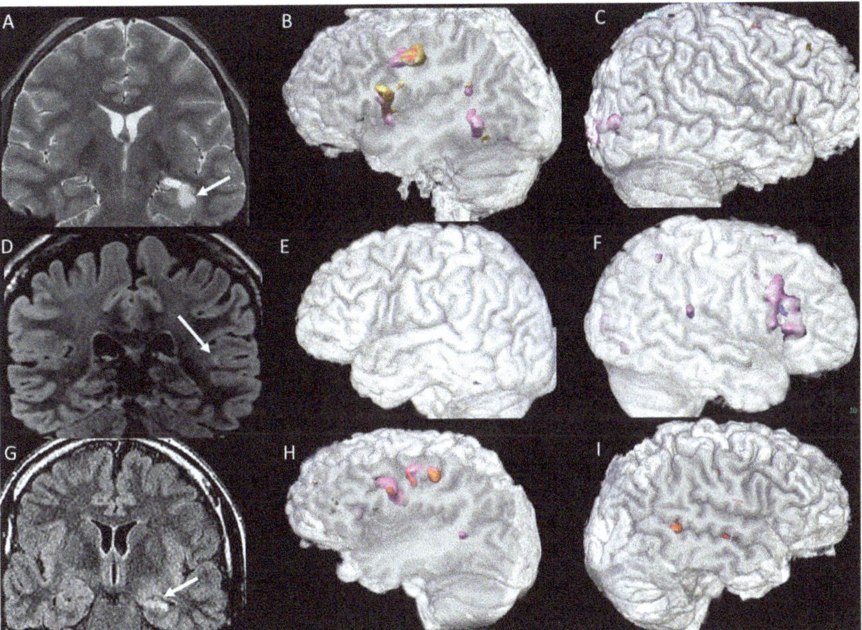

Fig. 6.21 Examples of language lateralization in three patients with temporal epilepsy: Patient 1 (**a–c**) with temporo-mesial DNET on T2 coronal view (**a**). Left (**b**) and right (**c**) 3D reformat of a language fMRI shows a clear left lateralization. Patient 2 (**d–f**) with a left temporal FCD on FLAIR coronal view (**d**). Language fMRI (**e**, **f**) shows an atypical right lateralization resulting from a reorganization of the language network. Patient 3 (**g–i**) with a left hippocampal sclerosis on FLAIR coronal view (**g**). Language fMRI (**h**, **i**) shows an interhemispheric dissociation of left frontal and right temporal language responses leading to a very atypical crossed language dominance

Furthermore, fMRI can be used to map language networks using various specific language tasks that activate frontal (fluency tasks) and temporal (comprehension tasks) language areas and thus determine patterns of language network, whether the lateralization is typical or not. Preoperative fMRI activations near the epileptogenic zone that is to be resected is a predictor of long-term postoperative language deficiency [72]. It is thus essential to choose the best set of paradigms in order to activate the cortical regions, frontal or temporal, that are targeted by the surgery. However, contrary to sensory-motor fMRI, language fMRI is to date not adequate to guide resection because of a lack of sensitivity when compared to direct cortical stimulation [73] even if specificity is high.

Episodic Memory

Temporal lobe surgery, especially in patients with hippocampal sclerosis, is associated with a risk of visual and/or verbal memory postsurgical complications. New paradigms are tailored for imaging the episodic memory network with encoding and retrieval tasks in order to visualize activations in mesial temporal structures [74]. Verbal event-related memory task seems to show the best reliability to distinguish between left-onset and right-onset patients [75]. As for language structures, an asymmetry index derived from activations in both hippocampi can evaluate compensatory mechanisms of the normal entorhinal cortex to counterweigh the impaired function of the sclerotic hippocampus. The aim of memory fMRI would thus be to predict the effect of resection of the sclerotic temporal structure on the postoperative memory decline [76–79]. Depending on the size of the HS, the category of memory decline risk will differ: verbal memory decline will be more frequent in patients undergoing left temporal lobe resection and visual memory decline in those with right temporal lobe surgery [80, 81]. Memory activation patterns before surgery seems to be the strongest predictor of verbal and visual memory loss as a result of anterior temporal lobe resection, and preserved function in the ipsilateral posterior hippocampus seems to help to maintain memory encoding after anterior temporal lobe resection [76].

Mapping Connectivity

As cortical eloquent area may affect one or several cognitive functions, white matter connections and functional connectivity between eloquent areas may also cause cognitive impairments when damaged. Although focal epilepsy is traditionally considered as a cortical and regional disorder based on the epileptogenic zone model from which seizures originate, recent studies suggest that widespread network alterations extend beyond this zone and may be correlated to cognitive impairments and surgical outcome prediction. Thus, presurgical imaging workup in epilepsy may require an evaluation of functional networks.

Resting State

Focal epilepsy, whether temporal or extratemporal, is associated with modifications of the connectivity that can be observed in regions directly connected to the epileptic zone. These modifications may also widespread well beyond the seizure onset area [82]. Regional connectivity modifications may be related to the lateralization of the hemisphere of seizure onset, and thus resting state fMRI could predict laterality of the epileptogenic hemisphere [83]. In patients with TLE, most studies suggest an increased connectivity between hippocampus and other ipsilateral limbic structures (including thalamus) involved in seizure propagation, compared to controls [84, 85]. In the same way, patients with frontal lobe epilepsy shows increased connectivity in the neighborhood of the epileptic zone [86, 87]. This increase of the regional connectivity, when included in the resection area, seems to be related to a better surgical outcome. Therefore, preoperative resting state fMRI can help localize the global epileptic zone (EZ) that should be targeted by surgery [88]. Moreover, this phenomenon of higher regional connectivity near the EZ is in most of the cases associated with a diminished connectivity in widespread distant regions throughout the brain, including those involved in cognition. In resting state fMRI analysis, the evaluation of the default mode network represents a good illustration of this remote effect of connectivity disturbance. Thus, patients with TLE and extratemporal epilepsy when compared to controls show a decrease of functional connectivity among default mode regions contrasting with increased connectivity within functional networks near the seizure onset [84, 87, 88]. Other distant cognitive networks may be affected such as frontoparietal association and primary sensorimotor networks. These local and widespread connectivity disturbances evaluated by resting state fMRI may also be related to cognitive impairments as studied for verbal and nonverbal episodic memory, language, working memory, and attentional functions [89]. Thereby, resting state fMRI seems to provide important clues for the understanding of pathophysiology related to focal epilepsy. However, the clinical benefit for individual patients is not established yet.

Diffusion Tensor Imaging and Tractography

Corticospinal Tract

Motor responses authenticated with fMRI may also help to define white matter projections with DTI tractography and to provide a reliable guide for the surgeon to avoid permanent motor deficit (Fig. 6.22). As for glioma surgery in which

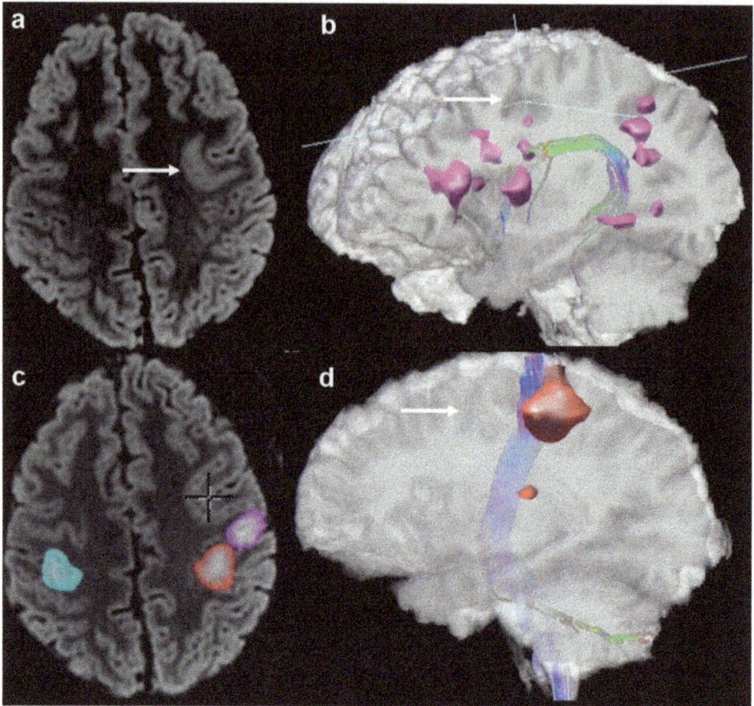

Fig. 6.22 Patient with a focal cortical dysplasia of the left precentral sulcus on axial FLAIR sequence (**a**). In this case both language and motor networks need to be analyzed. Language fMRI responses during a semantic association paradigm (clusters in pink) and arcuate fasciculus tractography (**b**) show relationship between this network and the lesion (arrow). Motor responses (**c**) during left hand (blue), right hand (red), and then mouth (violet) movements show that the lesion is near motor functions in the left hemisphere, without any functional reorganization. Pyramidal tract (**d**) is also traced from the cortical right hand motor response and the brain stem

intraoperative direct cortical stimulation supports tractography as a reliable method for showing the relationship between a glioma and the corticospinal tract [90], similar results were obtained to predict the risk of postoperative motor deficits in patients with frontal epilepsy [91].

Arcuate Fasciculus and Language Tracts

Left-right asymmetry of anisotropy along the two main language pathways, arcuate fasciculus and inferior occipitofrontal fasciculus, are observed in controls and reflect the language specialization and lateralization. In patients with left TLE, this asymmetry of anisotropy along the arcuate fasciculus lowers compared to patients with right TLE and is correlated with fMRI-based lateralization indices [92].

Visual Pathway

Temporal lobe epilepsy surgery exposes to the risk of damaging the inferior optic radiation during resection and can cause contralateral upper quadrantanopia. The preoperative tractography is predictive of the risk of a visual field alteration [93] The optic radiation can be accurately delineated by tractography and can help surgical planning and guide intraoperative procedure [94]. Correction for brain shift using intraoperative MRI also improves the accuracy of the technique [95].

Conclusion

MR diagnosis in patients with epilepsy may be difficult because of subtle cortical lesions which can be hardly distinguishable with "standard" MRI protocol. An appropriate MRI epilepsy protocol is essential in the assessment epileptogenic lesions. In this specific indication, high field MRIs and advanced sequences such as ASL, diffusion tensor, double inversion recovery, and functional MRI are especially appreciated to comfort challenging cases, to appraise the limits of the epileptic zone, and to assess its consequences on cortical network and connectivity. In all cases, MR abnormalities are valuable only when considered together with clinical, electroclinical data, and other imaging techniques.

References

1. Pitkänen A, Löscher W, Vezzani A, Becker AJ, Simonato M, Lukasiuk K, Gröhn O, Bankstahl JP, Friedman A, Aronica E, Gorter JA, Ravizza T, Sisodiya SM, Kokaia M, Beck H (2016) Advances in the development of biomarkers for epilepsy. Lancet Neurol 15:843–856. https://doi.org/10.1016/S1474-4422(16)00112-5
2. Gaillard WD, Chiron C, Helen Cross J, Simon Harvey A, Kuzniecky R, Hertz-Pannier L, Gilbert Vezina L (2009) Guidelines for imaging infants and children with recent-onset epilepsy. Epilepsia 50:2147–2153
3. Kwan P, Arzimanoglou A, Berg AT, Brodic MJ, Allen Hauser W, Mathern G, Moshé SL, Perucca E, Wiebe S, French J (2010) Definition of drug resistant epilepsy: consensus proposal by the ad hoc task force of the ILAE commission on therapeutic strategies. Epilepsia 51:1069–1077. https://doi.org/10.1111/j.1528-1167.2009.02397.x
4. de Tisi J, Bell GS, Peacock JL, McEvoy AW, Harkness WF, Sander JW, Duncan JS (2011) The long-term outcome of adult epilepsy surgery, patterns of seizure remission, and relapse: a cohort study. Lancet 378:1388–1395
5. Téllez-Zenteno JF, Ronquillo LH, Moien-Afshari F, Wiebe S (2010) Surgical outcomes in lesional and non-lesional epilepsy: a systematic review and meta-analysis. Epilepsy Res 89:310–318
6. von Oertzen J, Urbach H, Jungbluth S, Kurthen M, Reuber M, Fernández G, Elger CE (2002) Standard magnetic resonance imaging is inadequate for patients with refractory focal epilepsy. J Neurol Neurosurg Psychiatry 73:643–647. https://doi.org/10.1136/jnnp.73.6.643

7. Cendes F (2013) Neuroimaging in investigation of patients with epilepsy. Continuum Lifelong Learning Neurol 19:623–642
8. Wellmer J, Quesada CM, Rothe L, Elger CE, Bien CG, Urbach H (2013) Proposal for a magnetic resonance imaging protocol for the detection of epileptogenic lesions at early outpatient stages. Epilepsia 54:1977–1987
9. Saini J, Singh A, Kesavadas C, Thomas B, Rathore C, Bahuleyan B, Radhakrishnan A, Radhakrishnan K (2010) Role of three-dimensional fluid-attenuated inversion recovery (3D FLAIR) and proton density magnetic resonance imaging for the detection and evaluation of lesion extent of focal cortical dysplasia in patients with refractory epilepsy. Acta Radiol 51:218–225
10. Li Q, Zhang Q, Sun H, Zhang Y, Bai R (2011) Double inversion recovery magnetic resonance imaging at 3 T: diagnostic value in hippocampal sclerosis. J Comput Assist Tomogr 35:290–293
11. Rugg-Gunn FJ, Boulby PA, Symms MR, Barker GJ, Duncan JS (2006) Imaging the neocortex in epilepsy with double inversion recovery imaging. NeuroImage 31:39–50
12. Mellerio C, Labeyrie M-A, Chassoux F, Roca P, Alami O, Plat M, Naggara O, Devaux B, Meder J-F, Oppenheim C (2014) 3T MRI improves the detection of transmantle sign in type 2 focal cortical dysplasia. Epilepsia 55:117–122. https://doi.org/10.1111/epi.12464
13. Phal PM, Usmanov A, Nesbit GM, Anderson JC, Spencer D, Wang P, Helwig JA, Roberts C, Hamilton BE (2008) Qualitative comparison of 3-T and 1.5-T MRI in the evaluation of epilepsy. Am J Roentgenol 191:890–895
14. Rubinger L, Chan C, D'Arco F, Moineddin R, Muthaffar O, Rutka JT, Snead OC, Smith ML, Widjaja E (2016) Change in presurgical diagnostic imaging evaluation affects subsequent pediatric epilepsy surgery outcome. Epilepsia 57:32–40. https://doi.org/10.1111/epi.13229
15. Zijlmans M, de Kort GA, Witkamp TD, Huiskamp GM, Seppenwoolde J-H, van Huffelen AC, Leijten FS (2009) 3T versus 1.5 T phased-array MRI in the presurgical work-up of patients with partial epilepsy of uncertain focus. J Magn Reson Imaging 30:256–262
16. Coras R, de Boer OJ, Armstrong D, Becker A, Jacques TS, Miyata H, Thom M, Vinters HV, Spreafico R, Oz B et al (2012) Good interobserver and intraobserver agreement in the evaluation of the new ILAE classification of focal cortical dysplasias. Epilepsia 53:1341–1348
17. Wiggins GC, Polimeni JR, Potthast A, Schmitt M, Alagappan V, Wald LL (2009) 96-channel receive-only head coil for 3 tesla: design optimization and evaluation. Magn Reson Med 62:754–762. https://doi.org/10.1002/mrm.22028
18. Oppenheim C, Dormont D, Biondi A, Lehéricy S, Hasboun D, Clémenceau S, Baulac M, Marsault C (1998) Loss of digitations of the hippocampal head on high-resolution fast spin-echo MR: a sign of mesial temporal sclerosis. Am J Neuroradiol 19:457–463
19. Kim DW, Lee SK, Nam H, Chu K, Chung CK, Lee S-Y, Choe G, Kim HK (2010) Epilepsy with dual pathology: surgical treatment of cortical dysplasia accompanied by hippocampal sclerosis. Epilepsia 51:1429–1435. https://doi.org/10.1111/j.1528-1167.2009.02403.x
20. Spencer S, Huh L (2008) Outcomes of epilepsy surgery in adults and children. Lancet Neurol 7:525–537
21. Maccotta L, Moseley ED, Benzinger TL, Hogan RE (2015) Beyond the CA1 subfield: local hippocampal shape changes in MRI-negative temporal lobe epilepsy. Epilepsia 56:780–788. https://doi.org/10.1111/epi.12955
22. Kosior RK, Lauzon ML, Frayne R, Federico P (2009) Single-subject voxel-based relaxometry for clinical assessment of temporal lobe epilepsy. Epilepsy Res 86:23–31. https://doi.org/10.1016/j.eplepsyres.2009.04.001
23. Lim Y-M, Cho Y-W, Shamim S, Solomon J, Birn R, Luh WM, Gaillard WD, Ritzl EK, Theodore WH (2008) Usefulness of pulsed arterial spin labeling MR imaging in mesial temporal lobe epilepsy. Epilepsy Res 82:183–189
24. Pendse N, Wissmeyer M, Altrichter S, Vargas M, Delavelle J, Viallon M, Federspiel A, Seeck M, Schaller K, Lövblad KO (2010) Interictal arterial spin-labeling MRI perfusion in intractable epilepsy. J Neuroradiol 37:60–63. https://doi.org/10.1016/j.neurad.2009.05.006

25. Wolf RL, Alsop DC, Levy-Reis I, Meyer PT, Maldjian JA, Gonzalez-Atavales J, French JA, Alavi A, Detre JA (2001) Detection of mesial temporal lobe hypoperfusion in patients with temporal lobe epilepsy by use of arterial spin labeled perfusion MR imaging. AJNR Am J Neuroradiol 22:1334–1341

26. Eberhardt KE, Stefan H, Buchfelder M, Pauli E, Hopp P, Huk W, Tomandl BF (2000) The significance of bilateral CSI changes for the postoperative outcome in temporal lobe epilepsy. J Comput Assist Tomogr 24:919–926

27. Blumcke I, Vinters HV, Armstrong D (2009) Malformations of cortical development and epilepsies: neuropathological findings with emphasis on focal cortical dysplasia. Epileptic Disord 11:181–193

28. Colombo N, Salamon N, Raybaud C (2009) Imaging of malformations of cortical development. Epileptic Disord 11:194–205

29. Mellerio C, Labeyrie M-A, Chassoux F, Daumas-Duport C, Landre E, Turak B, Roux F-X, Meder J-F, Devaux B, Oppenheim C (2012) Optimizing MR imaging detection of type 2 focal cortical dysplasia: best criteria for clinical practice. Am J Neuroradiol 33:1932–1938

30. Widdess-Walsh P, Kellinghaus C, Jeha L (2005) Electro-clinical and imaging characteristics of focal cortical dysplasia: correlation with pathological subtypes. Epilepsy Res 67:25

31. Wang DD, Deans AE, Barkovich AJ, Tihan T, Barbaro NM, Garcia PA, Chang EF (2013) Transmantle sign in focal cortical dysplasia: a unique radiological entity with excellent prognosis for seizure control: clinical article. J Neurosurg 118:337–344. https://doi.org/10.3171/2012.10.JNS12119

32. Besson P, Andermann F, Dubeau F, Bernasconi A (2008) Small focal cortical dysplasia lesions are located at the bottom of a deep sulcus. Brain 131:3246–3255. https://doi.org/10.1093/brain/awn224

33. Mellerio C, Roca P, Chassoux F, Danière F, Cachia A, Lion S, Naggara O, Devaux B, Meder J-F, Oppenheim C (2015) The power button sign: a newly described central Sulcal pattern on surface rendering MR images of type 2 focal cortical dysplasia. Radiology 274:500–507. https://doi.org/10.1148/radiol.14140773

34. Colombo N, Tassi L, Deleo F, Citterio A, Bramerio M, Mai R, Sartori I, Cardinale F, Russo GL, Spreafico R (2012) Focal cortical dysplasia type IIa and IIb: MRI aspects in 118 cases proven by histopathology. Neuroradiology 54:1065–1077. https://doi.org/10.1007/s00234-012-1049-1

35. Kim DW, Lee SK, Chu K, Park KI, Lee SY, Lee CH, Chung CK, Choe G, Kim JY (2009) Predictors of surgical outcome and pathologic considerations in focal cortical dysplasia. Neurology 72:211–216

36. Chapman K, Wyllie E, Najm I, Ruggieri P, Bingaman W, Lüders J, Kotagal P, Lachhwani D, Dinner D, Lüders HO (2005) Seizure outcome after epilepsy surgery in patients with normal preoperative MRI. J Neurol Neurosurg Psychiatry 76:710–713

37. So EL, Lee RW (2014) Epilepsy surgery in MRI-negative epilepsies. Curr Opin Neurol 27:206–212

38. Kim H, Harrison A, Kankirawatana P, Rozzelle C, Blount J, Torgerson C, Knowlton R (2013) Major white matter fiber changes in medically intractable neocortical epilepsy in children: a diffusion tensor imaging study. Epilepsy Res 103:211–220

39. Altrichter S, Pendse N, Wissmeyer M, Jägersberg M, Federspiel A, Viallon M, Seeck M, Lövblad K-O (2009) Arterial spin-labeling demonstrates ictal cortical hyperperfusion in epilepsy secondary to hemimegalencephaly. J Neuroradiol 36:303–305. https://doi.org/10.1016/j.neurad.2009.04.001

40. Storti SF, Galazzo IB, Del Felice A, Pizzini FB, Arcaro C, Formaggio E, Mai R, Manganotti P (2014) Combining ESI, ASL and PET for quantitative assessment of drug-resistant focal epilepsy. NeuroImage 102:49–59. https://doi.org/10.1016/j.neuroimage.2013.06.028

41. Colliot O, Bernasconi N, Khalili N, Antel SB, Naessens V, Bernasconi A (2006) Individual voxel-based analysis of gray matter in focal cortical dysplasia. NeuroImage 29:162–171

42. Wagner J, Weber B, Urbach H, Elger CE, Huppertz H-J (2011) Morphometric MRI analysis improves detection of focal cortical dysplasia type II. Brain 134:2844–2854

43. Rugg-Gunn FJ, Boulby PA, Symms MR, Barker GJ, Duncan JS (2005) Whole-brain T2 mapping demonstrates occult abnormalities in focal epilepsy. Neurology 64:318–325

44. Roca P, Mellerio C, Chassoux F, Rivière D, Cachia A, Charron S, Lion S, Mangin J-F, Devaux B, Meder J-F, Oppenheim C (2015) Sulcus-based MR analysis of focal cortical dysplasia located in the central region. PLoS One 10:e0122252. https://doi.org/10.1371/jour nal.pone.0122252

45. Devaux B, Chassoux F, Landré E, Turak B, Laurent A, Zanello M, Mellerio C, Varlet P (2017) Surgery for dysembryoplastic neuroepithelial tumors and gangliogliomas in eloquent areas. Functional results and seizure control. Neurochirurgie 63:227–234. https://doi.org/10.1016/j. neuchi.2016.10.009

46. Zhang D, Henning TD, Zou L-G, Hu L-B, Wen L, Feng X-Y, Dai S-H, Wang W-X, Sun Q-R, Zhang Z-G (2008) Intracranial ganglioglioma: clinicopathological and MRI findings in 16 patients. Clin Radiol 63:80–91. https://doi.org/10.1016/j.crad.2007.06.010

47. Kikuchi T, Kumabe T, Higano S, Watanabe M, Tominaga T (2009) Minimum apparent diffusion coefficient for the differential diagnosis of ganglioglioma. Neurol Res 31:1102–1107. https://doi.org/10.1179/174313209X382539

48. Law M, Meltzer DE, Wetzel SG, Yang S, Knopp EA, Golfinos J, Johnson G (2004) Conventional MR imaging with simultaneous measurements of cerebral blood volume and vascular permeability in ganglioglioma. Magn Reson Imaging 22:599–606. https://doi.org/10.1016/j. mri.2004.01.031

49. Chassoux F, Rodrigo S, Mellerio C, Landré E, Miquel C, Turak B, Laschet J, Meder J-F, Roux F-X, Daumas-Duport C (2012) Dysembryoplastic neuroepithelial tumors an MRI-based scheme for epilepsy surgery. Neurology 79:1699–1707

50. Campos AR, Clusmann H, von Lehe M, Niehusmann P, Becker AJ, Schramm J, Urbach H (2009) Simple and complex dysembryoplastic neuroepithelial tumors (DNT) variants: clinical profile, MRI, and histopathology. Neuroradiology 51:433–443. https://doi.org/10.1007/ s00234-009-0511-1

51. Chassoux F, Daumas-Duport C (2013) Dysembryoplastic neuroepithelial tumors: where are we now? Epilepsia 54:129–134. https://doi.org/10.1111/epi.12457

52. Bulakbasi N, Kocaoglu M, Sanal TH, Tayfun C (2007) Dysembryoplastic neuroepithelial tumors: proton MR spectroscopy, diffusion and perfusion characteristics. Neuroradiology 49:805–812. https://doi.org/10.1007/s00234-007-0263-8

53. Barkovich AJ, Guerrini R, Kuzniecky RI, Jackson GD, Dobyns WB (2012) A developmental and genetic classification for malformations of cortical development: update 2012. Brain 135:1348–1369. https://doi.org/10.1093/brain/aws019

54. Woodward KE, Gaxiola-Valdez I, Mainprize D, Grossi M, Goodyear BG, Federico P (2014) Recent seizure activity alters motor organization in frontal lobe epilepsy as revealed by task-based fMRI. Epilepsy Res 108:1286–1298

55. Janszky J, Ebner A, Kruse B, Mertens M, Jokeit H, Seitz RJ, Witte OW, Tuxhorn I, Woermann FG (2003) Functional organization of the brain with malformations of cortical development. Ann Neurol 53:759–767. https://doi.org/10.1002/ana.10545

56. Nikolova S, Bartha R, Parrent AG, Steven DA, Diosy D, Burneo JG (2015) Functional MRI of neuronal activation in epilepsy patients with malformations of cortical development. Epilepsy Res 116:1–7. https://doi.org/10.1016/j.eplepsyres.2015.06.012

57. Vitali P, Minati L, D'Incerti L, Maccagnano E, Mavilio N, Capello D, Dylgjeri S, Rodriguez G, Franceschetti S, Spreafico R, Villani F (2008) Functional MRI in malformations of cortical development: activation of dysplastic tissue and functional reorganization. J Neuroimaging 18:296–305. https://doi.org/10.1111/j.1552-6569.2007.00164.x

58. Achten E, Jackson GD, Cameron JA, Abbott DF, Stella DL, Fabinyi GCA (1999) Presurgical evaluation of the motor hand area with functional MR imaging in patients with tumors and dysplastic lesions. Radiology 210:529–538. https://doi.org/10.1148/radiology.210.2. r99ja31529

59. Lenge M, Barba C, Montanaro D, Aghakhanyan G, Frijia F, Guerrini R (2018) Relationships between morphologic and functional patterns in the polymicrogyric cortex. Cereb Cortex 28:1076–1086. https://doi.org/10.1093/cercor/bhx036

60. Christodoulou JA, Barnard ME, Del Tufo SN, Katzir T, Whitfield-Gabrieli S, Gabrieli JD, Chang BS (2013) Integration of gray matter nodules into functional cortical circuits in periventricular heterotopia. Epilepsy Behav 29:400–406

61. Dumoulin SO, Jirsch JD, Bernasconi A (2007) Functional organization of human visual cortex in occipital polymicrogyria. Hum Brain Mapp 28:1302–1312

62. Binder JR (2011) Functional MRI is a valid noninvasive alternative to Wada testing. Epilepsy Behav 20:214–222

63. Sabbah P, Chassoux F, Leveque C, Landre E, Baudoin-Chial S, Devaux B, Mann M, Godon-Hardy S, Nioche C, Aït-Ameur A, Sarrazin JL, Chodkiewicz JP, Cordoliani YS (2003) Functional MR imaging in assessment of language dominance in epileptic patients. NeuroImage 18:460–467. https://doi.org/10.1016/S1053-8119(03)00025-9

64. Thivard L, Hombrouck J, du Montcel ST, Delmaire C, Cohen L, Samson S, Dupont S, Chiras J, Baulac M, Lehéricy S (2005) Productive and perceptive language reorganization in temporal lobe epilepsy. NeuroImage 24:841–851

65. Wang A, Peters TM, de Ribaupierre S, Mirsattari SM (2012) Functional magnetic resonance imaging for language mapping in temporal lobe epilepsy. Epilepsy Res Treat 2012:198183

66. Benke T, Köylü B, Visani P, Karner E, Brenneis C, Bartha L, Trinka E, Trieb T, Felber S, Bauer G et al (2006) Language lateralization in temporal lobe epilepsy: a comparison between fMRI and the Wada test. Epilepsia 47:1308–1319

67. Janecek JK, Swanson SJ, Sabsevitz DS, Hammeke TA, Raghavan M, E Rozman M, Binder JR (2013) Language lateralization by fMRI and Wada testing in 229 patients with epilepsy: rates and predictors of discordance. Epilepsia 54:314–322. https://doi.org/10.1111/epi.12068

68. Fernandez G, Specht K, Weis S, Tendolkar I, Reuber M, Fell J, Klaver P, Ruhlmann J, Reul J, Elger CE (2003) Intrasubject reproducibility of presurgical language lateralization and mapping using fMRI. Neurology 60:969–975

69. Berl MM, Zimmaro LA, Khan OI, Dustin I, Ritzl E, Duke ES, Sepeta LN, Sato S, Theodore WH, Gaillard WD (2014) Characterization of atypical language activation patterns in focal epilepsy. Ann Neurol 75:33–42

70. Duke ES, Tesfaye M, Berl MM, Walker JE, Ritzl EK, Fasano RE, Conry JA, Pearl PL, Sato S, Theodore WH et al (2012) The effect of seizure focus on regional language processing areas. Epilepsia 53:1044–1050

71. Jensen EJ, Hargreaves IS, Pexman PM, Bass A, Goodyear BG, Federico P (2011) Abnormalities of lexical and semantic processing in left temporal lobe epilepsy: an fMRI study. Epilepsia 52:2013–2021

72. Rosazza C, Ghielmetti F, Minati L, Vitali P, Giovagnoli AR, Deleo F, Didato G, Parente A, Marras C, Bruzzone MG et al (2013) Preoperative language lateralization in temporal lobe epilepsy (TLE) predicts peri-ictal, pre-and post-operative language performance: an fMRI study. NeuroImage Clin 3:73–83

73. Austermuehle A, Cocjin J, Reynolds R, Agrawal S, Sepeta L, Gaillard WD, Zaghloul KA, Inati S, Theodore WH (2017) Language functional MRI and direct cortical stimulation in epilepsy preoperative planning. Ann Neurol 81:526–537. https://doi.org/10.1002/ana.24899

74. de Vanssay-Maigne A, Noulhiane M, Devauchelle AD, Rodrigo S, Baudoin-Chial S, Meder JF, Oppenheim C, Chiron C, Chassoux F (2011) Modulation of encoding and retrieval by recollection and familiarity: mapping the medial temporal lobe networks. NeuroImage 58:1131–1138

75. Towgood K, Barker GJ, Caceres A, Crum WR, Elwes RDC, Costafreda SG, Mehta MA, Morris RG, von Oertzen TJ, Richardson MP (2015) Bringing memory fMRI to the clinic: comparison of seven memory fMRI protocols in temporal lobe epilepsy. Hum Brain Mapp 36:1595–1608. https://doi.org/10.1002/hbm.22726

76. Bonelli SB, Powell RHW, Yogarajah M, Samson RS, Symms MR, Thompson PJ, Koepp MJ, Duncan JS (2010) Imaging memory in temporal lobe epilepsy: predicting the effects of temporal lobe resection. Brain 133:1186–1199. https://doi.org/10.1093/brain/awq006

77. Duncan JS, Winston GP, Koepp MJ, Ourselin S (2016) Brain imaging in the assessment for epilepsy surgery. Lancet Neurol 15:420–433

78. Dupont S, Duron E, Samson S, Denos M, Volle E, Delmaire C, Navarro V, Chiras J, Lehéricy S, Samson Y et al (2010) Functional MR imaging or Wada test: which is the better predictor of individual postoperative memory outcome? 1. Radiology 255:128–134

79. Sidhu MK, Stretton J, Winston GP, Symms M, Thompson PJ, Koepp MJ, Duncan JS (2015) Memory fMRI predicts verbal memory decline after anterior temporal lobe resection. Neurology 84:1512–1519. https://doi.org/10.1212/WNL.0000000000001461

80. Binder JR, Sabsevitz DS, Swanson SJ, Hammeke TA, Raghavan M, Mueller WM (2008) Use of preoperative functional MRI to predict verbal memory decline after temporal lobe epilepsy surgery. Epilepsia 49:1377–1394

81. Sidhu MK, Stretton J, Winston GP, Bonelli S, Centeno M, Vollmar C, Symms M, Thompson PJ, Koepp MJ, Duncan JS (2013) A functional magnetic resonance imaging study mapping the episodic memory encoding network in temporal lobe epilepsy. Brain 136:1868–1888

82. Englot DJ, Konrad PE, Morgan VL (2016) Regional and global connectivity disturbances in focal epilepsy, related neurocognitive sequelae, and potential mechanistic underpinnings. Epilepsia 57:1546–1557

83. Yang Z, Choupan J, Reutens D, Hocking J (2015) Lateralization of temporal lobe epilepsy based on resting-state functional magnetic resonance imaging and machine learning. Front Neurol 6:184

84. Haneef Z, Lenartowicz A, Yeh HJ, Levin HS, Engel J, Stern JM (2014) Functional connectivity of hippocampal networks in temporal lobe epilepsy. Epilepsia 55:137–145

85. Maccotta L, He BJ, Snyder AZ, Eisenman LN, Benzinger TL, Ances BM, Corbetta M, Hogan RE (2013) Impaired and facilitated functional networks in temporal lobe epilepsy. Neuroimage Clin 2:862–872. https://doi.org/10.1016/j.nicl.2013.06.011

86. Luo C, An D, Yao D, Gotman J (2014) Patient-specific connectivity pattern of epileptic network in frontal lobe epilepsy. Neuroimage Clin 4:668–675. https://doi.org/10.1016/j.nicl.2014.04.006

87. Pedersen M, Curwood EK, Vaughan DN, Omidvarnia AH, Jackson GD (2016) Abnormal brain areas common to the focal epilepsies: multivariate pattern analysis of fMRI. Brain Connect 6:208–215. https://doi.org/10.1089/brain.2015.0367

88. Englot DJ, Hinkley LB, Kort NS, Imber BS, Mizuiri D, Honma SM, Findlay AM, Garrett C, Cheung PL, Mantle M, Tarapore PE, Knowlton RC, Chang EF, Kirsch HE, Nagarajan SS (2015) Global and regional functional connectivity maps of neural oscillations in focal epilepsy. Brain 138:2249–2262. https://doi.org/10.1093/brain/awv130

89. Doucet GE, Rider R, Taylor N, Skidmore C, Sharan A, Sperling M, Tracy JI (2015) Presurgery resting-state local graph-theory measures predict neurocognitive outcomes after brain surgery in temporal lobe epilepsy. Epilepsia 56:517–526. https://doi.org/10.1111/epi.12936

90. Ohue S, Kohno S, Inoue A, Yamashita D, Harada H, Kumon Y, Kikuchi K, Miki H, Ohnishi T (2012) Accuracy of diffusion tensor magnetic resonance imaging-based tractography for surgery of gliomas near the pyramidal tract: a significant correlation between subcortical electrical stimulation and postoperative tractography. Neurosurgery 70:283–294. https://doi.org/10.1227/NEU.0b013e31823020e6

91. Jeong J-W, Asano E, Juhász C, Chugani HT (2014) Quantification of primary motor pathways using diffusion MRI tractography and its application to predict postoperative motor deficits in children with focal epilepsy. Hum Brain Mapp 35:3216–3226

92. Rodrigo S, Oppenheim C, Chassoux F, Hodel J, De Vanssay A, Baudoin-Chial S, Devaux B, Meder J-F (2008) Language lateralization in temporal lobe epilepsy using functional MRI and probabilistic tractography. Epilepsia 49:1367–1376. https://doi.org/10.1111/j.1528-1167.2008.01607.x

94. Winston GP, Daga P, Stretton J, Modat M, Symms MR, McEvoy AW, Ourselin S, Duncan JS (2012) Optic radiation tractography and vision in anterior temporal lobe resection. Ann Neurol 71:334–341. https://doi.org/10.1002/ana.22619
93. Yogarajah M, Focke NK, Bonelli S, Cercignani M, Acheson J, Parker GJM, Alexander DC, McEvoy AW, Symms MR, Koepp MJ et al (2009) Defining Meyer's loop–temporal lobe resections, visual field deficits and diffusion tensor tractography. Brain 132:1656–1668
95. Piper RJ, Yoong MM, Kandasamy J, Chin RF (2014) Application of diffusion tensor imaging and tractography of the optic radiation in anterior temporal lobe resection for epilepsy: a systematic review. Clin Neurol Neurosurg 124:59–65

Chapter 7
In Vivo Positron Emission Tomography of Extrastriatal Non-Dopaminergic Pathology in Parkinson Disease

Martijn L. T. M. Müller and Nicolaas I. Bohnen

Introduction

Motor and Non-motor Symptoms of Parkinson Disease

Parkinson disease (PD) is a progressive neurodegenerative disorder. It is estimated that the disease affects about 1–2% of the population [1]. Parkinson disease is a disease of later life, affecting mostly adults 65 years and older, and men are more affected by it than women [1]. Parkinson disease is characterized by the presence of both motor and non-motor symptoms. The cardinal motor symptoms of PD include slowness of movement (bradykinesia), tremor, rigidity, and postural instability and gait difficulties (PIGD). Non-motor symptoms are observed in nearly every patient and can be plentiful, such as cognitive dysfunction and dementia (PD with dementia, PDD), olfactory dysfunction (impaired sense of smell), apathy and depression, constipation, rapid eye movement sleep (REM) behavior disorder (RBD), and fatigue among other non-motor symptoms [2]. Neuropathological studies suggest that the multitude of these clinical features may be related to distinct neuropathological abnormalities [3, 4].

M. L. T. M. Müller (✉)
Department of Radiology, University of Michigan, Ann Arbor, MI, USA

Morris K. Udall Center of Excellence for Parkinson's Disease Research, University of Michigan, Ann Arbor, MI, USA
e-mail: mtmuller@med.umich.edu

N. I. Bohnen
Department of Radiology, University of Michigan, Ann Arbor, MI, USA

Department of Neurology, University of Michigan, Ann Arbor, MI, USA

Morris K. Udall Center of Excellence for Parkinson's Disease Research, University of Michigan, Ann Arbor, MI, USA

© Springer International Publishing AG, part of Springer Nature 2018
C. Habas (ed.), *The Neuroimaging of Brain Diseases*, Contemporary Clinical Neuroscience, https://doi.org/10.1007/978-3-319-78926-2_7

Neuropathology of PD

The hallmark neuropathology of PD is loss of dopamine in the striatum secondary to progressive degeneration of dopaminergic cells in the substantia nigra pars compacta due to the formation of Lewy bodies and Lewy neurites [5, 6]. Lewy bodies and Lewy neurites are composed of an aggregated form of the protein α-synuclein. Braak et al. proposed a temporal staging scheme describing the distribution of α-synuclein in a cohort of PD patients. According to this scheme, α-synuclein pathology is first deposited in the medulla and olfactory tubercle (Braak stage 1) and then ascends through the brainstem (Braak stage 2) to the midbrain and the substantia nigra (Braak stage 3). At this stage the clinical diagnosis of PD is often made [7]. α-Synuclein pathology subsequently propagates through the basal forebrain and limbic cortex (Braak stage 4) to ultimately culminate in widespread involvement of the neocortex (Braak stages 5 and 6) [5]. The Braak temporal staging system provides a conceptual and temporal framework to explain the heterogeneity of clinical symptom presentation [8]. The formation and propagation of Lewy bodies not only occur in dopaminergic neurons but extend to other neurotransmitter systems as well [3, 4]. For example, Friedrich Heinrich Lewy observed the eponymous Lewy bodies first in the cholinergic neurons of the nucleus basalis of Meynert (nbM) [9]. For decades, most of the research attention has initially focused on the clinical correlates of nigrostriatal dopaminergic system degeneration in PD; however, neuropathological reports have consistently reported degeneration of other neurotransmitter systems in addition to presence of Alzheimer disease (AD)-type pathology (β-amyloid (Aβ) amyloid plaques and neurofibrillary-*tau* tangles, NFT-*tau*). It has become clear that the heterogeneity of clinical motor and non-motor PD symptoms cannot be explained by nigrostriatal dopaminergic denervation alone [10, 11].

Positron emission tomography (PET) is a nuclear imaging technique that allows for in vivo examination of neurotransmission systems, energy metabolism, and pathology in the human brain using radioligands. Over the last 25 years, the introduction of novel radioligands has accelerated in vivo investigations of extrastriatal non-dopaminergic pathology on clinical symptoms of PD. In this chapter we will focus on the clinical sequelae of extrastriatal non-dopaminergic pathology in PD using in vivo PET, with an emphasis on cholinergic system degeneration and presence of AD-type pathology.

Positron Emission Tomography

Basic Principles

The basic principle of PET is the detection of pairs of γ-rays (high-energy photons) that are the result of positron annihilation by an isotope that is coupled with a biologically active molecule. In short, a biological active compound is labeled with a (low dose) decaying isotope (e.g., $[^{11}C]$ or $[^{18}F]$), the so-called radiotracer or radioligand. The radiotracer is injected in a subject to enter the blood stream and will eventually bind to the target of interest while emitting positrons. These emitted positrons will interact with a free electron and annihilation will occur. Annihilated positrons will produce two 511 keV photons simultaneously traveling in opposite direction. These pairs of photons are detected by the scintillator ("coincidence detection") in the PET scanner, and an image is reconstructed using tomographic techniques. This image will reflect the regional distribution of uptake of the radioligand by the target of interest. Single-photon emission computed tomography (SPECT) is similar to PET; however, the isotope emits a single photon rather than two photons through positron annihilation, as is the case in PET. As a result, PET allows a more precise spatial estimation of the sources of the emitted γ-rays. Therefore, PET allows for a higher spatial resolution compared to SPECT. For purposes of this chapter, we will focus on studies reporting PET results.

PET Radioligands

A multitude of different ligands usually exists to quantify the integrity of a particular neurotransmission system. These ligands are often based on naturally occurring substances and existing pharmaceuticals or find its origin in histological dyes used in neuropathological examinations. Ligands of neurotransmission systems can be markers of the actual neurotransmitter such as synthesis of precursors or analogs, vesicular transporters, reuptake transporters, enzymatic degradation activity, or receptors (Fig. 7.1). Ligands of brain proteinopathy, such as Aβ-amyloid and NFT-*tau*, mostly find their origin in histological dyes. It should be noted that at present, there is no validated PET radioligand for Lewy pathology or α-synuclein. For purposes of this chapter, we limit ourselves to studies with the most commonly used radioligands in PD.

Dopaminergic PET Imaging of PD

Pre- and Postsynaptic Dopamine PET Imaging in PD

Most of the PET imaging efforts in PD have focused on the dopaminergic system. A multitude of PET tracers have been developed that are constituents of the presynaptic or postsynaptic neuron (see also Fig. 7.1). For example, the first used dopaminergic ligand $[^{18}F]$-6-fluoroDOPA ($[^{18}F]$DOPA) is a precursor radioligand of dopamine synthesis [12]. Once dopamine is synthesized, it is transported by type 2 vesicular monoamine transporters (VMAT2) into synaptic vesicles from where it is released into the synaptic cleft. The PET radioligand $[^{11}C]$dihydrotetrabenazine ($[^{11}C]$DTBZ) and more recently a fluorinated ($[^{18}F]$) version of it are VMAT2 PET tracers

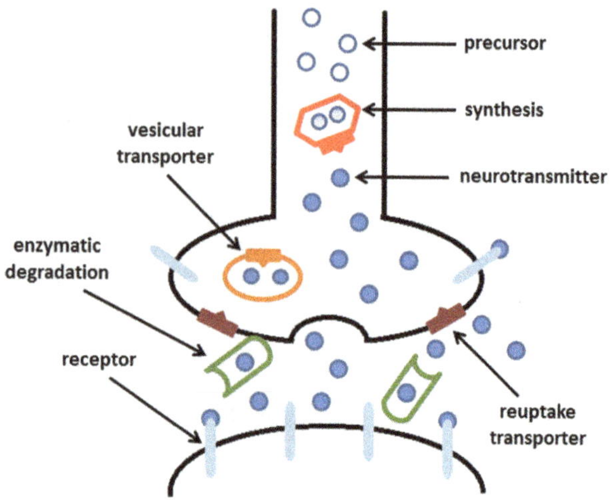

Fig. 7.1 Cartoon of presynaptic, synaptic cleft, and postsynaptic PET ligand targets

[13, 14]. While VMAT2 is a vesicular transporter for all monoaminergic neurotransmitters, in the striatum more than 95% of VMAT2 is utilized by dopamine [15]. Once dopamine is released in the synaptic cleft, it is recycled back by presynaptic dopamine reuptake transporters (DAT), which, for example, can be imaged with the cocaine analogs [11C]β-3beta-(4-fluorophenyl)tropane ([11C]β-CFT) or [11C]methylphenidate ([11C]MP). The [123I]ioflupane DAT SPECT ligand has clinical utility (DaTscan™ by GE Healthcare) for differentiating essential tremor from tremor due to parkinsonian syndromes. On the synaptic membrane, typically on the postsynaptic site of the synaptic cleft, PET ligands have been developed for dopamine receptors, such as [11C]SCH-23390 for the D1 receptor and [11C] raclopride for the D2 group of receptors. Generally there is reasonable good agreement between the presynaptic ligands, and they all show a nigrostriatal dopaminergic denervation pattern that is consistent with neuropathological reports of PD dopaminergic pathology, i.e., asymmetric and with the putamen affected more than the caudate (Fig. 7.2) [16–19]. However, Lee et al. showed in a multiligand study, using DOPA decarboxylase [18F]DOPA, VMAT2 [11C]DTBZ, and DAT [11C]MP, that DOPA decarboxylase may be upregulated and DAT may be downregulated in early PD patients, which may yield [11C]DTBZ as the most robust marker of nigrostriatal dopaminergic nerve terminal density [19, 20]. In vivo [11C]DTBZ VMAT2 PET studies found reductions of up to ~80% in VMAT2 binding in the posterior putamen contralateral to the most affected body side [17, 21]. Estimates for the annual rate of dopamine reduction in PD patients are around 12.5% [22, 23]. In vivo [18F]DOPA and [11C]DTBZ PET studies showed correlation of nigrostriatal dopaminergic denervation with clinical measures of PD including increased Hoehn and Yahr staging, increased ratings of bradykinesia and rigidity, increased disease duration, and decreased activities of daily living [21, 24].

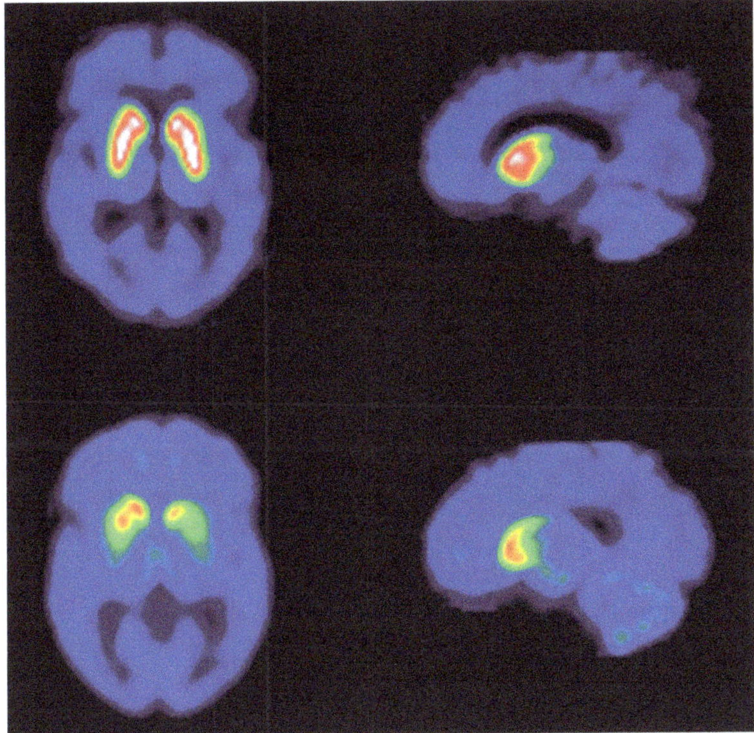

Fig. 7.2 Axial and sagittal [^{11}C]DTBZ VMAT2 PET in a normal non-PD subject (top row) and PD subject (bottom row). The normal non-PD subject shows more intense uptake in the putamen and caudate nucleus than the PD subject. The PD subject shows a typical pattern of striatal dopaminergic denervation with an asymmetrical pattern and the putamen affected more than the caudate

PET Imaging of Prodromal PD

Dopamine PET imaging may have utility in detection of prodromal PD. There is a significant body of literature demonstrating nigrostriatal dopaminergic denervation before the clinical diagnosis of PD. For example, several in vivo [^{18}F]DOPA PET studies identified healthy asymptomatic PD co-twins with abnormal striatal [^{18}F] DOPA uptake [25, 26]. Similarly, PET can identify at-risk asymptomatic family members in families with an increased incidence of parkinsonism [27, 28]. Assuming a linear relationship between decline in putaminal [^{18}F]DOPA uptake and disease duration, Morrish and coworkers estimated the preclinical window for PD to be about 6 ± 3 years and clinical symptoms to arise after a 30% or greater loss of [^{18}F] DOPA uptake [29]. Ultimately, the loss of dopamine in PD will follow a negative exponential progression pattern with most loss in the prodromal and early stages of the disease and a "floor" later in disease [30].

Olfactory loss (hyposmia) is a salient non-motor feature of PD. Impairment of olfactory function is highly prevalent in PD, and a recent in vivo [^{11}C]DTBZ

VMAT2 PET study showed that in PD patients whose clinical diagnosis of PD was confirmed by a PET pattern of nigrostriatal dopaminergic denervation, the loss of olfactory function was near universal [31, 32]. Olfactory loss can also predict the emergence of PD in asymptomatic older adults [33, 34]. In a series of studies, Ponsen and colleagues combined olfactory function assessments with in vivo [^{123}I]β-CIT DAT SPECT in asymptomatic adults. Their work showed that baseline hyposmia in subjects is associated with a 10–12% risk of developing a clinical diagnosis of PD within 2–4 years. Furthermore, they showed that all hyposmic subjects that converted to PD had evidence of reduced striatal DAT uptake [35–37]. The Parkinson Associated Risk Syndrome (PARS) study is a large multicenter longitudinal study that uses a two-tiered strategy of olfactory function screening (tier 1) and clinical and in vivo [^{123}I]β-CIT DAT SPECT (tier 2) to identify individuals at risk of PD. Results so far have shown that olfactory testing can be used to identify asymptomatic subjects with DAT deficit [38]. Furthermore, longitudinal assessments of this cohort of subjects showed that olfactory loss and [^{123}I]β-CIT DAT SPECT deficit were highly predictive of developing PD [39].

Cholinergic PET Imaging of PD

Cholinergic System Loss in PD

Acetylcholine is ubiquitous in the central nervous system and has widespread innervation in the cortex, subcortical structures, and the cerebellum. Cholinergic projections originate from three major sources in the brain. The nucleus basalis of Meynert (nbM), a basal forebrain (BF) nucleus, provides the cholinergic projections to the cerebral cortex and limbic regions [40, 41]. Cholinergic inputs to the thalamus but also connections to the cerebellum, several brainstem nuclei, some striatal fibers, and the spinal cord originate from the brainstem pedunculopontine nucleus-laterodorsal tegmental complex (PPN) [42]. Small populations of intrinsic cholinergic neurons have been observed in the hippocampus, striatum (cholinergic interneurons), cortex, the medial habenula, parts of the reticular formation, and cerebellum [43–46].

Neuropathological studies have shown that, in addition to dopaminergic loss, cholinergic loss also occurs in PD. According to Braak et al., Lewy bodies and neuronal accumulation of α-synuclein deposition in cholinergic neurons of the BF likely occur concurrently with the development of substantia nigra pathology, which indicates that cholinergic denervation may occur early in PD [5]. Neuropathological studies highlight the significance of cholinergic neuron loss in the etiology of PD dementia [47–51].

Cholinergic PET Ligands

Several cholinergic markers that are constituents of acetylcholine synthesis, storage, and recycling have been labeled for PET molecular imaging of cholinergic integrity. The two most widely used PET tracers are the acetylcholinesterase substrate analogs [^{11}C]methyl-4-piperidyl acetate ([^{11}C]MP4A) and [^{11}C]methyl-4-piperidinylpropionate ([^{11}C]PMP). These tracers are metabolized and then trapped by acetylcholinesterase (AChE) and as such reflect regional cerebral AChE distribution. Acetylcholinesterase is a reliable marker of brain cholinergic pathways as the regional distribution corresponds well with the regional distribution of choline transferase (the enzyme responsible for the biosynthesis of acetylcholine) [52]. As [^{11}C]PMP has a slower hydrolysis rate than [^{11}C]MP4A, this tracer allows for more precise estimates of AChE activity in regions of moderate to high AChE concentration with only slightly lower specificity for AChE [53]. Cholinergic innervation of the brain as measured with [^{11}C]PMP AChE PET shows highest activity in the basal ganglia, followed by the cerebellum and thalamus, and lowest activity in the cortex, similar to biodistribution findings in postmortem brains [53–55]. More recently, [^{18}F]FEOBV a cholinergic vesicular transporter (VAChT) PET ligand has been introduced, which is the PET analog of the VAChT SPECT ligand [^{123}I]IBVM [56, 57]. The advantage of this tracer is that, unlike the AChE PET ligands, uptake in the basal ganglia and cerebellum can be noninvasively estimated. Cholinergic receptor PET tracers are [^{18}F]flubatine and [^{18}F]2FA for α4β2 nicotinic acetylcholine receptors (nAChR) and [^{18}F]ASEM for α7 nAChR [58, 59]. Figure 7.3 shows a multiligand cholinergic PET study in a PD subject using [^{11}C]PMP AChE, [^{18}F]FEOBV VAChT, and [^{18}F]flubatine α4β2 nAChR PET.

Cholinergic Imaging of PD Pathology

In vivo PET studies of cholinergic denervation in PD are generally in agreement with histopathology reports. Acetylcholinesterase PET studies have reported cholinergic deficits in both PD without dementia and PDD patients [60–62]. PD patients without dementia show significant reductions in cortical AChE activity, and in PDD these are greater and more extensive [63]. The degree of cholinergic denervation in PD without dementia appears, however, to be more variable. In vivo PET imaging studies have reported cholinergic projection losses in the range from 5% to 25% in PD subjects without and with dementia [61, 63–65]. In a more detailed examination, variability in cortical (reflecting the integrity of nbM cholinergic nerve terminals) and thalamic cholinergic denervation (reflecting the integrity of PPN cholinergic nerve terminals), as assessed by in vivo [^{11}C]PMP AChE PET, was examined in PD patients without dementia and compared to healthy normal control subjects. Acetylcholinesterase activity in PD subjects was dichotomized as within-normal range or below-normal range based on a 5th percentile cutoff of AChE activity in the

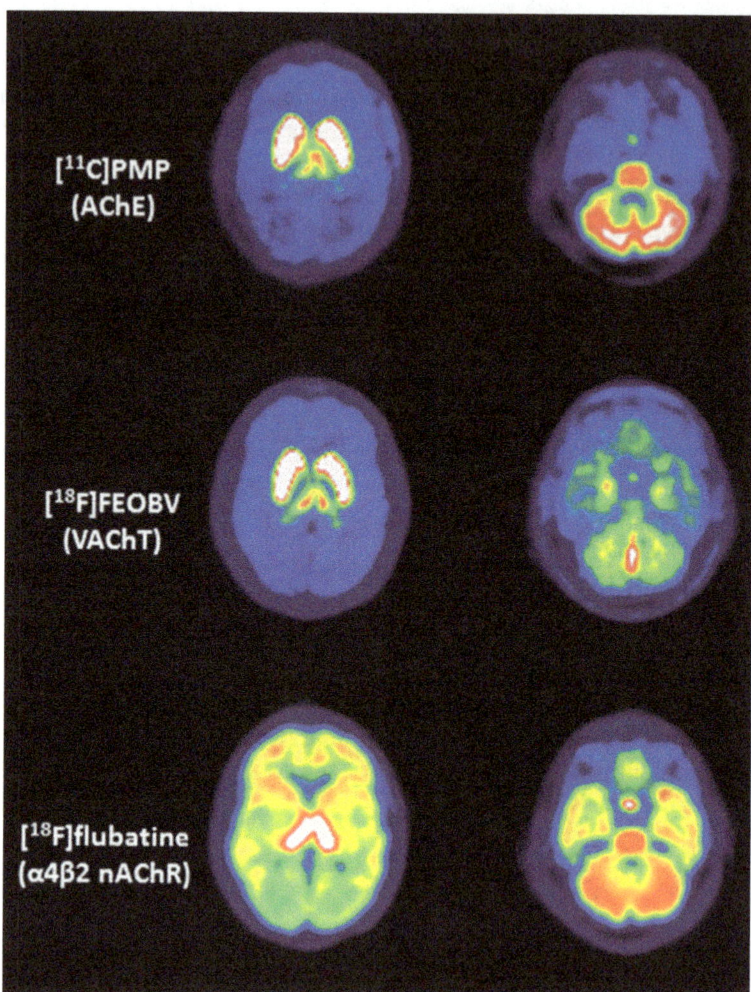

Fig. 7.3 Multiligand cholinergic PET study in a single PD subject using [¹¹C]PMP AChE, [¹⁸F] FEOBV VAChT, and [¹⁸F]flubatine α4β2 nAChR PET. Both [¹¹C]PMP and [¹⁸F]FEOBV show intense uptake in the striatum and to a lesser extent in the thalamus, while [¹⁸F]flubatine shows the most intense uptake in the thalamus but limited uptake in the striatum. [¹¹C]PMP shows intense uptake in the cerebellum, while [¹⁸F]FEOBV shows intense uptake in the cerebellar vermis especially. There is relative low [¹¹C]PMP uptake in the temporal cortex and hippocampus. [¹⁸F] FEOBV appears to have stronger uptake in the hippocampus, while [¹⁸F]flubatine seems to have stronger uptake in the temporal cortex. Note that all three tracers show, to a variable degree, cholinergic activity in the pituitary gland

healthy control subjects. Results showed within-normal range cortical and thalamic [¹¹C]PMP AChE activity for about 65% of the PD subjects. However, the remainder had combined neocortical and thalamic, isolated neocortical, or isolated thalamic cholinergic denervation [66].

Cholinergic muscarinic (mAChR) receptors and nAChR are another target for cholinergic PET imaging of PD. In vivo 2-[^{18}F]FA-85380 nAChR PET imaging of PD subjects without dementia has reported reductions of α4β2 nAChR binding in the striatum, substantia nigra, and cerebellum [67, 68]. In contrast, an mAChR PET study using [^{11}C]NMPB showed *increased* mAChR binding in the frontal cortex in PD patients without dementia [69].

Clinical Correlates of Cholinergic System Loss in PD

Cognition

Over the past several decades, there has been increasing interest in the functional sequelae of cholinergic system loss in PD. It has been associated especially with cognitive impairment and postural instability and gait difficulties (PIGD) in PD [70]. Cognitive impairment and dementia are common in PD, and approximately 75% of the PD patients will develop dementia after 10 years [71]. Even in PD patients without dementia, some cognitive impairment, in particular executive dysfunction, can be observed. Neuropathological studies implicate a role for the basal forebrain and its cholinergic cortical projection in the etiology of cognitive impairment in PD [72, 73]. Findings from AChE PET studies correspond with these observations and show that cortical cholinergic denervation associates with decreased performance on cognitive tasks [66, 74]. These findings, however, do not preclude a role for the dopaminergic system in the etiology of cognitive impairment in PD [75]. Autopsy findings showed that dopaminergic loss in the caudate nucleus was associated with cognitive impairment in PD patients [73]. In vivo dopamine PET imaging studies have also found negative effects of striatal dopaminergic denervation on working memory and overall cognition [66, 76].

Cognitive dysfunction in PD also appears to be associated with decreased nAChR expression. For example, evidence from neuropathological studies shows that cortical nAChR expression is reduced in PDD [77, 78]. An in vivo [^{18}F]2FA nAChR PET showed that cognitive symptoms correlated, albeit weakly, with reduced nAChR expression in the thalamus, midbrain, temporal cortex, hippocampus, and cerebellum [67].

Postural Instability and Gait Difficulties

Postural instability and gait difficulties is a disabling motor feature of PD and responds poorly to dopaminergic replacement therapy, especially with advancing disease [79, 80]. Degeneration of the cholinergic system is likely to be a major contributor to PIGD features in PD [62]. In vivo AChE PET imaging findings support a role for the PPN cholinergic projections in the etiology of PIGD features in PD. Brainstem PPN neurons project mainly to the thalamus, and loss of thalamic

AChE activity likely reflects PPN neuron dysfunction. In vivo [^{11}C]PMP AChE PET studies have shown decreased thalamic AChE activity in PD patients with a history of falls compared to PD non-fallers; however, both groups did not differ in the degree of nigrostriatal dopaminergic denervation [66, 81]. These findings agree with postmortem findings showing more severe PPN cholinergic cell losses in PD fallers compared with PD non-fallers [82]. Findings of an in vivo [^{11}C]DTBZ VMAT2 and [^{11}C]PMP AChE PET study emphasize the putative role of the PPN in balance control [83]. In this study, the efficacy of sensory integration during standing balance was expressed as the degree of postural sway that a person generates during challenging balancing conditions. Results of the study showed that decreased thalamic, but not cortical, AChE activity was associated with increased postural sway in PD patients; however, there was no effect of striatal dopaminergic (VMAT2) terminal integrity on sway [83]. Decreased gait speed in PD patients also appears to be associated with cholinergic system loss. In an in vivo [^{11}C]DTBZ VMAT2 and [^{11}C]PMP AChE PET study, dopaminergic and cholinergic correlates of gait speed in PD and normal control subjects were assessed. Results of the study showed that cortical cholinergic denervation was a more robust determinant of slow gait speed than nigrostriatal denervation alone in PD patients. Interestingly, the study also showed that gait speed was not significantly slower than gait speed of healthy non-PD control subjects in PD patients with relatively isolated nigrostriatal denervation [84].

Other Non-motor Symptoms

The cholinergic system has also been implicated in non-motor symptoms of PD. Parkinson disease patients frequently report depressive mood, major depression, and apathy [85]. An in vivo [^{11}C]PMP AChE PET study showed that, in PD patients with and without dementia, the presence of depressive symptoms and apathy was associated with the severity of cortical cholinergic denervation [86]. An in vivo [^{18}F] 2FA nAChR PET reported strong correlations of depression with nAChR expression in the anterior cingulate cortex, putamen, midbrain, and occipital cortex [67]. Although an in vivo [^{11}C]RTI-32, a marker of both dopamine and noradrenaline transporter binding, PET study has linked monoaminergic system changes to depression, the previous presented evidence suggests that cholinergic system dysfunction may play a role in the etiology of depression in PD as well [87].

Cholinergic system dysfunction may also play a role in the etiology of olfactory dysfunction in PD. Olfactory function is modulated by the cholinergic system, and an in vivo [^{11}C]PMP AChE PET study has shown that cholinergic system loss is associated with olfactory dysfunction [88–90]. While in vivo [^{11}C]β-CFT DAT PET studies suggest a role for decreased striatal and hippocampal dopaminergic activity in the etiology of olfactory dysfunction, limbic cholinergic denervation was the strongest predictor of olfactory dysfunction in PD [90, 91].

The loss of atonia during REM sleep (RBD) results in abnormal motor manifestations ("acting out of dreams"). RBD is common in PD patients, and some PD patients (retrospectively) report RBD as one of the earliest prodromal non-motor manifestations of the disease. Indeed, an in vivo [^{11}C]DTBZ VMAT2 PET study found reduced posterior putamen VMAT2 activity in six elderly adults with chronic idiopathic RBD. The observed pattern of dopaminergic loss in this region is typical of PD, and results of this study suggest that RBD is a possible herald of PD [92, 93]. As RBD is associated with cognitive impairment and dementia in PD, cholinergic system dysfunction may be a shared pathology between these two clinical symptoms [94–97]. An in vivo [^{11}C]DTBZ VMAT2 and [^{11}C]PMP AChE PET study showed that RBD was associated with cholinergic denervation in PD, even in the absence of dementia. In this study, about one-third of non-demented PD subjects had a history of RBD symptoms and exhibited decreased neocortical, limbic cortical, and thalamic cholinergic innervation compared to those without RBD symptoms, independent of the degree of striatal dopaminergic denervation [98].

Aβ-Amyloid and NFT-*tau* PET Imaging of PD

Proteinopathies in PD

Neuropathological studies of PD have shown that proteinopathies, other than α-synuclein aggregates, may also be present in PD, especially in PDD. These studies reveal the presence of AD-type Aβ-amyloid plaques and to a lesser degree NFT-*tau* accumulation in significant subsets of PD and PDD patients [7, 99–108]. In AD, the early development and progression over time of these proteinopathies have been associated with the development of cognitive impairment and dementia [109, 110]. A similar role for AD-type proteinopathies may be expected in cognitive symptomatology of PD and PPD. Indeed, neuropathological studies also suggest a putative role for AD-type pathology in the etiology of cognitive impairment in PD [111–116]. Since the introduction of Aβ-amyloid and NFT-*tau* PET ligands, the number of studies that examine in vivo the effect of AD-type pathology (in the context of α-synuclein aggregates) on cognitive impairment in PD has proliferated.

Aβ-Amyloid PET Imaging of PD

The relatively recent development of radioligands that visualize Aβ-amyloid plaques and NFT-*tau* offers a novel opportunity to study in vivo these protein aggregations in PD (Fig. 7.4). Klunk, Mathis, and colleagues from the University of Pittsburgh were the first to report the in vivo use of the thioflavin-T derivative [N-methyl-[^{11}C]2-(4′-methylaminophenyl)-6-hydroxybenzothiazole (also known as Pittsburgh compound B, [^{11}C]PiB), as a selective ligand to visualize Aβ-amyloid in the human brain [117 119]. This ligand is selective for Aβ-amyloid as it binds avidly to both fibrillar and diffuse amyloid plaques, but not significantly to Lewy bodies

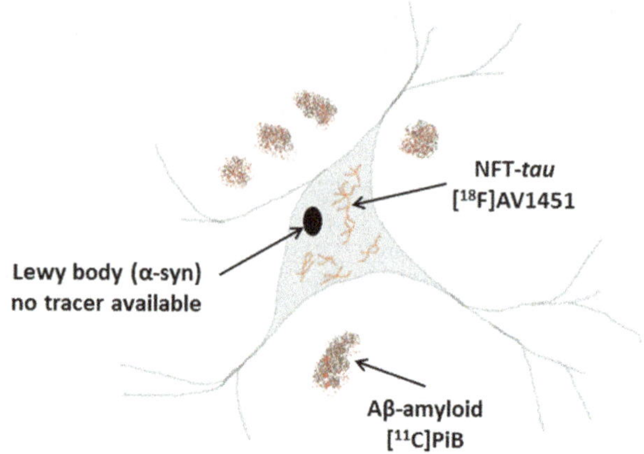

Fig. 7.4 Cartoon of inter- and extra-neuronal proteinopathy PET targets

[120–122]. Since the introduction of [11C]PiB, PET imaging with the [11C]PiB ligand and [18F]-labeled related successors ([18F]flutemetamol, [18F]florbetaben, and [18F]florbetapir) has been performed extensively in patients with mild cognitive impairment (MCI) and AD. Studies show that [11C]PiB binding in AD follows a pattern that matches the stages of AD as described in histopathological reports of amyloidopathy by Braak et al. [123–125].

Compared to Aβ-amyloid levels typically observed in AD, in vivo PET imaging studies of Aβ-amyloid in subjects with PD or PDD generally show lower and more variable levels of Aβ-amyloid in the neocortex [126–136]. For example, levels of Aβ-amyloid deposition that would be considered a "positive" Aβ-amyloid PET scan in AD are uncommon in PD(D) [137–139]. In fact, a recent meta-analysis of in vivo Aβ-amyloid PET studies suggested that PD patients both without and with MCI have a lower incidence of AD-range Aβ-amyloid deposition compared to elderly normal subjects [140].

Cognitive Correlates of Aβ-Amyloid Depositions in PD

The effect of Aβ-amyloid deposition on cognition is frequently examined in PD(D). There is some evidence that Aβ-amyloid deposition detrimentally affects cognition in PD [141]. For example, an in vivo [11C]PiB Aβ-amyloid PET study in PD patients at risk for dementia showed that increased cortical [11C]PiB binding was associated in a linear manner with worse overall performance on neuropsychological test performance as well as decreased performance on the Wechsler Adult Intelligence Scale [138]. Despite this finding, recent reviews concluded inconsistent correlations between cortical Aβ-amyloid deposition and cognitive function in PD [141, 142]. More recent evidence, however, suggests that region-specific cortical and subcortical Aβ-amyloid deposition may be more relevant to cognitive function in PD. For example, in vivo [11C]PiB Aβ-amyloid PET prospective cohort studies have shown that precuneus amyloidopathy was associated with faster progression to cognitive impairment and dementia in PD [133, 143]. Neuropathological studies show that striatal Aβ-amyloid deposition is significantly higher in patients with PDD than in patients with PD without dementia [144, 145]. Findings of a recent in vivo [^{11}C]PiB Aβ-amyloid PET study suggest that the combined presence of striatal and cortical Aβ-amyloid depositions is associated with cognitive impairment in PD rather than cortical Aβ-amyloid depositions alone [146]. Altogether these PET imaging results suggest a role for Aβ-amyloid, especially precuneus and striatal amyloidopathy, in the etiology of cognitive impairment in PD. Other AD-type pathology, such as NFT-*tau* tangles, may also play a role in the etiology of cognitive impairment in PD.

NFT-tau *PET Imaging of PD*

The development of tau PET tracers allows for the in vivo visualization and quantification of NFT-*tau* tangles in the brain. Several potential candidates have been developed and tested; however, [^{18}F]7-(6-fluoropyridin-3-yl)-5H-pyrido[4,3-b] indole ([^{18}F]AV1451, formerly designated as [^{18}F]T807), seems to have especially favorable characteristics and appears now to be most commonly used in in vivo human studies [147, 148]. [^{18}F]AV1451 crosses the blood-brain barrier easily, is highly selective for tau, and has no plasma metabolites entering the brain [149]. Most importantly, it is selective for detection of brain tau pathology in the form of tangles and paired helical filament-tau-containing neurites typical for Alzheimer pathology, but does not seem to bind to Aβ-amyloid or α-synuclein [150]. In vivo PET imaging of [^{18}F]AV1451 uptake in normal older adults showed age-related pattern of increasing binding mainly in the hippocampal formation [151, 152]. Gomperts et al. performed one of the first in vivo [^{18}F]AV1451 NFT-*tau* PET studies to examine [^{18}F]AV1451 uptake in patients with dementia with Lewy bodies (DLB) and cognitively impaired and cognitively normal PD patients. Normal control subjects were also included in the study. Results showed that compared to control subjects, [^{18}F]AV1451 uptake was mildly increased in the inferior and lateral temporal lobe and the precuneus regions for DLB subjects. Furthermore, greater [^{18}F]AV1451 uptake in the inferior temporal gyrus and precuneus was associated with greater cognitive impairment in the combined group of DLB and cognitively impaired PD patients [153]. Contradictory results, however, were found by Hansen et al. who performed in vivo [^{18}F]AV1451 NFT-*tau* PET imaging in a slightly larger group of PD subjects, PD subjects with mild cognitive impairment, and controls. Results showed no differences between these groups, and within the patient group, there was no association with [^{18}F]AV1451 uptake and cognitive task performance [154]. A recent [^{18}F]AV1451 PET study reported similar findings [155]. However, both studies had no subjects with more severe dementia who would be more likely to show tau pathology. These preliminary findings show that [^{18}F]AV1451 may have utility in neuroimaging of NFT-*tau* in PD. However, results of these studies also suggest that there is a gradient and that only more severe cognitive impairment and dementia in PD associate with increased NFT-*tau* [156].

Serotonergic PET Imaging of PD

Serotonergic System Loss in PD

Serotonin is widely distributed in the brain, with most serotonin-producing cell bodies located in the raphe nuclei of the brainstem, in particular in the dorsal and median raphe nuclei of the caudal brainstem. These neurons project widely to the thalamus, hypothalamus, basal ganglia, forebrain, and the neocortex [157–159]. The role of serotonin is recognized in the etiology of depressive symptomatology; however, serotonin appears also to play a role in cognition, motor behavior, and regulation of circadian rhythm [159–162]. It has been suggested that serotonin mostly behaves as modulator of other neurotransmitter systems, thereby indirectly affecting behavior, rather than have a direct influence on these behaviors [163–165].

Neuropathological reports have reported the loss of serotonergic neurons in PD. In the aforementioned temporal staging scheme by Braak et al., Lewy pathology occurs during stage 2 in caudal brainstem nuclei containing serotoninergic neurons, most importantly the raphe nuclei [5]. This suggests that caudal brainstem serotoninergic neurons may be affected by α-synucleinopathy even before dopaminergic midbrain neurons [166]. Neuronal loss as well as Lewy body pathology has been reported in the serotonin-producing raphe nuclei in several neuropathological studies [4, 167–171]. Widespread cerebral loss of serotonin can be observed secondary to the loss of serotonin-producing raphe nuclei in PD. For example, significant loss of serotonin has been reported in PD in the striatum, with the caudate effected more than the putamen, globus pallidus, substantia nigra, hypothalamus, and thalamus [172–177]. Cortical reductions of serotonin can also be observed in PD, including frontal cortex, cingulate cortex, entorhinal cortex, and the hippocampus [175, 178].

Serotonin Transporter PET Imaging of PD

Several PET tracers have been developed to assess different constituents of the serotonergic system in vivo; however, a majority of studies have focused on the serotonin (reuptake) transporter (SERT). In particular, [^{11}C]N,N-dimethyl-2-(2-amino-4-cyanophenylthio) benzylamine ([^{11}C]DASB) is a radioligand with high sensitivity and specificity for SERT (Fig. 7.5) [179]. Several in vivo [^{11}C] DASB PET studies have assessed brain regional SERT innervation in PD. For example, Gutmann et al. found widespread serotonergic loss in non-depressed PD patients compared to controls, with most pronounced losses in the orbitofrontal cortex, striatum, and midbrain [180]. In an in vivo [^{11}C]DASB PET study in PD

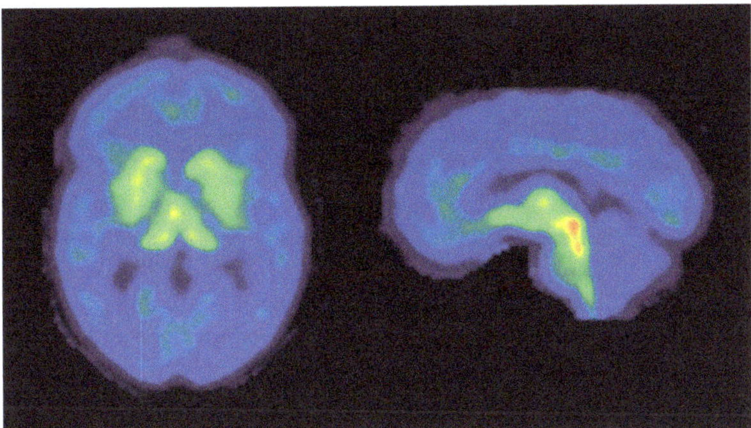

Fig. 7.5 Example of a [^{11}C]DASB SERT PET scan. Intense uptake can be observed in the caudal brainstem and moderate uptake in the thalamus and striatum

patients with less advanced disease severity, similar widespread losses were reported [166]. However, interestingly, SERT was relatively preserved in the caudal brainstem where serotonergic neurons project to brainstem nuclei and the spinal cord [158]. This result could contradict the temporal staging scheme by Braak et al. and its description of early Lewy body pathology in these nuclei, although Braak et al. may have documented Lewy body pathology in non-serotonergic neurons [5, 166]. Strecker et al. reported a similar preservation of SERT binding in the brainstem; however, they did not find widespread loss in other brain areas [181]. Serotonergic loss extends with advancing disease. In an in vivo [^{11}C]DASB SERT PET study, Politis et al. reported decreased [^{11}C]DASB uptake in the caudate, thalamus, and anterior cingulate across all stages of PD symptom severity. With increasing disease duration, the putamen, insular cortex, posterior cingulate, and prefrontal cortex became involved, while ventral striatal, extra-neuronal caudal, and rostral raphe nuclei and amygdala [^{11}C]DASB uptake were only reduced in the later stages of the disease [182]. None of these studies showed any strong correlation with the clinical motor phenotype of PD (e.g., Hoehn and Yahr staging or UPDRS part III subscore) [182].

Clinical Correlates of SERT Loss in PD

The degree of brain regional SERT availability appears to affect some clinical symptoms of PD. A small-scale in vivo [^{11}C]DASB SERT PET study in early PD patients with depression reported a widespread *increased* [^{11}C]DASB uptake, in particular in the dorsolateral cortex and prefrontal cortex [183]. A larger in vivo [^{11}C]DASB SERT PET study in depressed PD patients reported increased [^{11}C]DASB uptake in the raphe nuclei and limbic structures also [184]. An in vivo [^{11}C]DASB SERT PET showed that reduced [^{11}C]DASB uptake in the basal ganglia and limbic structures was associated with fatigue in PD patients [185]. In another study, results from a polysomnography study (particularly measures of sleep-disordered breathing) in PD patients were correlated with [^{11}C]DASB uptake in selected brain regions. The study found no association between severity of sleep-disordered breathing and caudal brainstem serotoninergic innervation in PD patients, suggesting that brainstem serotonergic motor neurons do not play a role in the etiology of sleep-disordered breathing in PD [186]. In a small-scale in vivo [^{11}C] DASB SERT PET study, decreased [^{11}C]DASB uptake in the striatum and raphe nuclei significantly correlated with posture and action tremor severity, but not with resting tremor in PD patients. Motor-related circuitry such as the thalamus and Brodmann areas 4 and 10 also appeared to play a role in the etiology of tremor in PD [187]. A larger cohort study, using [^{123}I]FP-CIT SPECT as a marker of striatal dopaminergic and a proxy marker of raphe nuclei serotonergic transporter availability, however, did show association with rest tremor in PD patients. Raphe [^{123}I]FP-CIT uptake was associated with the amplitude, constancy, and severity of rest tremor [188]. In a series of studies, Politis, Piccini, Brooks, and colleagues looked at the role of SERT availability in the etiology of levodopa-induced dyskinesias (LID). They showed that, in the context of striatal dopaminergic denervation ([^{11}C]raclopride D2 receptor PET), PD patients with LID showed relative preservation of serotonergic terminal function ([^{11}C]DASB SERT PET) compared to PD patients without LID [189]. Relative preservation of serotonergic terminals as PD progresses could therefore be a risk factor for LID [190]. In the context of these observations, they also suggested that a ratio of serotonin/dopamine transporter, as assessed in vivo with [^{11}C]DASB SERT and [^{18}F]FP-CIT DAT, respectively, might be a potential marker of LID-risk in PD [191]. Another of their studies suggested also a role for increased [^{11}C]DASB uptake in the globus pallidus in the etiology of LID in PD [192]. These LID studies provide further evidence for the probable neuromodulatory role that serotonin plays in the central nervous system [163].

Network PET Imaging of PD

Resting-State Glucose Metabolism Network PET Imaging of PD

So far, in vivo PET imaging studies in PD were discussed in the context of the effect of regional innervation of single or multiple, yet distinct, neurotransmitter systems on clinical symptom presentation in PD. The brain, however, is a network of interconnected neurons, and regional neurotransmitter innervation should be viewed in the setting of the larger network it encompasses. An increasing number of studies approach human brain function from this network perspective. This has utility in functional magnetic resonance imaging (fMRI) studies especially. Using resting-state functional connectivity fMRI and statistical techniques, a topographical network of "functional connectivity" can be defined that describes statistical patterns of dynamic interactions among regions [193, 194]. In vivo PET studies have employed a similar statistical modeling approach in order to define topographical functional networks by quantifying resting-state glucose metabolism with the glucose analog [^{18}F]fluorodeoxyglucose ([^{18}F]FDG) PET tracer [195]. Eidelberg and colleagues have developed the scaled subprofile model (SSM) method, which uses principal component analysis (PCA) to identify regional covariance patterns in PD patients. These patterns reflect the covariation of increased or decreased glucose metabolism activity in PD patients relative to a normal population [196, 197]. As reviewed by Peng et al., a vast body of studies by Eidelberg and colleagues have repeatedly shown the existence of a PD-related covariance pattern (PDRP) that shows hypermetabolism of lentiform, thalamic, cerebellar, pontine, and sensorimotor regions, along with hypometabolism in the lateral premotor cortex, supplementary motor area, and parieto-occipital regions [196]. Expression of the PDRP is associated with clinical symptoms of PD such as Hoehn and Yahr staging, quantitative measures of rigidity and bradykinesia, and overall PD symptom severity [198, 199]. Likewise, a PD-related cognitive covariance pattern (PDCP) was identified that correlated with tests of memory and executive functioning. The PDCP is characterized by metabolic reductions in frontal and parietal association areas and relative increases in the cerebellar vermis and dentate nuclei [200]. A full discussion of anatomical and physiological features of the PD-related glucose metabolic networks and clinical correlates of these metabolic networks would go beyond the scope of this chapter and has been extensively reviewed by others [e.g., 196, 201]. These in vivo [^{18}F]FDG resting-state glucose metabolism PET study findings emphasize, however, that the expression of PD clinical symptoms occurs in the context of whole-brain network changes.

Cholinergic Network PET Imaging

Network imaging of the cholinergic system has been used to identify the distinct neural networks underlying cholinergic-mediated cognitive deficits in PD. As cholinergic receptors are widely expressed in the brain, nAChR and mAChR receptor radiotracers are ideal targets to extract spatial covariance patterns of the cholinergic network. For example, an in vivo [^{123}I]QNB M1/M4 subtype mAChR SPECT spatial covariance study in PD patients with dementia found relatively preserved or even increased receptor expression in prefrontal cortex, medial frontal cortex, orbitofrontal cortex, parietal cortex, and posterior cingulate areas with concomitant decreases in mAChR receptor expression in basal forebrain, striatal, temporal, insula, and anterior cingulate areas relative to control subjects [202]. In an [^{123}I] 5IA α4β2 nAChR SPECT study in PD patients without cognitive impairment, reduced regional [^{123}I]5IA uptake was observed in the caudate nucleus, orbitofrontal cortex, and the middle temporal gyrus along with higher uptake in the putamen, the supplementary motor area, and insular cortex compared to normal controls [203]. The relative decreases in regional receptor expression seen in these studies may reflect cholinergic neuronal loss. However, the relative increases in regional cholinergic receptor expression can be interpreted as upregulation of the cholinergic system, either to compensate for cholinergic system loss in other brain regions or as a compensatory mechanism to maintain dopaminergic tone [202, 203].

Conclusions

Recent advances in PET imaging allow for specific in vivo molecular assessment of dopaminergic and non-dopaminergic neurotransmission systems and comorbid presence of AD-type proteinopathies in PD and PDD. With the proliferation of new PET probes, detailed assessment of individual biological determinants underlying the heterogeneity of motor and non-motor disease manifestations in PD becomes more viable. Furthermore, with these multiligand studies, both additive and interactive effects of neurotransmission and AD-type proteinopathy changes on clinical symptoms of PD(D) can now be assessed. Novel network analysis approaches allow for investigation of specific brain network changes that ultimately underlie clinical symptom manifestation in PD. All of these PET imaging techniques are poised to guide novel clinical trials in this field.

Acknowledgments Studies by our group reported in this chapter (Bohnen, Müller, Albin, Frey, and colleagues) were in part supported by NIH grants NS015655, NS099535, NS091856, and NS070856.

References

1. Alves G, Forsaa EB, Pedersen KF, Dreetz Gjerstad M, Larsen JP (2008) Epidemiology of Parkinson's disease. J Neurol 255 Suppl 5(5):18–32
2. Titova N, Qamar MA, Chaudhuri KR (2017) Chapter 3 – The nonmotor features of Parkinson's disease. In: Bhatia KP, Chaudhuri KR, Stamelou M (eds) International review of neurobiology, vol 132. Academic Press, pp 33–54
3. Hughes AJ, Daniel SE, Kilford L, Lees AJ (1992) Accuracy of clinical diagnosis of idiopathic Parkinson's disease: a clinico-pathological study of 100 cases. J Neurol Neurosurg Psychiatry 55(3):181–184
4. Paulus W, Jellinger K (1991) The neuropathologic basis of different clinical subgroups of Parkinson's disease. J Neuropathol Exp Neurol 50(6):743–755
5. Braak H, Del Tredici K, Rub U, de Vos RA, Jansen Steur EN, Braak E (2003) Staging of brain pathology related to sporadic Parkinson's disease. Neurobiol Aging 24(2):197–211
6. Hughes AJ, Daniel SE, Blankson S, Lees AJ (1993) A clinicopathologic study of 100 cases of Parkinson's disease. Arch Neurol 50(2):140–148

7. Hughes AJ, Ben-Shlomo Y, Daniel SE, Lees AJ (1992) What features improve the accuracy of clinical diagnosis in Parkinson's disease: a clinicopathologic study. Neurology 42 (6):1142–1146

8. Langston JW (2006) The Parkinson's complex: parkinsonism is just the tip of the iceberg. Ann Neurol 59(4):591–596

9. Holdorff B (2006) Fritz Heinrich Lewy (1885-1950). J Neurol 253(5):677–678

10. Hely MA, Morris JG, Reid WG, Trafficante R (2005) Sydney multicenter study of Parkinson's disease: non-L-dopa-responsive problems dominate at 15 years. Mov Disord 20(2):190–199

11. Hely MA, Reid WG, Adena MA, Halliday GM, Morris JG (2008) The Sydney multicenter study of Parkinson's disease: the inevitability of dementia at 20 years. Mov Disord 23 (6):837–844

12. Garnett ES, Firnau G, Chan PK, Sood S, Belbeck LW (1978) [18F]fluoro-dopa, an analogue of dopa, and its use in direct external measurements of storage, degradation, and turnover of intracerebral dopamine. Proc Natl Acad Sci U S A 75(1):464–467

13. Koeppe RA, Frey KA, Vander Borght TM, Karlamangla A, Jewett DM, Lee LC et al (1996) Kinetic evaluation of [11C]dihydrotetrabenazine by dynamic PET: measurement of vesicular monoamine transporter. J Cereb Blood Flow Metab 16(6):1288–1299

14. Lin SC, Lin KJ, Hsiao IT, Hsieh CJ, Lin WY, Lu CS et al (2014) In vivo detection of monoaminergic degeneration in early Parkinson disease by (18)F-9-fluoropropyl-(+)-dihydrotetrabenzazine PET. J Nucl Med 55(1):73–79

15. Vander Borght TM, Sima AA, Kilbourn MR, Desmond TJ, Kuhl DE, Frey KA (1995) [3H] methoxytetrabenazine: a high specific activity ligand for estimating monoaminergic neuronal integrity. Neuroscience 68(3):955–962

16. Kish SJ, Shannak K, Hornykiewicz O (1988) Uneven pattern of dopamine loss in the striatum of patients with idiopathic Parkinson's disease. Pathophysiologic and clinical implications. N Engl J Med 318(14):876–880

17. Frey KA, Koeppe RA, Kilbourn MR, Vander Borght TM, Albin RL, Gilman S et al (1996) Presynaptic monoaminergic vesicles in Parkinson's disease and normal aging. Ann Neurol 40 (6):873–884

18. Rakshi JS, Uema T, Ito K, Bailey DL, Morrish PK, Ashburner J et al (1999) Frontal, midbrain and striatal dopaminergic function in early and advanced Parkinson's disease A 3D [(18)F] dopa-PET study. Brain 122(Pt 9):1637–1650

19. Lee CS, Samii A, Sossi V, Ruth TJ, Schulzer M, Holden JE et al (2000) In vivo positron emission tomographic evidence for compensatory changes in presynaptic dopaminergic nerve terminals in Parkinson's disease. Ann Neurol 47(4):493–503

20. Bohnen NI, Frey KA (2003) The role of positron emission tomography imaging in movement disorders. Neuroimaging Clin N Am 13(4):791–803

21. Bohnen NI, Albin RL, Koeppe RA, Wernette KA, Kilbourn MR, Minoshima S et al (2006) Positron emission tomography of monoaminergic vesicular binding in aging and Parkinson disease. J Cereb Blood Flow Metab 26(9):1198–1212

22. Morrish PK, Sawle GV, Brooks DJ (1996) An [18F]dopa-PET and clinical study of the rate of progression in Parkinson's disease. Brain 119(Pt 2):585–591

23. Nurmi E, Ruottinen HM, Kaasinen V, Bergman J, Haaparanta M, Solin O et al (2000) Progression in Parkinson's disease: a positron emission tomography study with a dopamine transporter ligand [18F]CFT. Ann Neurol 47(6):804–808

24. Antonini A, Vontobel P, Psylla M, Gunther I, Maguire PR, Missimer J et al (1995) Complementary positron emission tomographic studies of the striatal dopaminergic system in Parkinson's disease. Arch Neurol 52(12):1183–1190

25. Burn DJ, Mark MH, Playford ED, Maraganore DM, Zimmerman TR Jr, Duvoisin RC et al (1992) Parkinson's disease in twins studied with 18F-dopa and positron emission tomography. Neurology 42(10):1894–1900

26. Holthoff VA, Vieregge P, Kessler J, Pietrzyk U, Herholz K, Bonner J et al (1994) Discordant twins with Parkinson's disease: positron emission tomography and early signs of impaired cognitive circuits. Ann Neurol 36(2):176–182

27. Sawle GV, Wroe SJ, Lees AJ, Brooks DJ, Frackowiak RS (1992) The identification of presymptomatic parkinsonism: clinical and [18F]dopa positron emission tomography studies in an Irish kindred. Ann Neurol 32(5):609–617

28. Piccini P, Morrish PK, Turjanski N, Sawle GV, Burn DJ, Weeks RA et al (1997) Dopaminergic function in familial Parkinson's disease: a clinical and 18F-dopa positron emission tomography study. Ann Neurol 41(2):222–229

29. Morrish PK, Rakshi JS, Bailey DL, Sawle GV, Brooks DJ (1998) Measuring the rate of progression and estimating the preclinical period of Parkinson's disease with [18F]dopa PET. J Neurol Neurosurg Psychiatry 64(3):314–319

30. Kaasinen V, Vahlberg T (2017) Striatal dopamine in Parkinson disease: A meta-analysis of imaging studies. Ann Neurol 82(6):873–882

31. Haugen J, Muller ML, Kotagal V, Albin RL, Koeppe RA, Scott PJ et al (2016) Prevalence of impaired odor identification in Parkinson disease with imaging evidence of nigrostriatal denervation. J Neural Transm (Vienna) 123(4):421–424

32. Haehner A, Boesveldt S, Berendse HW, Mackay-Sim A, Fleischmann J, Silburn PA et al (2009) Prevalence of smell loss in Parkinson's disease--a multicenter study. Parkinsonism Relat Disord 15(7):490–494

33. Ross GW, Petrovitch H, Abbott RD, Tanner CM, Popper J, Masaki K et al (2008) Association of olfactory dysfunction with risk for future Parkinson's disease. Ann Neurol 63(2):167–173

34. Haehner A, Hummel T, Hummel C, Sommer U, Junghanns S, Reichmann H (2007) Olfactory loss may be a first sign of idiopathic Parkinson's disease. Mov Disord 22(6):839–842

35. Berendse HW, Ponsen MM (2009) Diagnosing premotor Parkinson's disease using a two-step approach combining olfactory testing and DAT SPECT imaging. Parkinsonism Relat Disord 15(Suppl 3):S26–S30

36. Ponsen MM, Stoffers D, Booij J, van Eck-Smit BL, Wolters E, Berendse HW (2004) Idiopathic hyposmia as a preclinical sign of Parkinson's disease. Ann Neurol 56(2):173–181

37. Ponsen MM, Stoffers D, Wolters E, Booij J, Berendse HW (2010) Olfactory testing combined with dopamine transporter imaging as a method to detect prodromal Parkinson's disease. J Neurol Neurosurg Psychiatry 81(4):396–399

38. Jennings D, Siderowf A, Stern M, Seibyl J, Eberly S, Oakes D et al (2014) Imaging prodromal Parkinson disease: the Parkinson Associated Risk Syndrome Study. Neurology 83 (19):1739–1746

39. Jennings D, Siderowf A, Stern M, Seibyl J, Eberly S, Oakes D et al (2017) Conversion to Parkinson disease in the PARS hyposmic and dopamine transporter-deficit prodromal cohort. JAMA Neurol 74(8):933–940

40. Mesulam MM, Geula C (1988) Nucleus basalis (Ch4) and cortical cholinergic innervation in the human brain: observations based on the distribution of acetylcholinesterase and choline acetyltransferase. J Comp Neurol 275(2):216–240

41. Liu AK, Chang RC, Pearce RK, Gentleman SM (2015) Nucleus basalis of Meynert revisited: anatomy, history and differential involvement in Alzheimer's and Parkinson's disease. Acta Neuropathol 129(4):527–540

42. Heckers S, Geula C, Mesulam MM (1992) Cholinergic innervation of the human thalamus: dual origin and differential nuclear distribution. J Comp Neurol 325(1):68–82

43. de Lacalle S, Hersh LB, Saper CB (1993) Cholinergic innervation of the human cerebellum. J Comp Neurol 328(3):364–376

44. Fibiger HC (1982) The organization and some projections of cholinergic neurons of the mammalian forebrain. Brain Res 257(3):327–388

45. Lecourtier L, Kelly PH (2007) A conductor hidden in the orchestra? Role of the habenular complex in monoamine transmission and cognition. Neurosci Biobehav Rev 31(5):658–672

46. Mesulam MM, Mash D, Hersh L, Bothwell M, Geula C (1992) Cholinergic innervation of the human striatum, globus pallidus, subthalamic nucleus, substantia nigra, and red nucleus. J Comp Neurol 323(2):252–268

47. Candy JM, Perry RH, Perry EK, Irving D, Blessed G, Fairbairn AF et al (1983) Pathological changes in the nucleus of Meynert in Alzheimer's and Parkinson's diseases. J Neurol Sci 59 (2):277–289
48. Nakano I, Hirano A (1984) Parkinson's disease: neuron loss in the nucleus basalis without concomitant Alzheimer's disease. Ann Neurol 15(5):415–418
49. Rogers JD, Brogan D, Mirra SS (1985) The nucleus basalis of Meynert in neurological disease: a quantitative morphological study. Ann Neurol 17(2):163–170
50. Tagliavini F, Pilleri G, Bouras C, Constantinidis J (1984) The basal nucleus of Meynert in idiopathic Parkinson's disease. Acta Neurol Scand 70(1):20–28
51. Whitehouse PJ, Hedreen JC, White CL 3rd, Price DL (1983) Basal forebrain neurons in the dementia of Parkinson disease. Ann Neurol 13(3):243–248
52. Mesulam MM, Geula C (1992) Overlap between acetylcholinesterase-rich and choline acetyltransferase-positive (cholinergic) axons in human cerebral cortex. Brain Res 577 (1):112–120
53. Koeppe RA, Frey KA, Snyder SE, Meyer P, Kilbourn MR, Kuhl DE (1999) Kinetic modeling of N-[11C]methylpiperidin-4-yl propionate: alternatives for analysis of an irreversible positron emission tomography trace for measurement of acetylcholinesterase activity in human brain. J Cereb Blood Flow Metab 19(10):1150–1163
54. Kuhl DE, Koeppe RA, Minoshima S, Snyder SE, Ficaro EP, Foster NL et al (1999) In vivo mapping of cerebral acetylcholinesterase activity in aging and Alzheimer's disease. Neurology 52(4):691–699
55. Atack JR, Perry EK, Bonham JR, Candy JM, Perry RH (1986) Molecular forms of acetylcholinesterase and butyrylcholinesterase in the aged human central nervous system. J Neurochem 47(1):263–277
56. Petrou M, Frey KA, Kilbourn MR, Scott PJ, Raffel DM, Bohnen NI et al (2014) In vivo imaging of human cholinergic nerve terminals with (−)-5-(18)F-fluoroethoxybenzovesamicol: biodistribution, dosimetry, and tracer kinetic analyses. J Nucl Med 55(3):396–404
57. Kuhl DE, Koeppe RA, Fessler JA, Minoshima S, Ackermann RJ, Carey JE et al (1994) In vivo mapping of cholinergic neurons in the human brain using SPECT and IBVM. J Nucl Med 35 (3):405–410
58. Hillmer AT, Esterlis I, Gallezot JD, Bois F, Zheng MQ, Nabulsi N et al (2016) Imaging of cerebral alpha4beta2* nicotinic acetylcholine receptors with (−)-[(18)F]Flubatine PET: Implementation of bolus plus constant infusion and sensitivity to acetylcholine in human brain. NeuroImage 141:71–80
59. Gao Y, Mease RC, Olson TT, Kellar KJ, Dannals RF, Pomper MG et al (2015) [(125)I]Iodo-ASEM, a specific in vivo radioligand for alpha7-nAChR. Nucl Med Biol 42(5):488–493
60. Shinotoh H, Namba H, Yamaguchi M, Fukushi K, Nagatsuka S, Iyo M et al (1999) Positron emission tomographic measurement of acetylcholinesterase activity reveals differential loss of ascending cholinergic systems in Parkinson's disease and progressive supranuclear palsy. Ann Neurol 46(1):62–69
61. Hilker R, Thomas AV, Klein JC, Weisenbach S, Kalbe E, Burghaus L et al (2005) Dementia in Parkinson disease: functional imaging of cholinergic and dopaminergic pathways. Neurology 65(11):1716–1722
62. Gilman S, Koeppe RA, Nan B, Wang CN, Wang X, Junck L et al (2010) Cerebral cortical and subcortical cholinergic deficits in parkinsonian syndromes. Neurology 74(18):1416–1423
63. Bohnen NI, Kaufer DI, Ivanco LS, Lopresti B, Koeppe RA, Davis JG et al (2003) Cortical cholinergic function is more severely affected in parkinsonian dementia than in Alzheimer disease: an in vivo positron emission tomographic study. Arch Neurol 60(12):1745–1748
64. Shinotoh H, Aotsuka A, Fukushi K, Nagatsuka S, Tanaka N, Ota T et al (2001) Effect of donepezil on brain acetylcholinesterase activity in patients with AD measured by PET. Neurology 56(3):408–410
65. Shimada H, Hirano S, Shinotoh H, Aotsuka A, Sato K, Tanaka N et al (2009) Mapping of brain acetylcholinesterase alterations in Lewy body disease by PET. Neurology 73(4):273–278

66. Bohnen NI, Muller ML, Kotagal V, Koeppe RA, Kilbourn MR, Gilman S et al (2012) Heterogeneity of cholinergic denervation in Parkinson's disease without dementia. J Cereb Blood Flow Metab 32(8):1609–1617

67. Meyer PM, Strecker K, Kendziorra K, Becker G, Hesse S, Woelpl D et al (2009) Reduced alpha4beta2*-nicotinic acetylcholine receptor binding and its relationship to mild cognitive and depressive symptoms in Parkinson disease. Arch Gen Psychiatry 66(8):866–877

68. Kas A, Bottlaender M, Gallezot JD, Vidailhet M, Villafane G, Gregoire MC et al (2009) Decrease of nicotinic receptors in the nigrostriatal system in Parkinson's disease. J Cereb Blood Flow Metab 29(9):1601–1608

69. Asahina M, Suhara T, Shinotoh H, Inoue O, Suzuki K, Hattori T (1998) Brain muscarinic receptors in progressive supranuclear palsy and Parkinson's disease: a positron emission tomographic study. J Neurol Neurosurg Psychiatry 65(2):155–163

70. Muller ML, Bohnen NI (2013) Cholinergic dysfunction in Parkinson's disease. Curr Neurol Neurosci Rep 13(9):377

71. Aarsland D, Kurz MW (2010) The epidemiology of dementia associated with Parkinson disease. J Neurol Sci 289(1–2):18–22

72. Ruberg M, Rieger F, Villageois A, Bonnet AM, Agid Y (1986) Acetylcholinesterase and butyrylcholinesterase in frontal cortex and cerebrospinal fluid of demented and non-demented patients with Parkinson's disease. Brain Res 362(1):83–91

73. Mattila PM, Roytta M, Lonnberg P, Marjamaki P, Helenius H, Rinne JO (2001) Choline acetytransferase activity and striatal dopamine receptors in Parkinson's disease in relation to cognitive impairment. Acta Neuropathol 102(2):160–166

74. Bohnen NI, Kaufer DI, Hendrickson R, Ivanco LS, Lopresti BJ, Constantine GM et al (2006) Cognitive correlates of cortical cholinergic denervation in Parkinson's disease and parkinsonian dementia. J Neurol 253(2):242–247

75. Bohnen NI, Albin RL, Muller ML, Petrou M, Kotagal V, Koeppe RA et al (2015) Frequency of cholinergic and caudate nucleus dopaminergic deficits across the predemented cognitive spectrum of Parkinson disease and evidence of interaction effects. JAMA Neurol 72 (2):194–200

76. Sawamoto N, Piccini P, Hotton G, Pavese N, Thielemans K, Brooks DJ (2008) Cognitive deficits and striato-frontal dopamine release in Parkinson's disease. Brain 131 (Pt 5):1294–1302

77. Aubert I, Araujo DM, Cecyre D, Robitaille Y, Gauthier S, Quirion R (1992) Comparative alterations of nicotinic and muscarinic binding sites in Alzheimer's and Parkinson's diseases. J Neurochem 58(2):529–541

78. Whitehouse PJ, Martino AM, Wagster MV, Price DL, Mayeux R, Atack JR et al (1988) Reductions in [^{3}H]nicotinic acetylcholine binding in Alzheimer's disease and Parkinson's disease: an autoradiographic study. Neurology 38(5):720–723

79. Muslimovic D, Post B, Speelman JD, Schmand B, de Haan RJ, Group CS (2008) Determinants of disability and quality of life in mild to moderate Parkinson disease. Neurology 70 (23):2241–2247

80. Sethi K (2008) Levodopa unresponsive symptoms in Parkinson disease. Mov Disord 23(Suppl 3):S521–S533

81. Bohnen NI, Muller ML, Koeppe RA, Studenski SA, Kilbourn MA, Frey KA et al (2009) History of falls in Parkinson disease is associated with reduced cholinergic activity. Neurology 73(20):1670–1676

82. Karachi C, Grabli D, Bernard FA, Tande D, Wattiez N, Belaid H et al (2010) Cholinergic mesencephalic neurons are involved in gait and postural disorders in Parkinson disease. J Clin Invest 120(8):2745–2754

83. Muller ML, Albin RL, Kotagal V, Koeppe RA, Scott PJ, Frey KA et al (2013) Thalamic cholinergic innervation and postural sensory integration function in Parkinson's disease. Brain 136(Pt 11):3282–3289

84. Bohnen NI, Frey KA, Studenski S, Kotagal V, Koeppe RA, Scott PJ et al (2013) Gait speed in Parkinson disease correlates with cholinergic degeneration. Neurology 81(18):1611–1616
85. Kulisevsky J, Pagonabarraga J, Pascual-Sedano B, Garcia-Sanchez C, Gironell A, Trapecio Group Study (2008) Prevalence and correlates of neuropsychiatric symptoms in Parkinson's disease without dementia. Mov Disord 23(13):1889–1896
86. Bohnen NI, Kaufer DI, Hendrickson R, Constantine GM, Mathis CA, Moore RY (2007) Cortical cholinergic denervation is associated with depressive symptoms in Parkinson's disease and parkinsonian dementia. J Neurol Neurosurg Psychiatry 78(6):641–643
87. Remy P, Doder M, Lees A, Turjanski N, Brooks D (2005) Depression in Parkinson's disease: loss of dopamine and noradrenaline innervation in the limbic system. Brain 128 (Pt 6):1314–1322
88. Doty RL (2012) Olfactory dysfunction in Parkinson disease. Nat Rev Neurol 8(6):329–339
89. Bohnen NI, Muller ML (2013) In vivo neurochemical imaging of olfactory dysfunction in Parkinson's disease. J Neural Transm (Vienna). 120(4):571–576
90. Bohnen NI, Muller ML, Kotagal V, Koeppe RA, Kilbourn MA, Albin RL et al (2010) Olfactory dysfunction, central cholinergic integrity and cognitive impairment in Parkinson's disease. Brain 133(Pt 6):1747–1754
91. Bohnen NI, Gedela S, Kuwabara H, Constantine GM, Mathis CA, Studenski SA et al (2007) Selective hyposmia and nigrostriatal dopaminergic denervation in Parkinson's disease. J Neurol 254(1):84–90
92. Albin RL, Koeppe RA, Chervin RD, Consens FB, Wernette K, Frey KA et al (2000) Decreased striatal dopaminergic innervation in REM sleep behavior disorder. Neurology 55 (9):1410–1412
93. Postuma RB, Gagnon JF, Vendette M, Montplaisir JY (2009) Idiopathic REM sleep behavior disorder in the transition to degenerative disease. Mov Disord 24(15):2225–2232
94. Jozwiak N, Postuma RB, Montplaisir J, Latreille V, Panisset M, Chouinard S et al (2017) REM sleep behavior disorder and cognitive impairment in Parkinson's disease. Sleep 40(8)
95. Marion MH, Qurashi M, Marshall G, Foster O (2008) Is REM sleep behaviour disorder (RBD) a risk factor of dementia in idiopathic Parkinson's disease? J Neurol 255(2):192–196
96. Vendette M, Gagnon JF, Decary A, Massicotte-Marquez J, Postuma RB, Doyon J et al (2007) REM sleep behavior disorder predicts cognitive impairment in Parkinson disease without dementia. Neurology 69(19):1843–1849
97. Chahine LM, Xie SX, Simuni T, Tran B, Postuma R, Amara A et al (2016) Longitudinal changes in cognition in early Parkinson's disease patients with REM sleep behavior disorder. Parkinsonism Relat Disord 27:102–106
98. Kotagal V, Albin RL, Muller ML, Koeppe RA, Chervin RD, Frey KA et al (2012) Symptoms of rapid eye movement sleep behavior disorder are associated with cholinergic denervation in Parkinson disease. Ann Neurol 71(4):560–568
99. Boller F, Mizutani T, Roessmann U, Gambetti P (1980) Parkinson disease, dementia, and Alzheimer disease: clinicopathological correlations. Ann Neurol 7(4):329–335
100. Hakim AM, Mathieson G (1979) Dementia in Parkinson disease: a neuropathologic study. Neurology 29(9 Pt 1):1209–1214
101. Gaspar P, Gray F (1984) Dementia in idiopathic Parkinson's disease. A neuropathological study of 32 cases. Acta Neuropathol 64(1):43–52
102. Xuereb JH, Tomlinson BE, Irving D, Perry RH, Blessed G, Perry EK (1990) Cortical and subcortical pathology in Parkinson's disease: relationship to parkinsonian dementia. Adv Neurol 53:35–40
103. Brown DF, Dababo MA, Bigio EH, Risser RC, Eagan KP, Hladik CL et al (1998) Neuropathologic evidence that the Lewy body variant of Alzheimer disease represents coexistence of Alzheimer disease and idiopathic Parkinson disease. J Neuropathol Exp Neurol 57(1):39–46
104. Chui HC, Mortimer JA, Slager U, Zarow C, Bondareff W, Webster DD (1986) Pathologic correlates of dementia in Parkinson's disease. Arch Neurol 43(10):991–995

105. Mattila PM, Roytta M, Torikka H, Dickson DW, Rinne JO (1998) Cortical Lewy bodies and Alzheimer-type changes in patients with Parkinson's disease. Acta Neuropathol 95 (6):576–582

106. Hurtig HI, Trojanowski JQ, Galvin J, Ewbank D, Schmidt ML, Lee VM et al (2000) Alpha-synuclein cortical Lewy bodies correlate with dementia in Parkinson's disease. Neurology 54 (10):1916–1921

107. Apaydin H, Ahlskog JE, Parisi JE, Boeve BF, Dickson DW (2002) Parkinson disease neuropathology: later-developing dementia and loss of the levodopa response. Arch Neurol 59(1):102–112

108. Mastaglia FL, Johnsen RD, Byrnes ML, Kakulas BA (2003) Prevalence of amyloid-beta deposition in the cerebral cortex in Parkinson's disease. Mov Disord 18(1):81–86

109. Jack CR Jr, Knopman DS, Jagust WJ, Petersen RC, Weiner MW, Aisen PS et al (2013) Tracking pathophysiological processes in Alzheimer's disease: an updated hypothetical model of dynamic biomarkers. Lancet Neurol 12(2):207–216

110. Fleisher AS, Chen K, Quiroz YT, Jakimovich LJ, Gomez MG, Langois CM et al (2012) Florbetapir PET analysis of amyloid-beta deposition in the presenilin 1 E280A autosomal dominant Alzheimer's disease kindred: a cross-sectional study. Lancet Neurol 11 (12):1057–1065

111. Braak H, Braak E (1990) Cognitive impairment in Parkinson's disease: amyloid plaques, neurofibrillary tangles, and neuropil threads in the cerebral cortex. J Neural Transm Park Dis Dement Sect 2(1):45–57

112. Braak H, Rub U, Jansen Steur EN, Del Tredici K, de Vos RA (2005) Cognitive status correlates with neuropathologic stage in Parkinson disease. Neurology 64(8):1404–1410

113. Horvath J, Herrmann FR, Burkhard PR, Bouras C, Kovari E (2013) Neuropathology of dementia in a large cohort of patients with Parkinson's disease. Parkinsonism Relat Disord 19(10):864–868

114. Howlett DR, Whitfield D, Johnson M, Attems J, O'Brien JT, Aarsland D et al (2015) Regional multiple pathology scores are associated with cognitive decline in lewy body dementias. Brain Pathol 25(4):401–408

115. Ruffmann C, Calboli FC, Bravi I, Gveric D, Curry LK, de Smith A et al (2016) Cortical Lewy bodies and Abeta burden are associated with prevalence and timing of dementia in Lewy body diseases. Neuropathol Appl Neurobiol 42(5):436–450

116. Walker L, McAleese KE, Thomas AJ, Johnson M, Martin-Ruiz C, Parker C et al (2015) Neuropathologically mixed Alzheimer's and Lewy body disease: burden of pathological protein aggregates differs between clinical phenotypes. Acta Neuropathol 129(5):729–748

117. Klunk WE, Engler H, Nordberg A, Wang Y, Blomqvist G, Holt DP et al (2004) Imaging brain amyloid in Alzheimer's disease with Pittsburgh Compound-B. Ann Neurol 55(3):306–319

118. Mathis CA, Bacskai BJ, Kajdasz ST, McLellan ME, Frosch MP, Hyman BT et al (2002) A lipophilic thioflavin-T derivative for positron emission tomography (PET) imaging of amyloid in brain. Bioorg Med Chem Lett 12(3):295–298

119. Wang Y, Klunk WE, Huang GF, Debnath ML, Holt DP, Mathis CA (2002) Synthesis and evaluation of 2-(3′-iodo-4′-aminophenyl)-6-hydroxybenzothiazole for in vivo quantitation of amyloid deposits in Alzheimer's disease. J Mol Neurosci 19(1–2):11–16

120. Lockhart A, Lamb JR, Osredkar T, Sue LI, Joyce JN, Ye L et al (2007) PIB is a non-specific imaging marker of amyloid-beta (Abeta) peptide-related cerebral amyloidosis. Brain 130 (Pt 10):2607–2615

121. Fodero-Tavoletti MT, Smith DP, McLean CA, Adlard PA, Barnham KJ, Foster LE et al (2007) In vitro characterization of Pittsburgh compound-B binding to Lewy bodies. J Neurosci 27 (39):10365–10371

122. Ye L, Velasco A, Fraser G, Beach TG, Sue L, Osredkar T et al (2008) In vitro high affinity alpha-synuclein binding sites for the amyloid imaging agent PIB are not matched by binding to Lewy bodies in postmortem human brain. J Neurochem 105(4):1428–1437

123. Braak H, Braak E (1995) Staging of Alzheimer's disease-related neurofibrillary changes. Neurobiol Aging 16(3):271–278. discussion 8-84
124. Rowe CC, Ng S, Ackermann U, Gong SJ, Pike K, Savage G et al (2007) Imaging beta-amyloid burden in aging and dementia. Neurology 68(20):1718–1725
125. Li Y, Rinne JO, Mosconi L, Pirraglia E, Rusinek H, DeSanti S et al (2008) Regional analysis of FDG and PIB-PET images in normal aging, mild cognitive impairment, and Alzheimer's disease. Eur J Nucl Med Mol Imaging 35(12):2169–2181
126. Edison P, Rowe CC, Rinne JO, Ng S, Ahmed I, Kemppainen N et al (2008) Amyloid load in Parkinson's disease dementia and Lewy body dementia measured with [11C]PIB positron emission tomography. J Neurol Neurosurg Psychiatry 79(12):1331–1338
127. Gomperts SN, Rentz DM, Moran E, Becker JA, Locascio JJ, Klunk WE et al (2008) Imaging amyloid deposition in Lewy body diseases. Neurology 71(12):903–910
128. Maetzler W, Liepelt I, Reimold M, Reischl G, Solbach C, Becker C et al (2009) Cortical PIB binding in Lewy body disease is associated with Alzheimer-like characteristics. Neurobiol Dis 34(1):107–112
129. Maetzler W, Reimold M, Liepelt I, Solbach C, Leyhe T, Schweitzer K et al (2008) [^{11}C]PIB binding in Parkinson's disease dementia. NeuroImage 39(3):1027–1033
130. Johansson A, Savitcheva I, Forsberg A, Engler H, Langstrom B, Nordberg A et al (2008) [(11) C]-PIB imaging in patients with Parkinson's disease: preliminary results. Parkinsonism Relat Disord 14(4):345–347
131. Burack MA, Hartlein J, Flores HP, Taylor-Reinwald L, Perlmutter JS, Cairns NJ (2010) In vivo amyloid imaging in autopsy-confirmed Parkinson disease with dementia. Neurology 74 (1):77–84
132. Gomperts SN, Locascio JJ, Marquie M, Santarlasci AL, Rentz DM, Maye J et al (2012) Brain amyloid and cognition in Lewy body diseases. Mov Disord 27(8):965–973
133. Gomperts SN, Locascio JJ, Rentz D, Santarlasci A, Marquie M, Johnson KA et al (2013) Amyloid is linked to cognitive decline in patients with Parkinson disease without dementia. Neurology 80(1):85–91
134. Jokinen P, Scheinin N, Aalto S, Nagren K, Savisto N, Parkkola R et al (2010) [(11)C]PIB-, [(18)F]FDG-PET and MRI imaging in patients with Parkinson's disease with and without dementia. Parkinsonism Relat Disord 16(10):666–670
135. Shimada H, Shinotoh H, Hirano S, Miyoshi M, Sato K, Tanaka N et al (2013) Beta-Amyloid in Lewy body disease is related to Alzheimer's disease-like atrophy. Mov Disord 28(2):169–175
136. Villemagne VL, Ong K, Mulligan RS, Holl G, Pejoska S, Jones G et al (2011) Amyloid imaging with (18)F-florbetaben in Alzheimer disease and other dementias. J Nucl Med 52 (8):1210–1217
137. Mashima K, Ito D, Kameyama M, Osada T, Tabuchi H, Nihei Y et al (2017) Extremely low prevalence of amyloid positron emission tomography positivity in Parkinson's disease without dementia. Eur Neurol 77(5–6):231–237
138. Petrou M, Bohnen NI, Muller ML, Koeppe RA, Albin RL, Frey KA (2012) Abeta-amyloid deposition in patients with Parkinson disease at risk for development of dementia. Neurology 79(11):1161–1167
139. Klein JC, Eggers C, Kalbe E, Weisenbach S, Hohmann C, Vollmar S et al (2010) Neurotransmitter changes in dementia with Lewy bodies and Parkinson disease dementia in vivo. Neurology 74(11):885–892
140. Frey KA, Petrou M (2015) Imaging amyloidopathy in parkinson disease and parkinsonian dementia syndromes. Clin Transl Imaging 3(1):57–64
141. Petrou M, Dwamena BA, Foerster BR, MacEachern MP, Bohnen NI, Muller ML et al (2015) Amyloid deposition in Parkinson's disease and cognitive impairment: a systematic review. Mov Disord 30(7):928–935
142. Donaghy P, Thomas AJ, O'Brien JT (2015) Amyloid PET Imaging in Lewy body disorders. Am J Geriatr Psychiatry 23(1):23–37

143. Gomperts SN, Marquie M, Locascio JJ, Bayer S, Johnson KA, Growdon JH (2016) PET radioligands reveal the basis of dementia in parkinson's disease and dementia with lewy bodies. Neurodegener Dis 16(1–2):118–124

144. Kalaitzakis ME, Graeber MB, Gentleman SM, Pearce RK (2008) Striatal beta-amyloid deposition in Parkinson disease with dementia. J Neuropathol Exp Neurol 67(2):155–161

145. Kalaitzakis ME, Walls AJ, Pearce RK, Gentleman SM (2011) Striatal Abeta peptide deposition mirrors dementia and differentiates DLB and PDD from other parkinsonian syndromes. Neurobiol Dis 41(2):377–384

146. Shah N, Frey KA, Muller ML, Petrou M, Kotagal V, Koeppe RA et al (2016) Striatal and cortical beta-amyloidopathy and cognition in Parkinson's disease. Mov Disord 31(1):111–117

147. Dani M, Brooks DJ, Edison P (2016) Tau imaging in neurodegenerative diseases. Eur J Nucl Med Mol Imaging 43(6):1139–1150

148. Shah M, Catafau AM (2014) Molecular imaging insights into neurodegeneration: focus on tau PET radiotracers. J Nucl Med 55(6):871–874

149. Xia CF, Arteaga J, Chen G, Gangadharmath U, Gomez LF, Kasi D et al (2013) [(18)F]T807, a novel tau positron emission tomography imaging agent for Alzheimer's disease. Alzheimers Dement 9(6):666–676

150. Marquie M, Normandin MD, Vanderburg CR, Costantino IM, Bien EA, Rycyna LG et al (2015) Validating novel tau positron emission tomography tracer [F-18]-AV-1451 (T807) on postmortem brain tissue. Ann Neurol 78(5):787–800

151. Johnson KA, Schultz A, Betensky RA, Becker JA, Sepulcre J, Rentz D et al (2016) Tau positron emission tomographic imaging in aging and early Alzheimer disease. Ann Neurol 79 (1):110–119

152. Scholl M, Lockhart SN, Schonhaut DR, O'Neil JP, Janabi M, Ossenkoppele R et al (2016) PET imaging of tau deposition in the aging human brain. Neuron 89(5):971–982

153. Gomperts SN, Locascio JJ, Makaretz SJ, Schultz A, Caso C, Vasdev N et al (2016) Tau positron emission tomographic imaging in the lewy body diseases. JAMA Neurol 73 (11):1334–1341

154. Hansen AK, Damholdt MF, Fedorova TD, Knudsen K, Parbo P, Ismail R et al (2017) In Vivo cortical tau in Parkinson's disease using 18F-AV-1451 positron emission tomography. Mov Disord 32(6):922–927

155. Winer JR, Maass A, Pressman P, Stiver J, Schonhaut DR, Baker SL et al (2018) Associations between tau, beta-amyloid, and cognition in parkinson disease. JAMA Neurol 75:227–235

156. Kantarci K, Lowe VJ, Boeve BF, Senjem ML, Tosakulwong N, Lesnick TG et al (2017) AV-1451 tau and beta-amyloid positron emission tomography imaging in dementia with Lewy bodies. Ann Neurol 81(1):58–67

157. Graeff FG (1997) Serotonergic systems. Psychiatr Clin North Am 20(4):723–739

158. Hornung JP (2003) The human raphe nuclei and the serotonergic system. J Chem Neuroanat 26(4):331–343

159. Huot P, Fox SH, Brotchie JM (2011) The serotonergic system in Parkinson's disease. Prog Neurobiol 95(2):163–212

160. Stockmeier CA (2003) Involvement of serotonin in depression: evidence from postmortem and imaging studies of serotonin receptors and the serotonin transporter. J Psychiatr Res 37 (5):357–373

161. Nobler MS, Mann JJ, Sackeim HA (1999) Serotonin, cerebral blood flow, and cerebral metabolic rate in geriatric major depression and normal aging. Brain Res Brain Res Rev 30 (3):250–263

162. Meltzer HY (1990) Role of serotonin in depression. Ann N Y Acad Sci 600:486–499; discussion 99–500

163. Ciranna L (2006) Serotonin as a modulator of glutamate- and GABA-mediated neurotransmission: implications in physiological functions and in pathology. Curr Neuropharmacol 4 (2):101–114

164. Robbins TW (2005) Chemistry of the mind: neurochemical modulation of prefrontal cortical function. J Comp Neurol 493(1):140–146
165. Brodie BB, Shore PA (1957) A concept for a role of serotonin and norepinephrine as chemical mediators in the brain. Ann N Y Acad Sci 66(3):631–642
166. Albin RL, Koeppe RA, Bohnen NI, Wernette K, Kilbourn MA, Frey KA (2008) Spared caudal brainstem SERT binding in early Parkinson's disease. J Cereb Blood Flow Metab 28 (3):441–444
167. Gai WP, Blessing WW, Blumbergs PC (1995) Ubiquitin-positive degenerating neurites in the brainstem in Parkinson's disease. Brain 118(Pt 6):1447–1459
168. Gibb WR (1986) Neuropathology in movement disorders. J Neurol Neurosurg Psychiatry Suppl:55–67
169. Halliday GM, Blumbergs PC, Cotton RG, Blessing WW, Geffen LB (1990) Loss of brainstem serotonin- and substance P-containing neurons in Parkinson's disease. Brain Res 510 (1):104–107
170. Halliday GM, Li YW, Blumbergs PC, Joh TH, Cotton RG, Howe PR et al (1990) Neuropathology of immunohistochemically identified brainstem neurons in Parkinson's disease. Ann Neurol 27(4):373–385
171. Ohama E, Ikuta F (1976) Parkinson's disease: distribution of Lewy bodies and monoamine neuron system. Acta Neuropathol 34(4):311–319
172. Fahn S, Libsch LR, Cutler RW (1971) Monoamines in the human neostriatum: topographic distribution in normals and in Parkinson's disease and their role in akinesia, rigidity, chorea, and tremor. J Neurol Sci 14(4):427–455
173. Kish SJ, Tong J, Hornykiewicz O, Rajput A, Chang LJ, Guttman M et al (2008) Preferential loss of serotonin markers in caudate versus putamen in Parkinson's disease. Brain 131 (Pt 1):120–131
174. Raisman R, Cash R, Agid Y (1986) Parkinson's disease: decreased density of 3H-imipramine and 3H-paroxetine binding sites in putamen. Neurology 36(4):556–560
175. Scatton B, Javoy-Agid F, Rouquier L, Dubois B, Agid Y (1983) Reduction of cortical dopamine, noradrenaline, serotonin and their metabolites in Parkinson's disease. Brain Res 275(2):321–328
176. Shannak K, Rajput A, Rozdilsky B, Kish S, Gilbert J, Hornykiewicz O (1994) Noradrenaline, dopamine and serotonin levels and metabolism in the human hypothalamus: observations in Parkinson's disease and normal subjects. Brain Res 639(1):33–41
177. Wilson JM, Levey AI, Rajput A, Ang L, Guttman M, Shannak K et al (1996) Differential changes in neurochemical markers of striatal dopamine nerve terminals in idiopathic Parkinson's disease. Neurology 47(3):718–726
178. D'Amato RJ, Zweig RM, Whitehouse PJ, Wenk GL, Singer HS, Mayeux R et al (1987) Aminergic systems in Alzheimer's disease and Parkinson's disease. Ann Neurol 22 (2):229–236
179. Houle S, Ginovart N, Hussey D, Meyer JH, Wilson AA (2000) Imaging the serotonin transporter with positron emission tomography: initial human studies with [11C]DAPP and [11C]DASB. Eur J Nucl Med 27(11):1719–1722
180. Guttman M, Boileau I, Warsh J, Saint-Cyr JA, Ginovart N, McCluskey T et al (2007) Brain serotonin transporter binding in non-depressed patients with Parkinson's disease. Eur J Neurol 14(5):523–528
181. Strecker K, Wegner F, Hesse S, Becker GA, Patt M, Meyer PM et al (2011) Preserved serotonin transporter binding in de novo Parkinson's disease: negative correlation with the dopamine transporter. J Neurol 258(1):19–26
182. Politis M, Wu K, Loane C, Kiferle L, Molloy S, Brooks DJ et al (2010) Staging of serotonergic dysfunction in Parkinson's disease: an in vivo 11C-DASB PET study. Neurobiol Dis 40 (1):216–221
183. Boileau I, Warsh JJ, Guttman M, Saint-Cyr JA, McCluskey T, Rusjan P et al (2008) Elevated serotonin transporter binding in depressed patients with Parkinson's disease: a preliminary PET study with [11C]DASB. Mov Disord 23(12):1776–1780

184. Politis M, Wu K, Loane C, Turkheimer FE, Molloy S, Brooks DJ et al (2010) Depressive symptoms in PD correlate with higher 5-HTT binding in raphe and limbic structures. Neurology 75(21):1920–1927
185. Pavese N, Metta V, Bose SK, Chaudhuri KR, Brooks DJ (2010) Fatigue in Parkinson's disease is linked to striatal and limbic serotonergic dysfunction. Brain 133(11):3434–3443
186. Lelieveld IM, Muller ML, Bohnen NI, Koeppe RA, Chervin RD, Frey KA et al (2012) The role of serotonin in sleep disordered breathing associated with Parkinson disease: a correlative [11C]DASB PET imaging study. PLoS One 7(7):e40166
187. Loane C, Wu K, Bain P, Brooks DJ, Piccini P, Politis M (2013) Serotonergic loss in motor circuitries correlates with severity of action-postural tremor in PD. Neurology 80 (20):1850–1855
188. Qamhawi Z, Towey D, Shah B, Pagano G, Seibyl J, Marek K et al (2015) Clinical correlates of raphe serotonergic dysfunction in early Parkinson's disease. Brain 138(Pt 10):2964–2973
189. Politis M, Wu K, Loane C, Brooks DJ, Kiferle L, Turkheimer FE et al (2014) Serotonergic mechanisms responsible for levodopa-induced dyskinesias in Parkinson's disease patients. J Clin Invest 124(3):1340–1349
190. Roussakis AA, Politis M, Towey D, Piccini P (2016) Serotonin-to-dopamine transporter ratios in Parkinson disease: relevance for dyskinesias. Neurology 86(12):1152–1158
191. Lee JY, Seo S, Lee JS, Kim HJ, Kim YK, Jeon BS (2015) Putaminal serotonergic innervation: monitoring dyskinesia risk in Parkinson disease. Neurology 85(10):853–860
192. Smith R, Wu K, Hart T, Loane C, Brooks DJ, Bjorklund A et al (2015) The role of pallidal serotonergic function in Parkinson's disease dyskinesias: a positron emission tomography study. Neurobiol Aging 36(4):1736–1742
193. Sporns O (2011) The human connectome: a complex network. Ann N Y Acad Sci 1224:109–125
194. Sporns O (2013) Structure and function of complex brain networks. Dialogues Clin Neurosci 15(3):247–262
195. Mergenthaler P, Lindauer U, Dienel GA, Meisel A (2013) Sugar for the brain: the role of glucose in physiological and pathological brain function. Trends Neurosci 36(10):587–597
196. Peng S, Eidelberg D, Ma Y (2014) Brain network markers of abnormal cerebral glucose metabolism and blood flow in Parkinson's disease. Neurosci Bull 30(5):823–837
197. Spetsieris PG, Ko JH, Tang CC, Nazem A, Sako W, Peng S et al (2015) Metabolic resting-state brain networks in health and disease. Proc Natl Acad Sci U S A 112(8):2563–2568
198. Eidelberg D, Moeller JR, Dhawan V, Spetsieris P, Takikawa S, Ishikawa T et al (1994) The metabolic topography of parkinsonism. J Cereb Blood Flow Metab 14(5):783–801
199. Kaasinen V, Maguire RP, Hundemer HP, Leenders KL (2006) Corticostriatal covariance patterns of 6-[18F]fluoro-L-dopa and [18F]fluorodeoxyglucose PET in Parkinson's disease. J Neurol 253(3):340–348
200. Huang C, Mattis P, Tang C, Perrine K, Carbon M, Eidelberg D (2007) Metabolic brain networks associated with cognitive function in Parkinson's disease. NeuroImage 34 (2):714–723
201. Fitzpatrick T, Mattis P, Eidelberg D (2010) Functional imaging of cognitive impairment in Parkinson's disease. Clin EEG Neurosci 41(3):119–126
202. Colloby SJ, McKeith IG, Burn DJ, Wyper DJ, O'Brien JT, Taylor JP (2016) Cholinergic and perfusion brain networks in Parkinson disease dementia. Neurology 87(2):178–185
203. Isaias IU, Spiegel J, Brumberg J, Cosgrove KP, Marotta G, Oishi N et al (2014) Nicotinic acetylcholine receptor density in cognitively intact subjects at an early stage of Parkinson's disease. Front Aging Neurosci 6:213

Chapter 8
Structural and Functional Neuroimaging in Multiple Sclerosis: From Atrophy, Lesions to Global Network Disruption

Prejaas Tewarie, Menno Schoonheim, and Arjan Hillebrand

Abbreviations

BOLD	Blood-oxygenation level dependent
CIS	Clinically isolated syndrome
DTI	Diffusion tensor imaging
EAE	Experimental allergic encephalomyelitis
EE	Gelectroencephalogram
fMRI	Functional magnetic resonance
MAGNIMS	Magnetic resonance imaging in multiple sclerosis
MEG	Magnetoencephalography
MRI	Magnetic resonance imaging
MS	Multiple sclerosis
MST	Minimum spanning tree
MTI	Magnetisation transfer imaging
NMSS	National MS society
PASAT	Paced auditory serial addition test
PP	Primary progressive
RR	Relapsing remitting
SP	Secondary progressive

P. Tewarie (✉) · A. Hillebrand
Department of Clinical Neurophysiology and MEG Center, VU University Medical Center, Amsterdam, The Netherlands
e-mail: p.tewarie@vumc.nl

M. Schoonheim
Department of Anatomy & Neurosciences, VU University Medical Center, Amsterdam, The Netherlands

© Springer International Publishing AG, part of Springer Nature 2018
C. Habas (ed.), *The Neuroimaging of Brain Diseases*, Contemporary Clinical Neuroscience, https://doi.org/10.1007/978-3-319-78926-2_8

Introduction

Multiple sclerosis (MS) is an inflammatory and degenerative disease that affects the brain and the spinal cord. The disease is one of the most common central nervous system disorders, affects twice as many women as men and generally starts at the age of 20–40 [1, 2]. The pathological hallmark seems to be myelin damage, although it is currently unknown whether the initial damage to myelin is the result of a primary auto-immune response (i.e. outside-in) or initially degenerates on its own due to other factors, leading to a secondary immune response (i.e. inside-out) [3]. Symptoms due to MS can range from balance problems, paresis, autonomic dysfunction, eye sight defects to cognitive problems in various cognitive domains. Classical MRI findings used for diagnosis are characteristic white matter lesions that can be found adjacent to the ventricles, juxtacortical, infratentorial and in the spinal cord. More recently, extensive grey matter lesions, diffuse brain changes and atrophy patterns have also been identified (see below). The disease expresses itself in three main disease phenotypes [4]: the relapsing-remitting disease type (RR), the secondary progressive disease type (SP) and the primary progressive disease type (PP) (Fig. 8.1). The RR disease type is the most common, characterised by episodes (usually days/weeks) with complaints, the relapses, alternated with periods with normal functioning, the remissions, which can be accompanied with residual complaints, slowly rising the level of disability over time. After 10 years, 50% of RRMS patients progress into SPMS, which is characterised by the strongly reduced amount of relapses, with somewhat faster disease progression and a failing efficiency of immunomodulatory treatment [5]. PPMS is the most uncommon disease type, which appears similar to SPMS both in disease progression as in average age of onset, however without a previous history of RRMS and a higher male prevalence [6]. Current treatment options strongly focus on the dampening of the immune response [7, 8], effectively lowering the duration and amount of relapses in RRMS. Unfortunately, no treatment options have shown to be effective in reducing progression in SPMS or PPMS, as almost all treatment options for MS target the immune response related to relapses, both of which are less clearly involved in progressive MS. In fact, current therapies are very much focused on traditional lesion-based measures of tissue damage in MS, which have a poor correlation with patient outcomes and disease progression, and do not address the other forms of structural (and functional) brain changes. Therefore there is a strong need for new, more advanced, structural and functional neuroimaging markers. Given the numerous types of structural pathology that can appear in MS, we argue that related functional changes may also strongly vary between patients, which could in turn explain the large heterogeneity in cognitive and physical disability in MS.

In the current chapter, we discuss the recent developments in structural and functional neuroimaging in MS and how these have enhanced our understanding of the clinico-radiological paradox (see section "Structural Imaging in Multiple Sclerosis") in MS. We will start with structural imaging in MS, as damage to anatomical structure has classically been the focus in the field (section "Structural

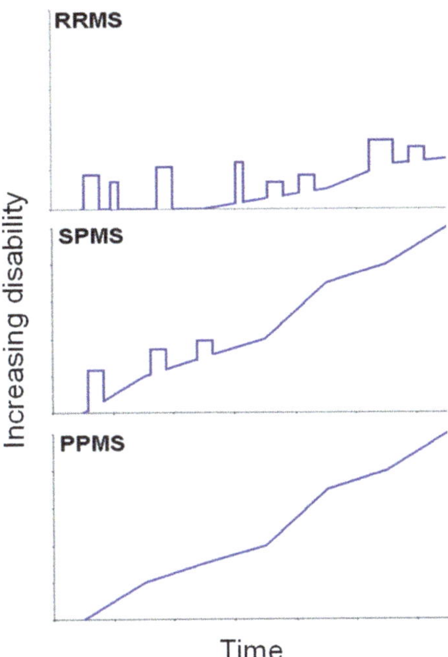

Fig. 8.1 Progression of physical disability in MS phenotypes. In relapsing-remitting MS (*RRMS*), disability remains relatively low and is marked by periods of exacerbations (relapses) that can completely remit. Over time, however, some level of physical disability will accumulate. Around half of patients suffering from RRMS will convert to secondary progressive MS (*SPMS*). This disease phenotype is marked by the disappearance of relapses, but a sudden accelerated increase in the accumulation of grey matter damage and disability. In primary progressive MS (*PPMS*), physical disability accumulates steadily over time, without the presence of relapses

Imaging in Multiple Sclerosis"). We will discuss how focal lesions can be visualised and how they are used in diagnosing MS patients. We will then highlight the clinico-radiological paradox in MS, as these lesions do not relate well to clinical symptoms, and how studying cortical lesions, atrophy patterns and diffuse changes in normal-appearing brain tissues have enhanced our understanding of the heterogeneity in MS. Section "Functional Imaging in Multiple Sclerosis" then discusses how functional imaging studies have been essential in explaining the clinico-radiological paradox. We highlight studies that have used either functional MRI (fMRI) or magnetoencephalography (MEG) to study neuronal activity, functional connectivity and functional network topology in MS. Section "Network Theory" argues how functional metrics could be used as a bridge between structural pathology and cognitive and clinical disability in MS, forming a more holistic view of the brain. We close the chapter with perspectives for future studies (section "Multimodal Imaging in MS: Integrating Structure and Function").

Structural Imaging in Multiple Sclerosis

A Brief History of Detecting MS Pathology

Multiple sclerosis has a long history [9]. In 1838, Jean-Martin Charcot described one of his patients, a woman with symptoms common for MS, such as changes in eye movements and slurred speech. When she died, he performed an autopsy and described multiple sclerotic plaques in the woman's brain. In the 1870's physicians agreed that MS was indeed a neurological disease. The disease mechanism remained unclear, as knowledge on the immune system and neurobiology was poor. In fact, myelin was not discovered until 1878, while the oligodendrocytes that produce it were not discovered until 1928. In 1916 the inflammatory process around blood vessels and damage to the myelin in multiple sclerosis plaques was described. In 1935 it was discovered that immune cells can induce myelin loss, by injecting myelin into animals, creating the animal model experimental allergic encephalomyelitis (EAE). The link between the immune system and MS, and not only its animal model, was demonstrated much later by the discovery of oligoclonal bands in the cerebrospinal fluid of MS patients in 1947. This finding, together with the formation of the national MS society (NMSS) in the United States, sparked research into MS like never before. The NMSS funded a panel of experts to come up with the first diagnostic criteria for MS in 1960, mostly based on clinical symptoms. The first studies looking at treatment strategies targeting the immune systems (using steroids, interferons and glatiramer acetate), were performed in the 1970s. Then, in 1981, we entered the era of brain imaging in MS, as the first magnetic resonance imaging (MRI) scan of an MS patient was performed. This era not only revolutionised the way we see MS today [10, 11] but also drastically improved the diagnostic process and clinical trials.

Conventional Imaging of White Matter Lesions

The classic multiple sclerosis diagnostic imaging protocol will usually contain two types of sequences, firstly a T2-based sequence such as a dual echo sequence or FLAIR, which is a T2 scan with suppressed CSF signal for improved contrast in order to identify almost all white matter lesions for diagnosis, and secondly a T1-weighted scan using contrast to classify these lesions further. The latter provides additional information, as it is impossible to see whether or not a lesion is new, old, mild or severely destructive using only T2 scans. Using a conventional 2D proton-density fast spin-echo sequence (left panel in Fig. 8.2), most white matter lesions appear hyperintense, while the surrounding white matter appears dark. This strongly facilitates the detection of MS lesions in the brain, with subsequent manual or (semi-)automated delineation of total T2 lesion volume. It is also possible to see the so-called diffusely abnormal white matter on such a standard T2 image, frequently

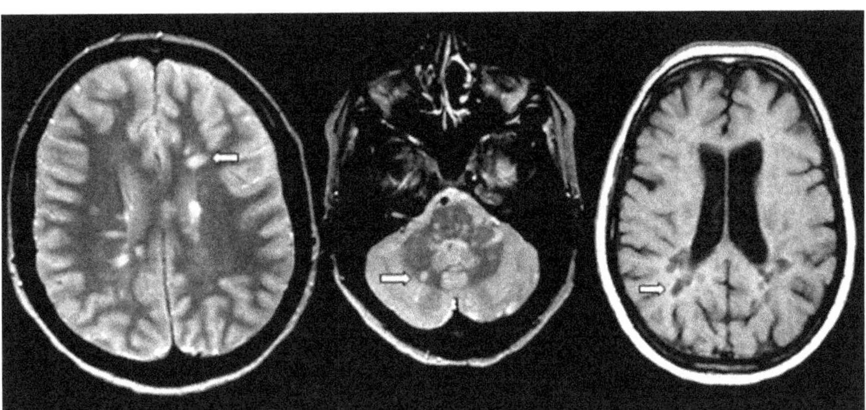

Fig. 8.2 Conventional MRI imaging of white matter damage in MS. Left: So-called 2D PD/T2-based MR image, which is used most commonly for diagnosis; axial slice through the top section of the lateral ventricles, featuring the stereotypical periventricular ovoid-shaped white matter lesions. One example is demarcated with an arrow. Middle: Axial slice through the cerebellum, showing infratentorial white matter lesions. Right: 2D T1-based MR image used most commonly for lesion classification; axial slice through the ventricles showing especially destructive "black hole" MS-lesions

observed as a slightly cloudy abnormal region, thought to possibly reflect the Wallerian degeneration ("dying back") of axons from more severely damaged lesions [12, 13]. Once a lesion has been identified on T2, it is possible to classify lesions further on a 2D T1-weighted spin-echo sequence (right panel in Fig. 8.2). Using T1 imaging it is possible to see whether or not the lesion is new, i.e. hyperintense if a contrast agent such as gadolinium is administered to the patient [14], or especially destructive, i.e. a black hole [15], hypointense on T1. By following T1 lesions over time, it was discovered that MS lesions are dynamic, i.e. they appear as contrast-enhancing when they are new, but this effect wears off when the lesion gets older and reappears in newer lesions [16]. Using this information it is also possible to determine the outcome of such a lesion, for example, whether or not the lesion becomes a "persistent black hole" [17], which can be used in clinical trials looking at neurodegeneration [18]. The measurement of total T1 and T2 lesion counts and volumes has been the imaging hallmarks of most classic placebo-controlled therapeutic trials [19], with a strong focus on preventing new, contrast-enhancing T1 lesions [20].

Diagnostic Criteria

For a long time, multiple sclerosis was thought to be purely immune-mediated and only characterised by focal white matter lesions. The reason for this rather limited view was primarily due to the very limited visibility of the full scale of pathology using the initial MRI technology in the 1980s and 1990s, during the first major clinical trials. Nonetheless, imaging these white matter lesions (see Fig. 8.2) has proven to be excellent for diagnosing MS [21]. The diagnostic criteria, which have been revised several times, are based on the concept of dissemination of lesions in space and time. In other words, they are based on the detection of lesions in multiple regions of the central nervous system and an increase in lesion numbers over time. If patients do not yet fulfil these criteria but do show clear clinical signs of having had an MS-like relapse, they are characterised as having a "clinically isolated syndrome", or CIS, which frequently later converts to so-called clinically definite MS (CDMS), i.e. fulfilling the diagnostic criteria. The most recent diagnostic criteria [22] allow for a diagnosis of MS using only one MRI scan in some RRMS patients, using contrast-enhanced imaging to distinguish asymptomatic active (i.e. new) and inactive (i.e. older) lesions, more easily fulfilling the criterion of dissemination over time. The criterion of dissemination in space centres around the presence of more than one lesion on a T2-weighted scan in periventricular, juxtacortical, infratentorial and/or spinal cord regions. In 2016, the magnetic resonance imaging in multiple sclerosis (MAGNIMS) collaborative research network published new recommendations [23]. The main changes were in the requirement for "dissemination in space", which now also includes optic neuritis and intra-cortical lesions. While abovementioned criteria work relatively well for RRMS, diagnosing PPMS has always been challenging, given the lack of relapses, focus on spinal cord pathology and low amount of lesions in the brain. The latest criteria aim to improve this by an even stronger focus on the spinal cord pathology, sustained disability progression and oligoclonal bands. Using these latest criteria, most MS patients can be diagnosed with relatively high sensitivity and specificity.

The Clinico-Radiological Paradox

Interestingly, it was soon noted that markers such as the gadolinium-enhanced appearance of new lesions on T1-weighted MR images are strong predictors of the occurrence of relapses [24], but not of the accumulation of disability [25]. In RRMS, classifying T1 lesions further results in a better correlation with clinic, such as specifically looking at black holes [26] or by adding magnetisation transfer imaging (MTI) information, assessing the severity of damage [27]. Unfortunately, lesion-based (i.e. neuroinflammatory) markers still correlate rather poorly with clinic [28, 29] especially in PPMS [30, 31], while stronger correlations are found using more advanced (neurodegenerative) measures such as brain atrophy [32]. Similarly, while initial studies suggested a strong relationship between white matter lesions and cognitive dysfunction in MS [33], it was soon noted that other measures such as brain atrophy [34] relate much more strongly to cognitive dysfunction than lesion volumes. These findings were thus summarised as a clinico-radiological paradox in MS [35], with some patients showing very few lesions, but a poor clinical status, and vice versa, regardless of disease duration and severity [29]. Most studies up to that point in time were limited to whole-brain lesion counts and volumes. This led some research groups to specifically study the lesion location to solve this paradox [36–40], showing some added value of such an approach. Unfortunately, the paradox remained, highlighting the need for more advanced imaging measures in MS looking beyond the effects of focal white matter lesions.

Cortical Lesions

As the clinico-radiological paradox remained unsolved, the search was on for newer ways of detecting pathology in MS, leading to the discovery of grey matter lesions, which are regions of cortical demyelination with much less immune activity than in the white matter [41] and poorly correlated to the number of white matter lesions [42]. These lesions are very common in MS [43], but not frequently monitored in the setting of a clinical trial [44]. Most of these lesions affect both the grey and white matter (type 1, mixed lesions), although lesions also appear within the cortex (type 2), on the pial surface (type 3), or can span the entire surface of the cortex, without affecting the white matter (type 4) [45]. The clinical relevance of cortical lesions is quite high, with a strong history of correlations with physical disability [46–48], cognitive dysfunction [49–53] and epilepsy [54, 55]. Interestingly, the relationship between cortical lesions and cortical atrophy is not so clear [56, 57].The visualisation of cortical lesions has been highly problematic on standard T2 imaging but has improved somewhat by the application of the FLAIR sequence [58]. Using FLAIR, more lesions are visible than on T2, especially in the deep grey matter, but most cortical lesions are still missed [59, 60]. It is possible to also suppress the white matter signal on top of the CSF suppression of FLAIR, thus creating a double-inversion recovery sequence (DIR). This sequence, an example result of which is shown in Fig. 8.3, is especially sensitive to intra-cortical lesions in MS [61], although most, especially smaller, lesions are missed [62, 63]. The scoring of cortical lesions on DIR can be difficult, however, given the rather noisy appearance and presence of specific artefacts, which resulted in scoring guidelines [64]. More recent explorations into high-field strengths have highlighted that especially cortical lesions, and not white matter lesions, are more easily visualised at 7T [65] and that FLAIR, and not DIR, is currently the best sequence to visualise cortical lesions at 7T [66]. Unfortunately, even at 7T, most lesions are missed, as especially those deep within the cortex or on the pial surface continue to be difficult to visualise [67].

Basics and Limitations of Global Brain Atrophy

Even before MRI scanning was available, it was noted that neurodegenerative processes such as atrophy are a common finding in MS brains [68]. Applying MRI technology (see Fig. 8.3) allowed for a much more precise quantification, frequently

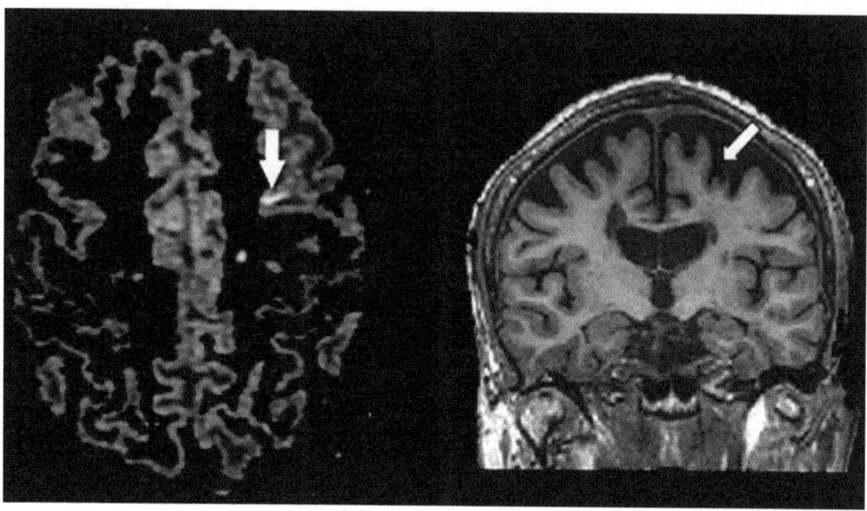

Fig. 8.3 Grey matter damage in MS. Left: 3D DIR-based MR image; axial slice, showing a lesion in the cortex that cannot be seen on T2-based sequences. Right: 3D T1-based MR image; coronal slice through the basal ganglia, showing extensive cortical atrophy

noting corpus callosum atrophy [69]. It was also shown that both brain [70] and spinal cord atrophy [71–73] progresses over time, which both relate to clinical status. Similarly, especially atrophy, and not the number of lesions, was related to cognitive dysfunction [74–76]. The nature of brain atrophy in MS was (and still remains) rather mysterious, as atrophy does not seem to be related to active lesions [77], can even appear without the presence of lesions [78] but is nonetheless related to neurodegenerative biomarkers [79] and neuro-axonal loss [80, 81]. Techniques to measure brain atrophy have evolved considerably over the years, although it should be noted that considerable variability between techniques remains [82, 83]. Initial explorations of corpus callosum and ventricular atrophy [84] involved manually delineating these structures on sagittal T1 images [85], which was also applied to other regions of the brain [86], an extremely time-consuming process. As most studies already used semiautomated techniques to delineate white matter lesion volumes, subsequent atrophy studies used such a local thresholding to delineate cortical outlines. By measuring the volume of several brain slices, combined with manual editing, an estimate could be made of whole-brain volumes in a relatively timely manner [87, 88]. Subsequent semiautomated techniques were able to segment out most of the CSF [89], again showing strong correlations of whole-brain volume with clinical scores, but not with T2 lesion volumes. During these technological developments, the scientific community started comparing methods, highlighting the need for normalising for head size and techniques able to reliably measure regional volumes and discussing histopathological correlates [90]. Longitudinal techniques to detect atrophic changes over time were then developed and applied to MS [91] and could be applied to conventional 2D-T1 images, which enabled large datasets to be analysed in this way. Using such a technique, it was shown that lesion volumes and atrophy actually contain complementary, independent, information on long-term disability progression [92, 93]. More current sophisticated techniques, such as voxel-based morphometry (as explored below), however, require more advanced techniques such as so-called 3D (and not 2D) T1-weighted imaging.

Regional Atrophy: Cortex

Automated techniques have resulted in many studies investigating cortical grey matter atrophy in MS [94, 95]. Separating grey and white matter became possible using automated software [96–98]. Although both grey and white matter volumes decrease over time, grey matter atrophy appears especially clinically relevant [99]. Extensive grey matter atrophy is worst in progressive MS [100], even accelerating when converting to SPMS [101]. Specific recommendations of standardised protocols [102–104] further aided in the detection of clinical relations. These explorations also revealed MS-specific problems with automatic segmentation, as white matter lesions often have a signal intensity proportional to that of grey matter or CSF. This not only negatively influences brain volumes but also registration pipelines. One possible solution to this issue is to fill all lesions on the 3D-T1 image [105] using, e.g. LEAP [106] with intensity levels of surrounding normal-appearing white matter (a technique also called inpainting). Explaining grey matter atrophy in MS has been very difficult [91], especially in progressive MS. Studies in RRMS indicate a central role for both lesional [107, 108] and normal-appearing tissue changes [109, 110] in the development of cortical atrophy. Most recently, regional cortical thickness estimation has become feasible in MS, showing global as well as slightly more focal cortical thinning, related to clinical symptoms [54, 111–116]. Using such a technique, it was also shown that most, but not all, cortical atrophy occurs in brain regions connected to damaged white matter tracts [117, 118]. Although it might seem that cortical atrophy occurs randomly, due to the largely random formation of lesions [36], there is most definitely structure in focal cortical thickness changes [119, 120], perhaps indicating a network-based propagation of cortical atrophy in MS. These findings could indicate that although cortical atrophy does in part occur due to white matter damage, other causes independent of white matter lesions are also present, especially in SPMS.

Regional Atrophy: Deep Grey Matter

While it might seem obvious to especially focus on cortical grey matter, newer techniques that are more sensitive to deep grey matter structures actually showed that atrophy in MS often begins in the deep grey matter [121, 122]. This had not been observed previously because most approaches that quantify atrophy rely on automated segmentation techniques that are relatively insensitive to deep grey matter structures [82], whereas newer techniques apply registration-based approaches in order to circumvent such contrast-dependent issues. Relations between deep grey matter atrophy and white matter lesions appear stronger than for cortical atrophy, although all forms of atrophy remain difficult to explain in progressive MS [109, 123]. Grey matter lesions, however, are less frequent in the deep grey matter than in the cortex and perhaps contribute less to deep grey matter atrophy than white matter pathology does [57]. Deep grey matter atrophy was previously indicated by the frequent observations of enlargement of the third ventricle in MS [84], which could be explained by thalamic atrophy [34], one of the strongest correlates of cognitive dysfunction in MS [124–131]. In fact, thalamic atrophy appears to be very important in MS [132], as it appears early on in RRMS [133], and in paediatric MS [134]. In fact, thalamic atrophy can even occur in clinically isolated syndrome, where it is predictive of conversion to clinically definite MS [135]. Although virtually all deep grey matter structures show atrophic changes, the thalamus shows the strongest clinical correlations [127, 136]. Together with the facts that (1) cortical atrophy is much more difficult to measure with current techniques and (2) cortical atrophy appears to develop later on in the disease, the thalamus has become a focal point of investigation in atrophy studies in MS. In fact, thalamic volume is now even monitored in newer clinical trials [137].

Normal-Appearing Brain Tissue

As mentioned above, neurodegenerative processes such as grey matter atrophy can occur without the presence of or connectedness to lesions, indicating that other processes might still be at play in MS, especially in progressive MS. Researchers have previously suggested that the so-called normal-appearing brain tissues are, in fact, not normal at all [35, 138]. Such regions are defined as not showing any focal lesions or diffusely abnormal regions visible on T2-weighted (or FLAIR/DIR) scans. Imaging studies, however, mostly featured relapsing-remitting MS patients, as these form the bulk of the patient population. Post-mortem studies, interestingly, have the reverse problem, as most patients were in the progressive phase of the disease at the time of death. In post-mortem studies, it was noted, as also seen in conventional imaging, that focal demyelination appears in all forms of MS. However, mainly in progressive MS patients, a lot of diffuse inflammation was also observed (i.e. the diffuse infiltration of inflammatory cells), combined with a diffuse loss of axons and myelin in the normal-appearing white matter [42, 139]. These diffuse changes in the normal-appearing tissues have also been observed in vivo, enabling the study of these changes in all MS phenotypes. Using advanced, more quantitative, MRI techniques, changes have been observed in the normal-appearing white and grey matter using diffusion tensor imaging (DTI) [109, 140–143], T1- and T2-relaxometry-based approaches [12, 144–146], MR spectroscopy [133, 147–150] and MTI [151–155], to name a few. These studies highlight the mostly subtle changes within these tissues, which may, for instance, explain the occurrence of brain atrophy in the absence of brain lesions in connected tracts, or without the presence of lesions within the affected grey matter regions. In fact, although such changes in the normal-appearing tissues are most severe close to lesions, they are still present in regions quite remote from focal pathology [156]. In summary, these studies show that there are many advanced (quantitative) MRI techniques that may play an essential role in elucidating the subtle pathology in the normal-appearing tissues [157], which appear to be a very strong structural correlate of clinical symptoms [52]. Of these techniques especially diffusion-weighted MR imaging (DWI) has gained attention [158], given its high sensitivity, and presence as a commonly used clinical protocol in diagnosing neurological diseases. A more advanced form of DWI that is commonly used in neuroscience is called diffusion tensor imaging (DTI). The grey matter and especially the thalamus were also highlighted in many diffusion studies, showing marked changes in diffusivity [159], even without the presence of lesions [160], which was later related to physical disability [123, 161, 162] and cognitive impairment [129, 163, 164]. As the thalamus is a major hub in the brain, these results sparked an interest to study structural networks in MS further using DTI.

Structural Networks

Although DTI has been used in many MS studies [52, 157, 165], precise voxel-wise comparisons of white matter voxels has been problematic for many years, given the highly variable neuroanatomy as well as effects of white matter atrophy that are difficult to deal with, limiting the interpretability of somewhat older DTI studies. Newer approaches, such as tract-based spatial statistics (TBSS) [166], focus on the major white matter bundles of the brain only, by linking voxels with the highest fractional anisotropy. Although this technique is highly reliable in investigating voxel-wise changes in diffusivity in MS, there are more sophisticated approaches to investigate structural pathways, namely, tractography. This technique has been applied to the healthy brain in many landmark studies, enabling scientists to study the topology of the structural brain network in great detail [167, 168]. In MS, tractography has also been used frequently to study tract disconnection and the connectivity of damaged brain regions [169], for example. Structural covariance is another structural network approach, which has also been applied in MS but is not as sensitive to lesional pathology. This technique assesses the covariance in cortical thickness across areas in the brain [170], based on the assumption that regions that wire together also tend to covary in cortical volumes. The results of such approaches will be discussed in more detail in the network section below (section "Network Theory"). In summary, these structural brain techniques have greatly enhanced our knowledge on MS, with more sophisticated techniques on the horizon. Bridging the gap between brain structure and clinic, however, still requires a last step, as we will argue below: the study of brain function in MS.

Functional Imaging in Multiple Sclerosis

Basics and Limitations of fMRI

Brain function in MS has mainly been studied using functional MRI. This technique is dependent on the so-called blood-oxygenation-level dependent (BOLD) response, which is a slightly delayed and convoluted haemodynamic response aimed to provide oxygen to regions of the brain that are active. As oxygenised and deoxygenised haemoglobin have different magnetic properties, this can be detected using MRI, providing a surrogate measure of brain function. Common processing steps involved include spatial smoothing, motion correction and registration to a common "standard" space, all of which have developed over time. Methodological issues specific to fMRI could also be at play, however. For instance, field strengths have increased considerably over the years. Additionally, motion correction is carried out in a much more stringent way as before [171, 172], which is known to impact connectivity values greatly, as higher residual motion in the data can induce false-positive increases in connectivity. To complicate matters further, false positives could also be the result of smoothing the data. Together, it is currently not clear how to optimally set these parameters for different scanner types/magnetic fields [171]. As multiple sclerosis features astrocyte dysfunction [173–175], this disease process may also influence fMRI measurements. Astrocytes are essential for the normal haemodynamic response [176–179], since astrocyte activity induces changes in arterial volumes involved in the BOLD response [177, 179]. The BOLD response is therefore of mixed origin; it is an indirect reflection of vascular, neuronal and astrocyte dynamics [180, 181]. In MS, astrocyte damage within and surrounding lesions may therefore lead to altered patterns of BOLD responses in the vicinity of these lesions that are not necessarily of neuronal origin. These astrocyte-induced BOLD responses could affect connectivity estimates. These effects are likely not able to induce connectivity itself, but it might blur correlated patterns of activities, which therefore might to erroneous conclusions of decreased connectivity in MS.

Functional Reorganisation

The first activation studies in MS using fMRI have mainly looked at activation patterns during the paced auditory serial addition test (PASAT), which is a test measuring information processing speed, one of the first affected cognitive domains in MS. A consistent finding was that frontal activation in cognitively preserved MS patients was stronger than in cognitive impaired patients [182–184]. The same kind of hyperactivation in preserved patients was then also observed during other tasks, such as during the well-known N-back task in the dorsolateral prefrontal cortex [185], or during an episodic memory task in the hippocampus [186].

Most of these studies have been conducted with task-based fMRI paradigms [187, 188], which collectively led to the rather controversial functional reorganisation concept [189], based on the hypothesis that increased brain function in cognitively preserved patients or in patients in the early stage of the disease was a compensatory mechanism to maintain (semi-)optimal cognitive performance. This concept initially included any form of increased brain function, i.e. both activation and connectivity, based on the observation of increased connectivity in clinically isolated syndrome CIS [190] and decreased connectivity in progressive MS [191]. Such functional connectivity is generally calculated using relatively simple correlation coefficients in fMRI, although directed (or effective) connectivity approaches are also possible using more sophisticated statistical models between regions of interest. Interestingly, the connectivity findings were not as consistent as the activation studies. For instance, connectivity of the dorsolateral prefrontal and medial frontal cortices was found to be reduced during an N-back study in patients that scored normally on the N-back task [192], while others have found increased connectivity in impaired patients between the dorsolateral prefrontal cortex and medial frontal cortices during the Go/No Go task, [193]. Differences in functional connectivity between patients and controls during the PASAT task have also been studied frequently, showing both increases and decreases in MS [194, 195]. In summary, these traditional approaches studying specific brain regions have been rather problematic in the context of functional reorganisation.

Resting-State Networks

Over the years, there has been a shift in methodology towards whole-brain approaches studying the so-called resting-state networks obtained with independent component analysis (ICA), as the field moved from task-based paradigms towards resting-state imaging [196]. One of the reasons for this shift is the idea that task-based connectivity restricts the analysis to a priori defined regions, such that other adaptive changes in the MS brain are potentially missed [197]. A number of studies using the popular ICA-based approach followed and observed connectivity change in several subnetworks [198–203]. Especially the default mode network (DMN) received attention in this context as it is especially active during rest, and because regions within this network are known to lie very central in the whole brain, and therefore likely also the most sensitive to network damage [204, 205]. Unfortunately, results on default mode network connectivity appeared to be just as contradictive to the concept of functional reorganisation. For instance, decreased DMN connectivity was related to cognitive dysfunction (i.e. in line with the functional reorganisation concept) [199, 200, 206]) but so was increased connectivity [203, 207]. Such processes were also observed in paediatric MS [208].

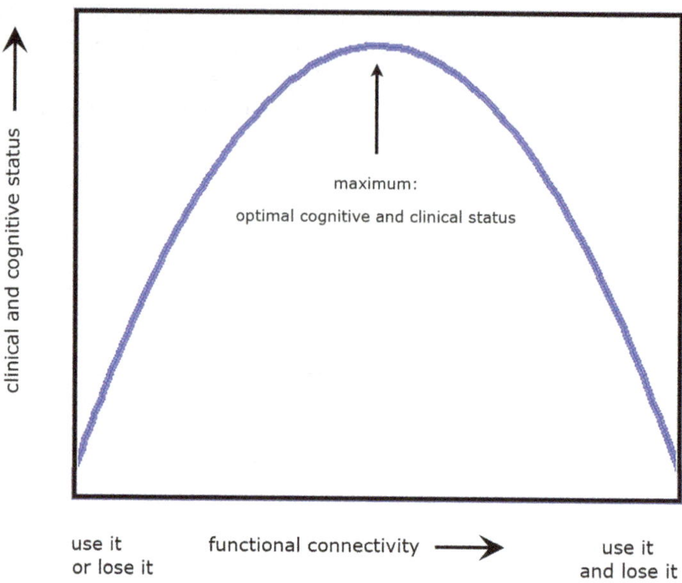

Fig. 8.4 The hypothesis that the relationship between functional connectivity and clinical and cognitive status may follow an inverted U-curve. Too weak connectivity (on the left) may lead to a loss of cognitive (physical) ability. Likewise connections that are too strong may also lead to loss of cognitive (physical) ability

Use It and/or Lose It

These findings thus far seemed to indicate that any change in connectivity could be related to poorer patient outcomes, i.e. forming an inverted U-curve (see Fig. 8.4), although it remains unclear why some patients would display increases and others decreases. This means that there is an optimal range for connectivity to uphold cognitive functioning and that any shift from this optimum might induce clinico-cognitive decline. So in terms of expressions frequently used in neuroscience [209], these changes appear to be a combination of the "use it and lose it" and "use it or lose it" principles, which were formerly regarded as mutually exclusive. This can be understood by examining our hypothesis in Fig. 8.4. On the extreme left, there is decreased connectivity that leads to cognitive and clinical dysfunction ("use it or lose it"), in line with the functional reorganisation hypothesis. Likewise, however, on the extreme right, there is an increase in connectivity which also leads to cognitive and clinical dysfunction ("use it and lose it"); see [210] for a modelling study that explains this. The maximum in between these extremes corresponds to the functional connectivity range that upholds optimal cognitive and clinical status. Unfortunately, the curve shape and position of the peak may be dependent on the region or connections of interest. It is argued that a combination of (i) sufficient intact tissue, (ii) degeneracy (i.e. other systems take over) and (iii) increased activation in unaffected areas (compensation) may be means in the brain to uphold cognitive and clinical status [197]. Therefore, mechanism (iii) may influence the working point for our inverted U-curve: if there is a disease-induced shift towards the left of the curve, then increased activation or connectivity may bring it back to the optimum.

Basics and Limitations of MEG

Abovementioned methodological considerations on fMRI has led to a search for more direct methods to measure brain function to study MS, such as magnetoen-cephalography (MEG). MEG has the advantage of measuring neuronal activity directly [211, 212], without the delay, mixed and convolved nature of the BOLD response that is seen with fMRI. It has a much higher temporal resolution but varying spatial resolution. Compared to fMRI, it is less sensitive for measuring activity from deeper (grey matter) structures such as the thalamus. MEG measures fluctuating magnetic fields originated from the brain, using special sensors that are sensitive to very small fluctuations in magnetic fields. These magnetic fields mainly originate from currents induced by postsynaptic potentials. The strength of the magnetic field induced by these (mainly pyramidal) currents falls off with $1/r^2$, where r is the distance from the source to the sensor. In contrast, action potentials are far shorter events and fall off with $1/r^3$, which therefore contribute significantly less to MEG signals (but see [213]). Unlike electric fields such as measured with EEG, magnetic fields are almost not distorted by the tissues surrounding the brain, such as the cerebral spinal fluid, skull and scalp and do not require a reference. Currently, the most used MEG systems contain on the order of several hundred sensors. Activity is generally quantified by the power within a frequency range. Power is usually quantified by the absolute value of the fast Fourier transform of a signal. Classical EEG frequency ranges are typically considered: delta (0.5–4 Hz), theta (4–8 Hz), alpha1 (8–10 Hz), alpha2 (10–13 Hz), beta (13–30 Hz) and gamma (30–48 Hz), although the definition for gamma may vary and may include much higher frequencies. A peak (peak frequency) in the power spectrum can usually be observed in the alpha range. The different frequency bands correspond to different computational roles in the brain, where higher frequency bands (e.g. gamma band) are more involved in bottom-up processing and lower frequency bands (especially alpha) are involved in top-down modulation of activity [214]. Unfortunately there is still a low availability of MEG scanners throughout the world, and the technology is now rapidly improving. Significant considerations also include the inverse problem and field spread effects (see below).

Source Reconstruction with MEG

Whereas in the early days of MEG, the analysis was limited to the locations of individual sensors, modern approaches often project to "source space", i.e. estimation of the location, orientation and strength of the original neuronal currents underlying the recorded magnetic fields using the information of all sensors [215, 216]. The location of the sources can also be defined by a predefined grid covering the cortical areas of the brain, where the challenge is to adequately sample anatomical space. One important ingredient for source reconstruction is an accurate head model to calculate the lead fields, which is usually based on the outline of the scalp segmented from a subject's co-registered MRI. Lead fields are defined as the magnetic field that would be measured if there was a unit current at a particular location (with a given orientation). Unfortunately, there is no unique solution for source reconstruction (an inverse problem). There are several reasons for attempting to solve the inverse problem: (i) it allows for the interpretation of the functional information in an anatomical context of, e.g. an atlas [218], which enables multi-modal studies; (ii) it increases the signal-to-noise ratio [217]; and (iii) it reduces field spread/mixing (see below) between signals, although it does not completely abolish it [218, 219]. In order to be able derive a unique solution, a priori assumptions about the neuronal activity need to be made, and this is where the various source recon-struction approaches differ [216, 220]. An accepted popular source reconstruction approach is beamforming, a spatial filtering approach that reconstructs the neuronal time series for a particular location on the basis of the weighted sum of sensor space time series. These beamformer weights are determined by the lead fields and the data covariance of the sensor space time series and are computed such that all activity for the target location is reconstructed while minimising the influence of other sources of activity (i.e. including noise) [221, 222]. The main assumption behind the beamformer approach is that spatially distinct neuronal sources are linearly inde-pendent [216]. Lastly, not only the location and strength but also the orientation of the sources has to be estimated. This can be achieved by performing a search for the orientation that optimises the normalised beamformer output, or alternatively by an eigendecomposition to determine the optimum source orientation [219].

Connectivity Measures in MEG

Functional connectivity measures in MEG are more complicated than in fMRI, given the richer information in MEG time courses. Furthermore, MEG measures have to deal with signal leakage and source spread (see below). One can roughly divide MEG measures into phase synchronisation, generalised synchronisation, envelope/amplitude correlations, mutual information and coherence. Every type of functional connectivity metric has its advantages and disadvantages in terms of required signal-to-noise ratio, sensitivity for genuine connections, bias or computational efficiency. However, probably the most important requirement for MEG/EEG data is that a functional connectivity metric should discard spurious connections that are due to field spread. Field spread is the phenomenon where the magnetic field from one source is picked up by many sensors with zero delay. This can then lead to spurious functional connections between sensors. Although the effect of field spread is reduced by source reconstruction, it is not completely resolved in source space [219], where it is often referred to as signal leakage. An in-depth explanation of these metrics can be found in the following review papers [223, 224], and here we restrict ourselves to a brief explanation on phase synchronisation, as this is most commonly used in MEG data. Phase synchronisation refers to the phenomenon where oscillators display stable phase relationships when they are weakly coupled. Weak coupling implies that there is an interaction between the phases of the two systems without causing the amplitudes to become identical. Phase is defined as "the fractional part of a period through which an oscillating signal has moved, as measured at any point in time from an arbitrary time origin" [225]. A popular used measure of phase synchronisation in MEG-based studies in MS is the phase lag index (PLI) [226], which is a measure that captures the asymmetry of the phase difference distribution and is relatively insensitive to leakage, although alternative approaches exist that aim to reduce the effects of signal leakage [227]. Another metric used in MEG-based studies in MS is the synchronisation likelihood, which is a measure for generalised synchronisation [228]. Generalised synchronisation refers to the phenomenon where the state of a response system A is considered as a function of the state of driving system B [228].

MEG Power Studies in MS

Before we will describe studies that have looked at changes in connectivity in MS, we will first describe results obtained with power analysis, because they were the oldest MEG studies in MS. The first study using MEG in MS dates back to 1992, when an elementary MEG system with 24 sensors was used to analyse somatosensory responses in MS patients and healthy controls [229], showing abnormally large amplitude responses, followed by a study showing focal slowing of brain activity around lesions [230]. EEG was subsequently explored in MS [231]. The first whole-head resting-state MEG study in RRMS was small ($N = 10$) and showed no changes in power [232], while a later, larger, source-space study [233] revealed slowing of rhythmic activity in early MS. Power in the alpha1 band was increased, while power in the alpha2 band was decreased. Peak frequency was lower in the patient group than in the control group, and this disease-induced effect was positively associated with information processing speed in MS patients, which is commonly impaired in MS. Interestingly, it is known that these electrophysiological parameters significantly correlate to white matter integrity in the healthy brain [234, 235], and therefore might mediate between structural damage to the white matter and clinical and cognitive problems in MS.

MEG Connectivity Studies in MS

The first study that also looked at MEG changes in functional connectivity in RRMS was the earlier mentioned smaller MEG study that also studied power, which showed a significant disease effect [232] on reduced interhemispheric connectivity in the alpha band (8–13 Hz) using coherence, which did not relate to clinical scores. In a subsequent study [236], it was hypothesised that sensorimotor deficits would be captured by decreases in connectivity between sensorimotor areas in RRMS. This study indeed showed that sensorimotor areas showed reduced functional connectivity in the patients, although no such association with disease status was found, possibly due to a rather homogeneous and small patient sample. In the first larger study, alterations in connectivity in several frequency bands were seen in early RRMS compared to controls [237], using the synchronisation likelihood metric. Specifically, this study showed lower connectivity in the alpha2 band and higher connectivity in the theta, alpha1 and beta band, which correlated with cognitive performance. These first few results from sensor-level MEG studies were promising and showed the potential of MEG in MS to electrophysiological abnormalities. It should be noted that the results from these studies should be interpreted with care since the metrics that were used varied greatly, were all sensitive to field spread, and analyses were performed at the sensor level. All subsequent MEG studies that we describe below, however, were performed in source space and used PLI to estimate functional connectivity. The first source-space connectivity study revealed lower functional connectivity in the alpha2 band and higher connectivity in the beta band in early RRMS patients compared to healthy controls [238]. These disease-induced effects in connectivity were stronger for literature-based resting-state subnetworks than for whole-brain connectivity patterns. Increased connectivity in the DMN was found in MS patients in the beta band, which correlated negatively with cognitive and motor performance. Therefore, these results suggested that a specific subset of connections or regions might be more sensitive or important in clinical deterioration, although only early RRMS patients were studied. Two larger MS studies, one with a relatively short disease duration ($N = 86$) and the other with a long disease duration ($N = 102$) confirmed, in particular, the lower functional connectivity in the alpha2 band at the whole-brain level [239, 240], which could mean that it is a general disease effect that occurs early, although this is still unclear. The MEG study in the long disease duration cohort also found higher connectivity in the delta and theta band, which was related to correlated patterns of atrophic brain regions [240]. Given the low number of studies, performed at only a few sites, however, it would be beneficial for the field if these MEG connectivity findings could be reproduced by other MEG centres.

Network Theory

Basics and Limitations of Network Analysis

The last decade has seen an incredible amount of research on functional networks in the healthy brain or in neurological and psychiatric disorders [241, 242]. The neuroimaging field is increasingly aware that tools and methods from basic network science are helpful in elucidating and understanding brain functioning from a network point of view. Whereas functional connectivity refers to the *strength* of the connections between brain regions, functional network topology corresponds to the *set (or pattern) of connections* (links) between *regions* (nodes) in the brain. Therefore, functional connectivity and functional networks can provide distinct information about the characteristics of the underlying system. However, the observed network topology depends on the analysis steps executed before reconstruction of the network, including functional connectivity estimation [243]. Therefore, if spurious connections (e.g. due to field spread) are included in reconstruction of the network, all subsequently calculated topological measures may be misleading and may lead to erroneous conclusions. Networks can be analysed using tools from graph theory. To construct a graph, the sensors/brain regions are depicted as nodes and the anatomical or functional connections as links. In the field of basic network theory, the term "graph" generally refers to a small-scale system and "network" to a large-scale system. For a graph, the general convention is to use the words vertices and edges instead of nodes and links. An unweighted network is reconstructed by applying a threshold to select the strongest connections in the brain and setting these to one, and the remaining connections to zero (i.e. a binary matrix is created), while a weighted network is obtained by usually considering all connections, including their strength (or link weight). Since the brain can be classified as a large-scale system, we use the "network" description in the remaining of the chapter.

Small-World and Scale-Free Networks

A thorough explanation of different measures that characterise network topology can be found in [242, 244], and see also Fig. 8.5. Briefly, the most basic measures are related to the concepts of segregation and integration. While segregation enables brain regions to evaluate information locally, integration enables the brain to merge information from, for example, different sensory areas, e.g. visual and auditory areas. A balance between segregation and integration is believed to be crucial to allow for efficient communication between brain regions and at the same time comes along with relatively low economical costs [245]. Several measures for segregation have been used in the field, ranging from a completely local level, such as the clustering (are a node's neighbouring nodes also each other's neighbours), to the level of subnetworks, i.e. modules (a set of nodes that show a more dense pattern of

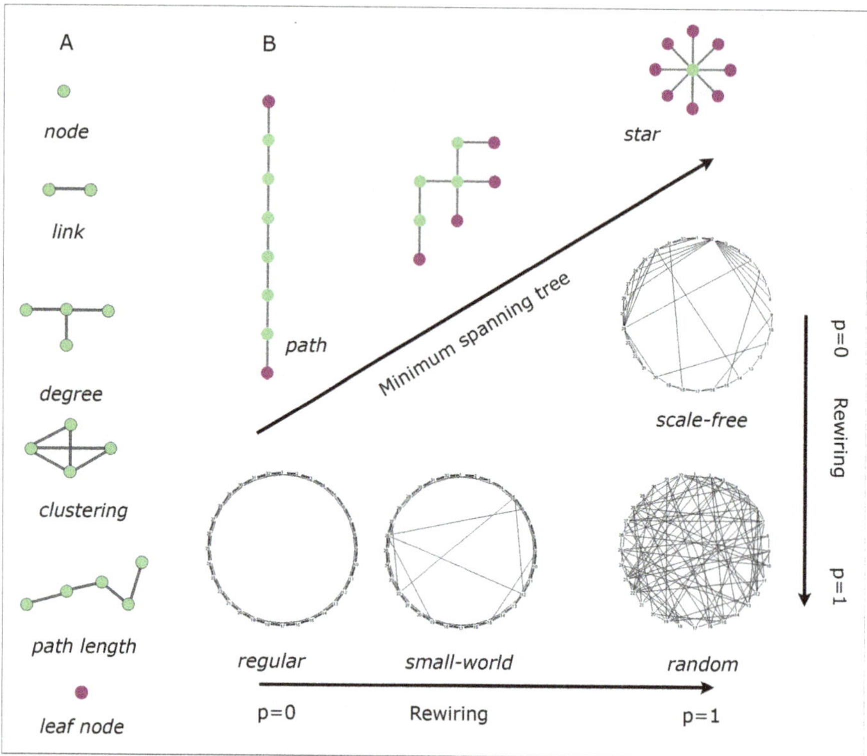

Fig. 8.5 Conventional and MST networks. Panel **A** illustrates the basic features of networks. Networks consist of nodes that are connected through links. The number of connections (links) of a node is called degree. The clustering coefficient of a node is the fraction of its neighbours that are also connected to each other. Path length corresponds to a route between any two nodes. The longest shortest path length in an MST network is called the diameter. Another characteristic of a MST is the number of leaf nodes, nodes with a degree of one. Panel **B** shows how the topologies of conventional networks are related to MST networks. An MST of a regular network corresponds to a path, whereas an MST of a scale-free network corresponds to a star network. Rewiring refers to swapping edges in a network randomly with a probability of p. The optimal MST configuration for the brain is likely to be somewhere between the two extremes (path and a star), such that there is network efficiency without the overload of hub nodes in the network. (This figure is based on Figure 9 in [251])

connections within this set than with nodes outside of this set). Popular measures for integration are the average characteristic path length (or global efficiency, which is its inverse). The characteristic path length is the length of the route between any two nodes that is minimised in terms of link weights (weighted networks) or number of links (unweighted networks). Based on clustering and path length, three network types can be identified: (i) regular networks, (ii) small-world networks, and (iii) random networks [246]. Regular networks are networks with high clustering and high path lengths, i.e. networks characterised by segregation, but with a lack of integration between clusters. Random networks are characterised by low clustering and low path length and thus characterised by integration but a lack of segregation. Lastly, small-world networks are characterised by high clustering and low path lengths, i.e. networks that are characterised by both segregation and integration. Another feature of networks is the importance of each node in the network, also called "hubness". Some nodes in the network can be considered more central than other nodes in the network, due to, e.g. their ubiquitous connections (high degree, where the degree indicates the number (or total weight) of connections to a node), due to the disproportional amount of traffic that trespasses these nodes (betweenness centrality) or due to the amount of connections to a node and also incorporating the connections to the neighbours of this node (eigenvector centrality (EC)). Networks for which the degree distribution can be described by a power law are also known as scale-free networks [247]. Functional brain networks have both small-world properties and scale-free properties, though both models cannot fully account for the network topology observed in brain networks as scale-free networks lack clustering and modularity while small-world networks lack modularity and hubness.

The realisation that the brain can be considered a complex network, and the availability of easy-to-use software tools, has led to extensive application of network analysis to neuroimaging data. The downside of this development is that this has unfortunately led to more confusion, rather than clarity, in the field. Many graph theoretical findings using these classical measures are contradictory, as has, for example, been shown in reviews of network studies in Alzheimer's disease and epilepsy [248, 249]. An important reason for these contradictory findings is that the observed network topology depends strongly on methodological choices [250], such as the number of nodes and links to include in the network, that is, the network size and density [243]. Therefore, for unweighted networks, different choices for the threshold might lead to entirely different inferences on network topology. For example, a threshold based on link weights may lead to different densities for networks for MS patients compared to healthy controls, simply because the groups differ in functional connectivity. Thus, observed group differences in network measures may not be due to differences in network topology but purely due to a difference in density. Usage of weighted networks does not solve the problem either, since measures computed on weighted networks depend on the average strength of connectivity, and differences in signal-to-noise between groups can lead to spurious results [251]. To correct for these biases, there is a tendency to normalise computed metrics with values obtained from appropriate surrogate networks. Unfortunately, this does not always solve the influence of the aforementioned biases and may even exaggerate them [243].

Minimum Spanning Tree

An approach that is able to avoid these methodological biases is the construction of the so-called minimum spanning tree. The minimum spanning tree (MST) is a subnetwork of the original weighted network that does not contain loops or cycles, but which does includes all nodes, and is unique for a given unique set of link weights. For any network one can extract a number of acyclic subnetworks (trees), but the MST is the one tree in which the sum of the link weight is minimised. Note that for functional and structural networks, we are interested in the strongest links; hence the inverse of the link weights are used as input for the MST algorithm. If the underlying network can be considered as a transportation network and if there is strong variability in link weights, then the MST forms the backbone of the underlying network over which most of the traffic flows [252]. Unfortunately, this last condition is not always met for brain networks [253]. Also note that the MST is a special case of a tree network among the union of shortest path trees, with the MST being the sparsest configuration among the union [253]. A denser network can be obtained by considering higher-order configurations, which can also be used to characterise brain networks, although with the necessity of a user-defined parameter. For now, we consider the MST, since this has mostly been used in MS studies. Just as for the original network, the MST can be characterised by topological measures. These include the diameter, leaf fraction, node degree and degree divergence. The diameter is the longest shortest path in the network [251]. Leaf fraction is computed as the number of nodes with a degree of one, divided by network size. Lastly, the degree divergence is related to synchronisation properties of the network and is a measure related to the width of the degree distribution (hubness). These MST measures capture the information of the underlying network and are related to conventional measures computed on the original networks [251]. The two extreme configurations of the MST are the path (MST equivalent of a regular network) and the star (MST equivalent of a scale-free network) (see Fig. 8.5). The tree equivalent of the well-known conventional random- and small-world configurations can be found in between these two extrema.

Functional Network Topology in MS

The first study on functional network topology in MS computed conventional graph metrics based on an unweighted network obtained from fMRI BOLD time courses [254]. Normalised versions of the clustering coefficient and path length were calculated in a group of 30 MS patients, who were mostly diagnosed with RRMS and some with SPMS. MS patients were mildly disabled. Results showed no significant difference with respect to the normalised clustering coefficient, but the normalised path length was significantly higher in MS patients for both males and females, indicating a less efficient network in MS patients or in other words, a shift towards a more regular topology. The second graph theoretical study in MS used MEG instead of fMRI [237]. The same authors computed conventional graph measures on weighted networks, where connections were estimated using synchronisation likelihood, in a cohort of 34 RRMS patients with relatively short disease duration. Results showed higher normalised clustering coefficient and path lengths in MS patients in the alpha1 band, where the former was found to be negatively associated with cognitive performance. Another fMRI study supported the notion that functional networks are shifted towards regular network topology in MS, i.e. the authors found a higher path length in unweighted networks in MS patients [255]. However, results from this study come with a difficulty in interpretation since the clustering coefficient in the same group was lower in MS patients, which is suggestive of a shift towards a more random network topology. Two recent MEG studies, making use of relatively large patient cohorts, computed in addition to MST measures (see below) also conventional graph metrics on weighted networks [239, 240]. Both these studies, with patients with relatively short and long disease duration, found lower normalised clustering and path lengths in MS patients in especially the alpha2 band but also in other frequency bands [239]. The discrepancy with the previous MEG literature could be related to the fact that in the latter studies, the PLI was used to estimate connections, a metric that is insensitive to field spread, whereas generalised synchronisation used in the former study is sensitive to field spread, which might, in particular, induce spurious clustering. Furthermore, the link weights will have influenced the aforementioned results indirectly, and therefore topology and mean connectivity effects are likely to be mixed within these results.

Other network studies in MS have also analysed other properties of functional networks, such as subnetworks (modularity) or hubness. In a standard PASAT task during fMRI acquisition, modularity was assessed in a group of 16 early MS patients [256]. Higher modularity was observed in MS patients compared to healthy subjects, which was associated with cognitive dysfunction (defined here performance during the task). Higher modularity was also related to white matter lesion load, which suggests that damage to white matter tracts may result in a splitting of the network in more segregate components. Two studies that have analysed hubness in terms of eigenvector centrality (EC) have confirmed each other's findings. An MEG study in a group of 34 RRMS patients showed that EC was higher in MS patients over parietal regions in the theta band compared to healthy controls [257]. The alpha2 and beta band showed lower EC over temporal regions, whereas the gamma band showed lower EC over parietal regions in MS patients. However, these analyses were performed in sensor space and employed the synchronisation likelihood metric that is sensitive to field spread, warranting some caution. An fMRI study investigated many more subjects from this same cohort [258], and found lower EC values in the occipito-parieto-temporal areas making up the ventral stream, in line with the alpha2, beta and gamma band results from the MEG study, and higher EC values for the thalami and posterior cingulate cortices. The latter was not found in the MEG study, nor was the equivalent of the increased theta parietal EC found in fMRI, which is surprising given that this frequency band is more likely to contribute to fMRI networks than the gamma band [259]. This could be due to the basic difference between fMRI and MEG or other aspects such as the possible inclusion of white matter voxels as nodes in the fMRI study. Additionally, since the eigenvector is normalised between one and zero, a genuine increase in hubness in specific regions is by definition accompanied with an artificial decrease in hubness elsewhere, although the latter is probably a more diffuse effect. Despite the fact that the hubness results were obtained with somewhat suboptimal approaches in the previously mentioned studies, they did correlate to cognitive dysfunction, making it worthwhile to investigate hubness in MS using these and other hub metrics in future studies.

Given the methodological hurdles with conventional network analyses, recent MEG studies have used the MST approach to study network alterations in MS. The first MST study in MS [260] showed that in 21 early RRMS patients, there was a shift towards a more path-like topology in especially the alpha2 band. This shift was characterised by a lower leaf fraction, larger diameter and lower degree divergence. Functional networks in relatively early MS patients thus become less efficient compared to those in healthy subjects, as any path between two nodes is longer for MS patients, which related to cognitive performance. Importantly, this shift in functional network topology towards a more path-like topology has been reproduced by two other studies with relative large sample sizes [239, 240], where one study also confirmed the relationship between the shift towards a path-like network and worse cognitive performance. MS patients that were included in these two studies had a relatively short disease duration and long disease duration, respectively. Therefore, for both short and long disease duration, networks in MS are less efficient. Note that the degree divergence was also found to be lower in all three MST studies, which indeed indicates a loss of hubness of nodes in the network. Longitudinal studies are required to analyse whether network efficiency worsens over time, or whether this stays stable.

Multimodal Imaging in MS: Integrating Structure and Function

The previous sections have highlighted the importance of functional disruption of activity, connectivity and topology in MS. The general network finding is that MS networks are characterised by a loss of efficiency and changes in hubness. These network alterations were associated with clinical and cognitive status. However, both structural damage and functional changes have been shown to relate with cognitive and physical problems in MS. Given the complex structural pathology in MS, it may be important to investigate if functional networks can form a bridge in the clinico-radiological paradox, in order to understand how structural pathology eventually may lead to clinical and cognitive dysfunction. Treating structural and functional damage in relation to each other could also help to elucidate if altered patterns of functional connectivity and network topology are induced by structural pathology or whether these disease-induced connectivity patterns are adaptive to uphold (sub)-optimal cognition. It can be argued that any structural damage of clinical significance (based on lesions, normal-appearing tissue changes and atrophy) should impact brain function as well. Therefore, functional disruption in MS might be an "integrator" of the different types of structural pathology in MS. Likewise, treating structure and function together might also give hints how functional damage may induce or aggravate structural pathology. There may even be a circular causality between the two, which has not been explored yet in MS. In order to assess causality, there is a pressing need for (multimodal) longitudinal studies combined with biophysical models in which such a causal claim can be tested.

The simplest way to relate structure to function in MS is to correlate lesion load with functional activation or connectivity. This has been done in several fMRI studies, which show that reduced connectivity may correlate to lesion load [198, 200, 255], but not always. It seems logical that white matter lesions disrupt structural connections between two nodes and therefore, affect functional connectivity as well. Unfortunately, this relationship between lesion load or lesion probability maps and functional connectivity has not yet been studied in great depth. This also holds for the relationship between other types of structural pathology and functional connectivity. Some studies have analysed the relationship between functional connectivity quantified with fMRI and white matter integrity [203, 261]. These studies showed that diffuse damage to white matter tracts is associated with lower functional connectivity between regions, which can occur for tracts and functional connections that are nonoverlapping. These preliminary studies therefore suggest that reduced connectivity might be an adverse effect induced by structural pathology, although definitive causal claims are impossible in such cross-sectional studies using correlative assessments.

Advancing Structural-Functional Concepts in MS

The real challenge in the field is to use more sophisticated ways to understand how structural and functional networks in MS are interrelated and how continued damage to either structural or functional networks may alter this relationship or how damage might lead to inefficiency of the functional network. The need to understand the relationship between structure and function in terms of networks is brought on by the observations that focal structural pathology can have global impact on a functional network. An example is the thalamus, which (as mentioned above) is one of the first brain structures to be hit with atrophic and microstructural changes in MS, in a landscape of random white matter damage. Using functional MRI, DTI and atrophy measures, a recent paper showed that functional connectivity increases only appear in more severely cognitively impaired MS patients. Interestingly, atrophy and diffusion changes already appeared in cognitively preserved or mildly impaired patient groups, indicating that a certain threshold of structural pathology has to be reached before functional network changes appear [130]. The consequence of such functional changes is still unclear, although it might be considered that these network changes may influence atrophy itself. In a multimodal fMR/MEG study, thalamic atrophy was shown to be related to a shift towards a less efficient configuration of the cortical functional network [239], highlighting that a rather focal and distant anatomical change is associated with diffuse functional changes within the cortex. The cortical network was also analysed in another multimodal imaging study in MS. Here, the relationship between structural covariance networks and functional networks with MEG was analysed [240]. It was shown that functional connection strengths in the theta band were correlated to covariation in thickness in MS patients. This relationship was not found for raw thickness itself, indicating that atrophy only has an effect on connectivity when there is also atrophy in another covarying region or the reverse, i.e. that atrophy is worsened by functional connections between structurally connected regions. Such a network-based propagation of atrophy was also indicated in a recent structural study in MS [120].

New insights in the relationship between structural and functional networks in MS could be gained from recent theory on this relationship in the healthy brain [262, 263]. It has been shown that there is a mapping between structural and functional brain networks, where the latter was based on both fMRI and MEG. The neurobiological interpretation of this mapping is that a functional connection between two regions is shaped by all possible walks in the structural network between these two regions. A walk is any path between two regions, which can also include loops. The contribution of the walks to a functional connection fell off with increasing number of included links (i.e. longer paths). Furthermore, fMRI-based functional networks were relatively more shaped by shorter walks in the structural network, whereas for MEG-based functional networks, this dominance for shorter walks was less dominant. The importance of this framework is that it could help to elucidate how white matter damage at the structural level might impact the functional networks. A lesion, for example, may cause disruption in the efficient shorter walks, which would therefore result that functional connections are more likely to be shaped by longer walks in the structural network, which could potentially explain the less efficient functional networks observed in MS. With this same structural to functional network mapping framework, the relationship between diffuse white matter damage and functional networks can be analysed. Hypotheses such as propagation of network damage throughout the network in terms of structural-functional networks can be analysed as well [263], which may be an important future topic in MS.

Apart from the structural to functional network relationships, multimodal studies in MS may also profit from advances in neuroimaging network methodology. Most of the network and connectivity studies in MS have either been conducted with fMRI or MEG. A lack of theory has prevented generalisation or interpretation of results between these modalities. Recent work in the neuroimaging field however points out that cortical fMRI networks can be predicted by MEG networks from the same subjects [259]. It seems that fMRI networks are a reflection of MEG networks obtained from multiple frequencies and from linear and non-linear neuronal interactions. In other words, a haemodynamic connection between two brain regions is a reflection of their direct neuronal connectivity (linear), but also a reflection of the overlap in connectivity profiles of these two brain regions (non-linear), i.e. shared connectivity. The non-linear component also extends to cross-frequency interactions. This mathematical understanding between MEG and fMRI networks may explain why a recent study found altered connectivity with MEG, and at the same time, a lack of cortical connectivity changes with fMRI [239]. Since the MEG connectivity findings were limited to a few frequency bands, the mathematical model informs us that the contribution of the other frequency bands (with non-significant results) might therefore blur any group differences at the haemodynamic level. Future neuroimaging studies will need to go beyond the relationship between fMRI and MEG networks at the connectivity level, such that network topology findings between the two modalities can be understood as well.

Future Perspectives

As mentioned in the first sections of this chapter, the clinico-radiological paradox in MS has driven the field to study the MS brain in a holistic, translational and multimodal approach. This approach has led to an expanding body of work, which has greatly enhanced our understanding of the brain and the disease, but also raised many additional questions. An important objective in the MS field is to understand the interrelationship between several types of functional and structural pathology. An understanding of the entire system could eventually lead to the identification of the "weak links" in the chain of pathological processes that may form objective measures for cognitive and clinical problems. But as stated earlier, before we can obtain a meaningful objective measure, we need to understand the pathophysiological processes at a systems level. The perspectives for such an objective measure for clinical and cognitive status would be very significant, as it could also help to judge treatment response in an individual patient. Furthermore, if one understands the dynamics of a system, one might also make better predictions about how the system will behave in the near future, and therefore it could help to give an individualised prognosis. So far, it seems that the heterogeneity within and between structural pathology, in the forms of focal lesions, diffuse changes and atrophy, leads to varied patterns of changes in brain function and network topology, which then results in clinico-cognitive decline. Such causal links, however, must be studied in a longitudinal setting in the future. Such studies could provide key explanations on clinical progression, cognitive deterioration and conversion to a progressive phenotype.

However, to understand the dynamics of the disease, we presumably also have to understand the dynamics of functional networks itself. Dynamics of functional networks in the healthy brain is increasingly being studied in the neuroimaging community [264] and allows to extract more information from the time series than only mean connectivity values. To achieve this in fMRI, it will be necessary to use considerably longer acquisition times in order to detect nonstationary connectivity [265], which could hold currently completely unknown types of information on brain function. In fact, these dynamic networks seem to correlate stronger to cognition than static networks [266]. MEG holds another key position in that it can study cortical networks directly, with a very high temporal resolution, with a lot of ground still to be covered given the low abundance of MEG literature in MS. Given the recent literature [259], spatial patterns of MEG-based dynamic networks can in fact be translated back to fMRI networks, which would allow for a translation between the two modalities. Initial MEG studies have indeed shown how MEG is also able to identify networks that fluctuate at sub-second (and longer) timescales [267–269]. Therefore, it may be insightful to analyse dynamic networks in MS pathology combined with longitudinal imaging and investigate whether there is any additional information in these dynamic networks that can explain the development of cognitive and clinical problems.

Two other new directions in the field of functional networks are directional connectivity and multilayer networks. Although directionality of connectivity has been assessed in the past [270], the limited information in fMRI time series and methodological hurdles for electrophysiological connectivity such as field spread have impeded advance in this field. Recent measures for MEG/EEG seem however to circumvent field spread and show front-to-back and back-to-front patterns of directed connectivity in resting-state MEG recordings for the theta and alpha2 band, respectively [271, 272]. Given the rich frequency content of MEG time series, studies have now began to analyse the relationship between networks obtained from different frequency bands in an integrative manner with the use of new advances in network science, the multilayer network approach. A multilayer network is a network of network and consists of several layers that are interacting with each other. Each layer is a network on its own characterised by its own topology. With a multilayer network approach, it has been shown that networks obtained from different frequency bands cannot be treated separately but are mutually influencing each other [273]. The importance for MS studies is that we could investigate how network alterations in different frequency bands are related to each other and how network disruption in one frequency band may cascade into the other frequency bands. It also allows us to combine structural and functional networks in the same hierarchical framework.

In summary, multiple sclerosis features structural pathology in the form of focal lesions, which are excellent for diagnosis but poor for clinical correlations, as well as diffuse changes throughout the brain and atrophic changes. Especially the latter two relate rather strongly to clinico-cognitive decline, although it is currently unclear how they arise. The clinico-radiological paradox in MS may only be unravelled using a holistic approach, i.e. when combining all measures of pathology in one network-based model. This model should also include changes in brain function, as these have been shown to relate strongly to clinical measures and provide relevant, additional information. Functional networks change drastically in MS, even with relatively mild structural damage, showing changes in network efficiency and hubness. The interplay between thalamic changes and broad-scale functional network pathology highlights the need to study the structural-functional interplay in MS further. There is a current lack of knowledge on advanced network changes and longitudinal imaging, especially in progressive MS, which we only understand poorly. Future studies should focus on a multimodal approach, following patients over time, in order to elucidate the complex network-based patterns of interplay between structure and function. Using such an approach, it might be possible to unravel the clinico-radiological paradox, possibly leading to a more powerful prediction of clinical progression in MS.

References

1. Kamm CP, Uitdehaag BM, Polman CH (2014) Multiple sclerosis: current knowledge and future outlook. Eur Neurol 72(3–4):132–141
2. Alonso A, Hernán MA (2008) Temporal trends in the incidence of multiple sclerosis a systematic review. Neurology 71(2):129–135
3. Stys PK, Zamponi GW, van Minnen J, Geurts JJ (2012) Will the real multiple sclerosis please stand up? Nat Rev Neurosci 13(7):507–514
4. Lublin FD, Reingold SC, Cohen JA, Cutter GR, Sørensen PS, Thompson AJ et al (2014) Defining the clinical course of multiple sclerosis the 2013 revisions. Neurology 83(3):278–286
5. La Mantia L, Vacchi L, Di Pietrantonj C, Ebers G, Rovaris M, Fredrikson S et al (2012) Interferon beta for secondary progressive multiple sclerosis. Cochrane Database Syst Rev 1: CD005181
6. Miller DH, Leary SM (2007) Primary-progressive multiple sclerosis. Lancet Neurol 6 (10):903–912
7. Filippini G, Del Giovane C, Vacchi L, D'Amico R, Di Pietrantonj C, Beecher D et al (2013) Immunomodulators and immunosuppressants for multiple sclerosis: a network meta-analysis. Cochrane Database Syst Rev 6:CD008933
8. Tramacere I, Del Giovane C, Salanti G, D'Amico R, Filippini G. Immunomodulators and immunosuppressants for relapsing-remitting multiple sclerosis: a network meta-analysis. Cochrane Database Syst Rev 2015;9:CD011381.
9. Rolak LA. The history of MS; the basic facts. http://www.nationalmssociety.org. National Multiple Sclerosis Society; 2016
10. Ormerod IE, Miller DH, McDonald WI, du Boulay EP, Rudge P, Kendall BE et al (1987) The role of NMR imaging in the assessment of multiple sclerosis and isolated neurological lesions. A quantitative study. Brain 110(Pt 6):1579–1616
11. Miller DH, Grossman RI, Reingold SC, McFarland HF (1998) The role of magnetic resonance techniques in understanding and managing multiple sclerosis. Brain 121(Pt 1):3–24
12. Neema M, Stankiewicz J, Arora A, Dandamudi VS, Batt CE, Guss ZD et al (2007) T1- and T2-based MRI measures of diffuse gray matter and white matter damage in patients with multiple sclerosis. J Neuroimaging 17(Suppl 1):16S–21S
13. Seewann A, Vrenken H, van der Valk P, Blezer EL, Knol DL, Castelijns JA et al (2009) Diffusely abnormal white matter in chronic multiple sclerosis: imaging and histopathologic analysis. Arch Neurol 66(5):601–609
14. Grossman RI, Gonzalez-Scarano F, Atlas SW, Galetta S, Silberberg DH (1986) Multiple sclerosis: gadolinium enhancement in MR imaging. Radiology 161(3):721–725
15. van Walderveen MA, Kamphorst W, Scheltens P, van Waesberghe JH, Ravid R, Valk J et al (1998) Histopathologic correlate of hypointense lesions on T1-weighted spin-echo MRI in multiple sclerosis. Neurology 50(5):1282–1288
16. Grossman RI, Braffman BH, Brorson JR, Goldberg HI, Silberberg DH, Gonzalez-Scarano F (1988) Multiple sclerosis: serial study of gadolinium-enhanced MR imaging. Radiology 169 (1):117–122
17. van Waesberghe JH, van Walderveen MA, Castelijns JA, Scheltens P, GJ L a N, Polman CH et al (1998) Patterns of lesion development in multiple sclerosis: longitudinal observations with T1-weighted spin-echo and magnetization transfer MR. AJNR Am J Neuroradiol 19 (4):675–683
18. van den Elskamp IJ, Boden B, Dattola V, Knol DL, Filippi M, Kappos L et al (2010) Cerebral atrophy as outcome measure in short-term phase 2 clinical trials in multiple sclerosis. Neuroradiology 52(10):875–881
19. Filippi M, Horsfield MA, Tofts PS, Barkhof F, Thompson AJ, Miller DH (1995) Quantitative assessment of MRI lesion load in monitoring the evolution of multiple sclerosis. Brain 118 (Pt 6):1601–1612

20. Miller DH, Barkhof F, Nauta JJ (1993) Gadolinium enhancement increases the sensitivity of MRI in detecting disease activity in multiple sclerosis. Brain 116(Pt 5):1077–1094

21. Barkhof F, Filippi M, Miller DH, Scheltens P, Campi A, Polman CH et al (1997) Comparison of MRI criteria at first presentation to predict conversion to clinically definite multiple sclerosis. Brain 120(Pt 11):2059–2069

22. Polman CH, Reingold SC, Banwell B, Clanet M, Cohen JA, Filippi M et al (2011) Diagnostic criteria for multiple sclerosis: 2010 revisions to the McDonald criteria. Ann Neurol 69 (2):292–302

23. Filippi M, Rocca MA, Ciccarelli O, De Stefano N, Evangelou N, Kappos L et al (2016) MRI criteria for the diagnosis of multiple sclerosis: MAGNIMS consensus guidelines. Lancet Neurol 15(3):292–303

24. Gonzalez-Scarano F, Grossman RI, Galetta S, Atlas SW, Silberberg DH (1987) Multiple sclerosis disease activity correlates with gadolinium-enhanced magnetic resonance imaging. Ann Neurol 21(3):300–306

25. Kappos L, Moeri D, Radue EW, Schoetzau A, Schweikert K, Barkhof F et al (1999) Predictive value of gadolinium-enhanced magnetic resonance imaging for relapse rate and changes in disability or impairment in multiple sclerosis: a meta-analysis. Gadolinium MRI Meta-analysis Group. Lancet 353(9157):964–969

26. Truyen L, van Waesberghe JH, van Walderveen MA, van Oosten BW, Polman CH, Hommes OR et al (1996) Accumulation of hypointense lesions ("black holes") on T1 spin-echo MRI correlates with disease progression in multiple sclerosis. Neurology 47(6):1469–1476

27. Rovaris M, Agosta F, Sormani MP, Inglese M, Martinelli V, Comi G et al (2003) Conventional and magnetization transfer MRI predictors of clinical multiple sclerosis evolution: a medium-term follow-up study. Brain 126(Pt 10):2323–2332

28. Barkhof F (1999) MRI in multiple sclerosis: correlation with expanded disability status scale (EDSS). Mult Scler 5(4):283–286

29. Thompson AJ, Kermode AG, MacManus DG, Kendall BE, Kingsley DP, Moseley IF et al (1990) Patterns of disease activity in multiple sclerosis: clinical and magnetic resonance imaging study. BMJ 300(6725):631–634

30. Montalban X (2005) Primary progressive multiple sclerosis. Curr Opin Neurol 18(3):261–266

31. Nijeholt GJ, van Walderveen MA, Castelijns JA, van Waesberghe JH, Polman C, Scheltens P et al (1998) Brain and spinal cord abnormalities in multiple sclerosis. Correlation between MRI parameters, clinical subtypes and symptoms. Brain 121(Pt 4):687–697

32. Khaleeli Z, Ciccarelli O, Manfredonia F, Barkhof F, Brochet B, Cercignani M et al (2008) Predicting progression in primary progressive multiple sclerosis: a 10-year multicenter study. Ann Neurol 63(6):790–793

33. Camp SJ, Stevenson VL, Thompson AJ, Miller DH, Borras C, Auriacombe S et al (1999) Cognitive function in primary progressive and transitional progressive multiple sclerosis: a controlled study with MRI correlates. Brain 122(Pt 7):1341–1348

34. Benedict RH, Weinstock-Guttman B, Fishman I, Sharma J, Tjoa CW, Bakshi R (2004) Prediction of neuropsychological impairment in multiple sclerosis: comparison of conventional magnetic resonance imaging measures of atrophy and lesion burden. Arch Neurol 61 (2):226–230

35. Barkhof F (2002) The clinico-radiological paradox in multiple sclerosis revisited. Curr Opin Neurol 15(3):239–245

36. Vellinga MM, Geurts JJ, Rostrup E, Uitdehaag BM, Polman CH, Barkhof F et al (2009) Clinical correlations of brain lesion distribution in multiple sclerosis. J Magn Reson Imaging 29(4):768–773

37. Hackmack K, Weygandt M, Wuerfel J, Pfueller CF, Bellmann-Strobl J, Paul F et al (2012) Can we overcome the 'clinico-radiological paradox' in multiple sclerosis? J Neurol 259 (10):2151–2160

38. Kincses ZT, Ropele S, Jenkinson M, Khalil M, Petrovic K, Loitfelder M et al (2011) Lesion probability mapping to explain clinical deficits and cognitive performance in multiple sclerosis. Mult Scler 17(6):681–689
39. Bodini B, Battaglini M, De Stefano N, Khaleeli Z, Barkhof F, Chard D et al (2011) T2 lesion location really matters: a 10 year follow-up study in primary progressive multiple sclerosis. J Neurol Neurosurg Psychiatry 82(1):72–77
40. Dalton CM, Bodini B, Samson RS, Battaglini M, Fisniku LK, Thompson AJ et al (2012) Brain lesion location and clinical status 20 years after a diagnosis of clinically isolated syndrome suggestive of multiple sclerosis. Mult Scler 18(3):322–328
41. Peterson JW, Bo L, Mork S, Chang A, Trapp BD (2001) Transected neurites, apoptotic neurons, and reduced inflammation in cortical multiple sclerosis lesions. Ann Neurol 50 (3):389–400
42. Kutzelnigg A, Lucchinetti CF, Stadelmann C, Bruck W, Rauschka H, Bergmann M et al (2005) Cortical demyelination and diffuse white matter injury in multiple sclerosis. Brain 128 (Pt 11):2705–2712
43. Kidd D, Barkhof F, McConnell R, Algra PR, Allen IV, Revesz T (1999) Cortical lesions in multiple sclerosis. Brain 122(Pt 1):17–26
44. Filippi M, Preziosa P, Rocca MA (2014) Magnetic resonance outcome measures in multiple sclerosis trials: time to rethink? Curr Opin Neurol 27(3):290–299
45. Bo L, Geurts JJ, Mork SJ, van der Valk P (2006) Grey matter pathology in multiple sclerosis. Acta Neurol Scand Suppl 183:48–50
46. Calabrese M, Rocca MA, Atzori M, Mattisi I, Bernardi V, Favaretto A et al (2009) Cortical lesions in primary progressive multiple sclerosis: a 2-year longitudinal MR study. Neurology 72(15):1330–1336
47. Pirko I, Lucchinetti CF, Sriram S, Bakshi R (2007) Gray matter involvement in multiple sclerosis. Neurology 68(9):634–642
48. Fisniku LK, Chard DT, Jackson JS, Anderson VM, Altmann DR, Miszkiel KA et al (2008) Gray matter atrophy is related to long-term disability in multiple sclerosis. Ann Neurol 64 (3):247–254
49. Calabrese M, Agosta F, Rinaldi F, Mattisi I, Grossi P, Favaretto A et al (2009) Cortical lesions and atrophy associated with cognitive impairment in relapsing-remitting multiple sclerosis. Arch Neurol 66(9):1144–1150
50. Calabrese M, Rinaldi F, Grossi P, Gallo P (2011) Cortical pathology and cognitive impairment in multiple sclerosis. Expert Rev Neurother 11(3):425–432
51. Rinaldi F, Calabrese M, Grossi P, Puthenparampil M, Perini P, Gallo P (2010) Cortical lesions and cognitive impairment in multiple sclerosis. Neurol Sci 31(Suppl 2):S235–S237
52. Rocca MA, Amato MP, De Stefano N, Enzinger C, Geurts JJ, Penner IK et al (2015) Clinical and imaging assessment of cognitive dysfunction in multiple sclerosis. Lancet Neurol 14 (3):302–317
53. Roosendaal SD, Moraal B, Vrenken H, Castelijns JA, Pouwels PJ, Barkhof F et al (2008) In vivo MR imaging of hippocampal lesions in multiple sclerosis. J Magn Reson Imaging 27 (4):726–731
54. Calabrese M, Grossi P, Favaretto A, Romualdi C, Atzori M, Rinaldi F et al (2012) Cortical pathology in multiple sclerosis patients with epilepsy: a 3 year longitudinal study. J Neurol Neurosurg Psychiatry 83(1):49–54
55. Uribe-San-Martin R, Ciampi-Diaz E, Suarez-Hernandez F, Vasquez-Torres M, Godoy-Fernandez J, Carcamo-Rodriguez C (2014) Prevalence of epilepsy in a cohort of patients with multiple sclerosis. Seizure 23(1):81–83
56. Calabrese M, Magliozzi R, Ciccarelli O, Geurts JJ, Reynolds R, Martin R (2015) Exploring the origins of grey matter damage in multiple sclerosis. Nat Rev Neurosci 16(3):147–158
57. van de Pavert SH, Muhlert N, Sethi V, Wheeler-Kingshott CA, Ridgway GR, Geurts JJ et al (2016) DIR-visible grey matter lesions and atrophy in multiple sclerosis: partners in crime? J Neurol Neurosurg Psychiatry 87(5):461–467

58. Bakshi R, Ariyaratana S, Benedict RH, Jacobs L (2001) Fluid-attenuated inversion recovery magnetic resonance imaging detects cortical and juxtacortical multiple sclerosis lesions. Arch Neurol 58(5):742–748

59. Geurts JJ, Bo L, Pouwels PJ, Castelijns JA, Polman CH, Barkhof F (2005) Cortical lesions in multiple sclerosis: combined postmortem MR imaging and histopathology. AJNR Am J Neuroradiol 26(3):572–577

60. Geurts JJ, Blezer EL, Vrenken H, van der Toorn A, Castelijns JA, Polman CH et al (2008) Does high-field MR imaging improve cortical lesion detection in multiple sclerosis? J Neurol 255(2):183–191

61. Geurts JJ, Pouwels PJ, Uitdehaag BM, Polman CH, Barkhof F, Castelijns JA (2005) Intracortical lesions in multiple sclerosis: improved detection with 3D double inversion-recovery MR imaging. Radiology 236(1):254–260

62. Seewann A, Vrenken H, Kooi EJ, van der Valk P, Knol DL, Polman CH et al (2011) Imaging the tip of the iceberg: visualization of cortical lesions in multiple sclerosis. Mult Scler 17 (10):1202–1210

63. Seewann A, Kooi EJ, Roosendaal SD, Pouwels PJ, Wattjes MP, van der Valk P et al (2012) Postmortem verification of MS cortical lesion detection with 3D DIR. Neurology 78 (5):302–308

64. Geurts JJ, Roosendaal SD, Calabrese M, Ciccarelli O, Agosta F, Chard DT et al (2011) Consensus recommendations for MS cortical lesion scoring using double inversion recovery MRI. Neurology 76(5):418–424

65. de Graaf WL, Kilsdonk ID, Lopez-Soriano A, Zwanenburg JJ, Visser F, Polman CH et al (2013) Clinical application of multi-contrast 7-T MR imaging in multiple sclerosis: increased lesion detection compared to 3 T confined to grey matter. Eur Radiol 23(2):528–540

66. Kilsdonk ID, de Graaf WL, Soriano AL, Zwanenburg JJ, Visser F, Kuijer JP et al (2013) Multicontrast MR imaging at 7T in multiple sclerosis: highest lesion detection in cortical gray matter with 3D-FLAIR. AJNR Am J Neuroradiol 34(4):791–796

67. Kilsdonk ID, Jonkman LE, Klaver R, van Veluw SJ, Zwanenburg JJ, Kuijer JP et al (2016) Increased cortical grey matter lesion detection in multiple sclerosis with 7 T MRI: a post-mortem verification study. Brain 8

68. McDonald WI (1974) Pathophysiology in multiple sclerosis. Brain 97(1):179–196

69. Simon JH, Holtas SL, Schiffer RB, Rudick RA, Herndon RM, Kido DK et al (1986) Corpus callosum and subcallosal-periventricular lesions in multiple sclerosis: detection with MR. Radiology 160(2):363–367

70. Losseff NA, Wang L, Lai HM, Yoo DS, Gawne-Cain ML, McDonald WI et al (1996) Progressive cerebral atrophy in multiple sclerosis. A serial MRI study. Brain 119 (Pt 6):2009–2019

71. Losseff NA, Webb SL, O'Riordan JI, Page R, Wang L, Barker GJ et al (1996) Spinal cord atrophy and disability in multiple sclerosis. A new reproducible and sensitive MRI method with potential to monitor disease progression. Brain 119(Pt 3):701–708

72. Lukas C, Knol DL, Sombekke MH, Bellenberg B, Hahn HK, Popescu V et al (2015) Cervical spinal cord volume loss is related to clinical disability progression in multiple sclerosis. J Neurol Neurosurg Psychiatry 86(4):410–418

73. Lukas C, Sombekke MH, Bellenberg B, Hahn HK, Popescu V, Bendfeldt K et al (2013) Relevance of spinal cord abnormalities to clinical disability in multiple sclerosis: MR imaging findings in a large cohort of patients. Radiology 269(2):542–552

74. Huber SJ, Paulson GW, Shuttleworth EC, Chakeres D, Clapp LE, Pakalnis A et al (1987) Magnetic resonance imaging correlates of dementia in multiple sclerosis. Arch Neurol 44 (7):732–736

75. Pozzilli C, Bastianello S, Padovani A, Passafiume D, Millefiorini E, Bozzao L et al (1991) Anterior corpus callosum atrophy and verbal fluency in multiple sclerosis. Cortex 27 (3):441–445

76. Pozzilli C, Fieschi C, Perani D, Paulesu E, Comi G, Bastianello S et al (1992) Relationship between corpus callosum atrophy and cerebral metabolic asymmetries in multiple sclerosis. J Neurol Sci 112(1–2):51–57

77. Saindane AM, Ge Y, Udupa JK, Babb JS, Mannon LJ, Grossman RI (2000) The effect of gadolinium-enhancing lesions on whole brain atrophy in relapsing-remitting MS. Neurology 55(1):61–65

78. Pelletier J, Habib M, Lyon-Caen O, Salamon G, Poncet M, Khalil R (1993) Functional and magnetic resonance imaging correlates of callosal involvement in multiple sclerosis. Arch Neurol 50(10):1077–1082

79. Eikelenboom MJ, Petzold A, Lazeron RH, Silber E, Sharief M, Thompson EJ et al (2003) Multiple sclerosis: Neurofilament light chain antibodies are correlated to cerebral atrophy. Neurology 60(2):219–223

80. Popescu V, Klaver R, Voorn P, Galis-de Graaf Y, Knol DL, Twisk JW et al (2015) What drives MRI-measured cortical atrophy in multiple sclerosis? Mult Scler 21(10):1280–1290

81. Klaver R, Popescu V, Voorn P, Galis-de Graaf Y, van der Valk P, de Vries HE et al (2015) Neuronal and axonal loss in normal-appearing gray matter and subpial lesions in multiple sclerosis. J Neuropathol Exp Neurol 74(5):453–458

82. Derakhshan M, Caramanos Z, Giacomini PS, Narayanan S, Maranzano J, Francis SJ et al (2010) Evaluation of automated techniques for the quantification of grey matter atrophy in patients with multiple sclerosis. NeuroImage 52(4):1261–1267

83. Popescu V, Schoonheim MM, Versteeg A, Chaturvedi N, Jonker M, Xavier de Menezes R et al (2016) Grey matter atrophy in multiple sclerosis: clinical interpretation depends on choice of analysis method. PLoS One 11(1):e0143942

84. Simon JH, Jacobs LD, Campion MK, Rudick RA, Cookfair DL, Herndon RM et al (1999) A longitudinal study of brain atrophy in relapsing multiple sclerosis. The multiple sclerosis collaborative research group (MSCRG). Neurology 53(1):139–148

85. Barkhof FJ, Elton M, Lindeboom J, Tas MW, Schmidt WF, Hommes OR et al (1998) Functional correlates of callosal atrophy in relapsing-remitting multiple sclerosis patients. A preliminary MRI study. J Neurol 245(3):153–158

86. Filippi M, Mastronardo G, Rocca MA, Pereira C, Comi G (1998) Quantitative volumetric analysis of brain magnetic resonance imaging from patients with multiple sclerosis. J Neurol Sci 158(2):148–153

87. Rovaris M, Bozzali M, Rodegher M, Tortorella C, Comi G, Filippi M (1999) Brain MRI correlates of magnetization transfer imaging metrics in patients with multiple sclerosis. J Neurol Sci 166(1):58–63

88. Molyneux PD, Kappos L, Polman C, Pozzilli C, Barkhof F, Filippi M et al (2000) The effect of interferon beta-1b treatment on MRI measures of cerebral atrophy in secondary progressive multiple sclerosis. European study group on interferon beta-1b in secondary progressive multiple sclerosis. Brain 123(Pt 11):2256–2263

89. Ge Y, Grossman RI, Udupa JK, Wei L, Mannon LJ, Polansky M et al (2000) Brain atrophy in relapsing-remitting multiple sclerosis and secondary progressive multiple sclerosis: longitudinal quantitative analysis. Radiology 214(3):665–670

90. Miller DH, Barkhof F, Frank JA, Parker GJ, Thompson AJ (2002) Measurement of atrophy in multiple sclerosis: pathological basis, methodological aspects and clinical relevance. Brain 125(Pt 8):1676–1695

91. Sastre-Garriga J, Ingle GT, Chard DT, Cercignani M, Ramio-Torrenta L, Miller DH et al (2005) Grey and white matter volume changes in early primary progressive multiple sclerosis: a longitudinal study. Brain 128(Pt 6):1454–1460

92. Popescu V, Agosta F, Hulst HE, Sluimer IC, Knol DL, Sormani MP et al (2013) Brain atrophy and lesion load predict long term disability in multiple sclerosis. J Neurol Neurosurg Psychiatry 84(10):1082–1091

93. Chard DT, Brex PA, Ciccarelli O, Griffin CM, Parker GJ, Dalton C et al (2003) The longitudinal relation between brain lesion load and atrophy in multiple sclerosis: a 14 year follow up study. J Neurol Neurosurg Psychiatry 74(11):1551–1554

94. van Munster CE, Jonkman LE, Weinstein HC, Uitdehaag BM, Geurts JJ (2015) Gray matter damage in multiple sclerosis: impact on clinical symptoms. Neuroscience 303:446–461

95. Grassiot B, Desgranges B, Eustache F, Defer G (2009) Quantification and clinical relevance of brain atrophy in multiple sclerosis: a review. J Neurol 256(9):1397–1412

96. Ge Y, Grossman RI, Udupa JK, Babb JS, Nyul LG, Kolson DL (2001) Brain atrophy in relapsing-remitting multiple sclerosis: fractional volumetric analysis of gray matter and white matter. Radiology 220(3):606–610

97. Chard DT, Griffin CM, Parker GJ, Kapoor R, Thompson AJ, Miller DH (2002) Brain atrophy in clinically early relapsing-remitting multiple sclerosis. Brain 125(Pt 2):327–337

98. Hardmeier M, Wagenpfeil S, Freitag P, Fisher E, Rudick RA, Kooijmans-Coutinho M et al (2003) Atrophy is detectable within a 3-month period in untreated patients with active relapsing remitting multiple sclerosis. Arch Neurol 60(12):1736–1739

99. Jacobsen C, Hagemeier J, Myhr KM, Nyland H, Lode K, Bergsland N et al (2014) Brain atrophy and disability progression in multiple sclerosis patients: a 10-year follow-up study. J Neurol Neurosurg Psychiatry 85(10):1109–1115

100. Lin X, Blumhardt LD, Constantinescu CS (2003) The relationship of brain and cervical cord volume to disability in clinical subtypes of multiple sclerosis: a three-dimensional MRI study. Acta Neurol Scand 108(6):401–406

101. Fisher E, Lee JC, Nakamura K, Rudick RA (2008) Gray matter atrophy in multiple sclerosis: a longitudinal study. Ann Neurol 64(3):255–265

102. Popescu V, Battaglini M, Hoogstrate WS, Verfaillie SC, Sluimer IC, van Schijndel RA et al (2012) Optimizing parameter choice for FSL-brain extraction tool (BET) on 3D T1 images in multiple sclerosis. NeuroImage 61(4):1484–1494

103. Vrenken H, Jenkinson M, Horsfield MA, Battaglini M, van Schijndel RA, Rostrup E et al (2013) Recommendations to improve imaging and analysis of brain lesion load and atrophy in longitudinal studies of multiple sclerosis. J Neurol 260(10):2458–2471

104. Ceccarelli A, Jackson JS, Tauhid S, Arora A, Gorky J, Dell'Oglio E et al (2012) The impact of lesion in-painting and registration methods on voxel-based morphometry in detecting regional cerebral gray matter atrophy in multiple sclerosis. AJNR Am J Neuroradiol 33(8):1579–1585

105. Popescu V, Ran NC, Barkhof F, Chard DT, Wheeler-Kingshott CA, Vrenken H (2014) Accurate GM atrophy quantification in MS using lesion-filling with co-registered 2D lesion masks. Neuroimage Clin 4:366–373

106. Chard DT, Jackson JS, Miller DH, Wheeler-Kingshott CA (2010) Reducing the impact of white matter lesions on automated measures of brain gray and white matter volumes. J Magn Reson Imaging 32(1):223–228

107. Gabilondo I, Martinez-Lapiscina EH, Martinez-Heras E, Fraga-Pumar E, Llufriu S, Ortiz S et al (2014) Trans-synaptic axonal degeneration in the visual pathway in multiple sclerosis. Ann Neurol 75(1):98–107

108. Bendfeldt K, Blumhagen JO, Egger H, Loetscher P, Denier N, Kuster P et al (2010) Spatio-temporal distribution pattern of white matter lesion volumes and their association with regional grey matter volume reductions in relapsing-remitting multiple sclerosis. Hum Brain Mapp 31 (10):1542–1555

109. Steenwijk MD, Daams M, Pouwels PJ, Balk LJ, Tewarie PK, Killestein J et al (2014) What explains gray matter atrophy in long-standing multiple sclerosis? Radiology 272(3):832–842

110. Bodini B, Khaleeli Z, Cercignani M, Miller DH, Thompson AJ, Ciccarelli O (2009) Exploring the relationship between white matter and gray matter damage in early primary progressive multiple sclerosis: an in vivo study with TBSS and VBM. Hum Brain Mapp 30(9):2852–2861

111. Tillema JM, Hulst HE, Rocca MA, Vrenken H, Steenwijk MD, Damjanovic D et al (2016) Regional cortical thinning in multiple sclerosis and its relation with cognitive impairment: a multicenter study. Mult Scler 22(7):901–909

112. Achiron A, Chapman J, Tal S, Bercovich E, Gil H, Achiron A (2013) Superior temporal gyrus thickness correlates with cognitive performance in multiple sclerosis. Brain Struct Funct 218 (4):943–950
113. Charil A, Dagher A, Lerch JP, Zijdenbos AP, Worsley KJ, Evans AC (2007) Focal cortical atrophy in multiple sclerosis: relation to lesion load and disability. NeuroImage 34(2):509–517
114. Geisseler O, Pflugshaupt T, Bezzola L, Reuter K, Weller D, Schuknecht B et al (2016) Cortical thinning in the anterior cingulate cortex predicts multiple sclerosis patients' fluency performance in a lateralised manner. Neuroimage Clin 10:89–95
115. Nygaard GO, Walhovd KB, Sowa P, Chepkoech JL, Bjornerud A, Due-Tonnessen P et al (2015) Cortical thickness and surface area relate to specific symptoms in early relapsing-remitting multiple sclerosis. Mult Scler 21(4):402–414
116. Sailer M, Fischl B, Salat D, Tempelmann C, Schonfeld MA, Busa E et al (2003) Focal thinning of the cerebral cortex in multiple sclerosis. Brain 126(Pt 8):1734–1744
117. Steenwijk MD, Daams M, Pouwels PJ, JB L, Tewarie PK, Geurts JJ et al (2015) Unraveling the relationship between regional gray matter atrophy and pathology in connected white matter tracts in long-standing multiple sclerosis. Hum Brain Mapp 36(5):1796–1807
118. Bergsland N, Lagana MM, Tavazzi E, Caffini M, Tortorella P, Baglio F et al (2015) Corticospinal tract integrity is related to primary motor cortex thinning in relapsing-remitting multiple sclerosis. Mult Scler 21(14):1771–1780
119. Parisi L, Rocca MA, Mattioli F, Riccitelli GC, Capra R, Stampatori C et al (2014) Patterns of regional gray matter and white matter atrophy in cortical multiple sclerosis. J Neurol 261 (9):1715–1725
120. Steenwijk MD, Geurts JJ, Daams M, Tijms BM, Wink AM, Balk LJ et al (2016) Cortical atrophy patterns in multiple sclerosis are non-random and clinically relevant. Brain 139 (Pt 1):115–126
121. Prinster A, Quarantelli M, Orefice G, Lanzillo R, Brunetti A, Mollica C et al (2006) Grey matter loss in relapsing-remitting multiple sclerosis: a voxel-based morphometry study. NeuroImage 29(3):859–867
122. Audoin B, Zaaraoui W, Reuter F, Rico A, Malikova I, Confort-Gouny S et al (2010) Atrophy mainly affects the limbic system and the deep grey matter at the first stage of multiple sclerosis. J Neurol Neurosurg Psychiatry 81(6):690–695
123. Cappellani R, Bergsland N, Weinstock-Guttman B, Kennedy C, Carl E, Ramasamy DP et al (2014) Subcortical deep gray matter pathology in patients with multiple sclerosis is associated with white matter lesion burden and atrophy but not with cortical atrophy: a diffusion tensor MRI study. AJNR Am J Neuroradiol 35(5):912–919
124. Houtchens MK, Benedict RH, Killiany R, Sharma J, Jaisani Z, Singh B et al (2007) Thalamic atrophy and cognition in multiple sclerosis. Neurology 69(12):1213–1223
125. Benedict RH, Ramasamy D, Munschauer F, Weinstock-Guttman B, Zivadinov R (2009) Memory impairment in multiple sclerosis: correlation with deep grey matter and mesial temporal atrophy. J Neurol Neurosurg Psychiatry 80(2):201–206
126. Riccitelli G, Rocca MA, Pagani E, Rodegher ME, Rossi P, Falini A et al (2011) Cognitive impairment in multiple sclerosis is associated to different patterns of gray matter atrophy according to clinical phenotype. Hum Brain Mapp 32(10):1535–1543
127. Schoonheim MM, Popescu V, Rueda Lopes FC, Wiebenga OT, Vrenken H, Douw L et al (2012) Subcortical atrophy and cognition: sex effects in multiple sclerosis. Neurology 79 (17):1754–1761
128. Batista S, Zivadinov R, Hoogs M, Bergsland N, Heininen-Brown M, Dwyer MG et al (2012) Basal ganglia, thalamus and neocortical atrophy predicting slowed cognitive processing in multiple sclerosis. J Neurol 259(1):139–146
129. Benedict RH, Hulst HE, Bergsland N, Schoonheim MM, Dwyer MG, Weinstock-Guttman B et al (2013) Clinical significance of atrophy and white matter mean diffusivity within the thalamus of multiple sclerosis patients. Mult Scler 19(11):1478–1484

130. Schoonheim MM, Hulst HE, Brandt RB, Strik M, Wink AM, Uitdehaag BM et al (2015) Thalamus structure and function determine severity of cognitive impairment in multiple sclerosis. Neurology 84(8):776–783

131. Benedict RH, Zivadinov R (2011) Risk factors for and management of cognitive dysfunction in multiple sclerosis. Nat Rev Neurol 7(6):332–342

132. Minagar A, Barnett MH, Benedict RH, Pelletier D, Pirko I, Sahraian MA et al (2013) The thalamus and multiple sclerosis: modern views on pathologic, imaging, and clinical aspects. Neurology 80(2):210–219

133. Wylezinska M, Cifelli A, Jezzard P, Palace J, Alecci M, Matthews PM (2003) Thalamic neurodegeneration in relapsing-remitting multiple sclerosis. Neurology 60(12):1949–1954

134. Mesaros S, Rocca MA, Absinta M, Ghezzi A, Milani N, Moiola L et al (2008) Evidence of thalamic gray matter loss in pediatric multiple sclerosis. Neurology 70(13 Pt 2):1107–1112

135. Zivadinov R, Havrdova E, Bergsland N, Tyblova M, Hagemeier J, Seidl Z et al (2013) Thalamic atrophy is associated with development of clinically definite multiple sclerosis. Radiology 268(3):831–841

136. Shiee N, Bazin PL, Zackowski KM, Farrell SK, Harrison DM, Newsome SD et al (2012) Revisiting brain atrophy and its relationship to disability in multiple sclerosis. PLoS One 7(5): e37049

137. Filippi M, Rocca MA, Pagani E, De Stefano N, Jeffery D, Kappos L et al (2014) Placebo-controlled trial of oral laquinimod in multiple sclerosis: MRI evidence of an effect on brain tissue damage. J Neurol Neurosurg Psychiatry 85(8):851–858

138. Vrenken H, Geurts JJ (2007) Gray and normal-appearing white matter in multiple sclerosis: an MRI perspective. Expert Rev Neurother 7(3):271–279

139. Evangelou N, Esiri MM, Smith S, Palace J, Matthews PM (2000) Quantitative pathological evidence for axonal loss in normal appearing white matter in multiple sclerosis. Ann Neurol 47 (3):391–395

140. Werring DJ, Clark CA, Barker GJ, Thompson AJ, Miller DH (1999) Diffusion tensor imaging of lesions and normal-appearing white matter in multiple sclerosis. Neurology 52 (8):1626–1632

141. Oreja-Guevara C, Rovaris M, Iannucci G, Valsasina P, Caputo D, Cavarretta R et al (2005) Progressive gray matter damage in patients with relapsing-remitting multiple sclerosis: a longitudinal diffusion tensor magnetic resonance imaging study. Arch Neurol 62(4):578–584

142. Rovaris M, Judica E, Gallo A, Benedetti B, Sormani MP, Caputo D et al (2006) Grey matter damage predicts the evolution of primary progressive multiple sclerosis at 5 years. Brain 129 (Pt 10):2628–2634

143. Vrenken H, Pouwels PJ, Geurts JJ, Knol DL, Polman CH, Barkhof F et al (2006) Altered diffusion tensor in multiple sclerosis normal-appearing brain tissue: cortical diffusion changes seem related to clinical deterioration. J Magn Reson Imaging 23(5):628–636

144. van Walderveen MA, van Schijndel RA, Pouwels PJ, Polman CH, Barkhof F (2003) Multislice T1 relaxation time measurements in the brain using IR-EPI: reproducibility, normal values, and histogram analysis in patients with multiple sclerosis. J Magn Reson Imaging 18 (6):656–664

145. Vrenken H, Geurts JJ, Knol DL, van Dijk LN, Dattola V, Jasperse B et al (2006) Whole-brain T1 mapping in multiple sclerosis: global changes of normal-appearing gray and white matter. Radiology 240(3):811–820

146. Steenwijk MD, Vrenken H, Jonkman LE, Daams M, Geurts JJ, Barkhof F et al (2015) High-resolution T1-relaxation time mapping displays subtle, clinically relevant, gray matter damage in long-standing multiple sclerosis. Mult Scler 12

147. Matthews PM, Arnold DL (2001) Magnetic resonance imaging of multiple sclerosis: new insights linking pathology to clinical evolution. Curr Opin Neurol 14(3):279–287

148. De Stefano N, Narayanan S, Francis GS, Arnaoutelis R, Tartaglia MC, Antel JP et al (2001) Evidence of axonal damage in the early stages of multiple sclerosis and its relevance to disability. Arch Neurol 58(1):65–70

149. De Stefano N, Iannucci G, Sormani MP, Guidi L, Bartolozzi ML, Comi G et al (2002) MR correlates of cerebral atrophy in patients with multiple sclerosis. J Neurol 249(8):1072–1077
150. Inglese M, Li BS, Rusinek H, Babb JS, Grossman RI, Gonen O (2003) Diffusely elevated cerebral choline and creatine in relapsing-remitting multiple sclerosis. Magn Reson Med 50 (1):190–195
151. Rovaris M, Bozzali M, Santuccio G, Ghezzi A, Caputo D, Montanari E et al (2001) In vivo assessment of the brain and cervical cord pathology of patients with primary progressive multiple sclerosis. Brain 124(Pt 12):2540–2549
152. Traboulsee A, Dehmeshki J, Peters KR, Griffin CM, Brex PA, Silver N et al (2003) Disability in multiple sclerosis is related to normal appearing brain tissue MTR histogram abnormalities. Mult Scler 9(6):566–573
153. Deloire MS, Salort E, Bonnet M, Arimone Y, Boudineau M, Amieva H et al (2005) Cognitive impairment as marker of diffuse brain abnormalities in early relapsing remitting multiple sclerosis. J Neurol Neurosurg Psychiatry 76(4):519–526
154. Khaleeli Z, Sastre-Garriga J, Ciccarelli O, Miller DH, Thompson AJ (2007) Magnetisation transfer ratio in the normal appearing white matter predicts progression of disability over 1 year in early primary progressive multiple sclerosis. J Neurol Neurosurg Psychiatry 78 (10):1076–1082
155. Vrenken H, Pouwels PJ, Ropele S, Knol DL, Geurts JJ, Polman CH et al (2007) Magnetization transfer ratio measurement in multiple sclerosis normal-appearing brain tissue: limited differences with controls but relationships with clinical and MR measures of disease. Mult Scler 13 (6):708–716
156. Vrenken H, Geurts JJ, Knol DL, Polman CH, Castelijns JA, Pouwels PJ et al (2006) Normal-appearing white matter changes vary with distance to lesions in multiple sclerosis. AJNR Am J Neuroradiol 27(9):2005–2011
157. Enzinger C, Barkhof F, Ciccarelli O, Filippi M, Kappos L, Rocca MA et al (2015) Nonconventional MRI and microstructural cerebral changes in multiple sclerosis. Nat Rev Neurol 11(12):676–686
158. Kacar K, Rocca MA, Copetti M, Sala S, Mesaros S, Stosic Opincal T et al (2011) Overcoming the clinical-MR imaging paradox of multiple sclerosis: MR imaging data assessed with a random forest approach. AJNR Am J Neuroradiol 32(11):2098–2102
159. Fabiano AJ, Sharma J, Weinstock-Guttman B, Munschauer FE, 3rd, Benedict RH, Zivadinov R, et al. Thalamic involvement in multiple sclerosis: a diffusion-weighted magnetic resonance imaging study. J Neuroimaging 2003;13(4):307–314.
160. Deppe M, Kramer J, Tenberge JG, Marinell J, Schwindt W, Deppe K et al (2016) Early silent microstructural degeneration and atrophy of the thalamocortical network in multiple sclerosis. Hum Brain Mapp 37(5):1866–1879
161. Mesaros S, Rocca MA, Pagani E, Sormani MP, Petrolini M, Comi G et al (2011) Thalamic damage predicts the evolution of primary-progressive multiple sclerosis at 5 years. AJNR Am J Neuroradiol 32(6):1016–1020
162. Hannoun S, Durand-Dubief F, Confavreux C, Ibarrola D, Streichenberger N, Cotton F et al (2012) Diffusion tensor-MRI evidence for extra-axonal neuronal degeneration in caudate and thalamic nuclei of patients with multiple sclerosis. AJNR Am J Neuroradiol 33(7):1363–1368
163. Schoonheim MM, Vigeveno RM, Rueda Lopes FC, Pouwels PJ, Polman CH, Barkhof F et al (2014) Sex-specific extent and severity of white matter damage in multiple sclerosis: implications for cognitive decline. Hum Brain Mapp 35(5):2348–2358
164. Hulst HE, Steenwijk MD, Versteeg A, Pouwels PJ, Vrenken H, Uitdehaag BM et al (2013) Cognitive impairment in MS: impact of white matter integrity, gray matter volume, and lesions. Neurology 80(11):1025–1032
165. Filippi M, Rocca MA, De Stefano N, Enzinger C, Fisher E, Horsfield MA et al (2011) Magnetic resonance techniques in multiple sclerosis: the present and the future. Arch Neurol 68(12):1514–1520

166. Roosendaal SD, Geurts JJ, Vrenken H, Hulst HE, Cover KS, Castelijns JA et al (2009) Regional DTI differences in multiple sclerosis patients. NeuroImage 44(4):1397–1403
167. Gong G, He Y, Concha L, Lebel C, Gross DW, Evans AC et al (2009) Mapping anatomical connectivity patterns of human cerebral cortex using in vivo diffusion tensor imaging tractography. Cereb Cortex 19(3):524–536
168. van den Heuvel MP, Sporns O (2011) Rich-club organization of the human connectome. J Neurosci 31(44):15775–15786
169. Ciccarelli O, Catani M, Johansen-Berg H, Clark C, Thompson A (2008) Diffusion-based tractography in neurological disorders: concepts, applications, and future developments. Lancet Neurol 7(8):715–727
170. Alexander-Bloch A, Giedd JN, Bullmore E (2013) Imaging structural co-variance between human brain regions. Nat Rev Neurosci 14(5):322–336
171. Bright MG, Murphy K (2015) Is fMRI "noise" really noise? Resting state nuisance regressors remove variance with network structure. NeuroImage 114:158–169
172. Bright MG, Murphy K (2013) Removing motion and physiological artifacts from intrinsic BOLD fluctuations using short echo data. NeuroImage 64:526–537
173. Nair A, Frederick TJ, Miller SD (2008) Astrocytes in multiple sclerosis: a product of their environment. Cell Mol Life Sci 65(17):2702–2720
174. Brosnan CF, Raine CS (2013) The astrocyte in multiple sclerosis revisited. Glia 61 (4):453–465
175. Correale J, Farez MF (2015) The role of astrocytes in multiple sclerosis progression. Front Neurol 6
176. Rossi DJ (2006) Another BOLD role for astrocytes: coupling blood flow to neural activity. Nat Neurosci 9(2):159–161
177. Haydon PG, Carmignoto G (2006) Astrocyte control of synaptic transmission and neurovascular coupling. Physiol Rev 86(3):1009–1031
178. Petzold GC, Murthy VN (2011) Role of astrocytes in neurovascular coupling. Neuron 71 (5):782–797
179. Bennett M, Farnell L, Gibson W (2008) Origins of the BOLD changes due to synaptic activity at astrocytes abutting arteriolar smooth muscle. J Theor Biol 252(1):123–130
180. Aquino KM, Robinson PA, Schira MM, Breakspear M (2014) Deconvolution of neural dynamics from fMRI data using a spatiotemporal hemodynamic response function. NeuroImage 94:203–215
181. Drysdale P, Huber J, Robinson P, Aquino K (2010) Spatiotemporal BOLD dynamics from a poroelastic hemodynamic model. J Theor Biol 265(4):524–534
182. Staffen W, Mair A, Zauner H, Unterrainer J, Niederhofer H, Kutzelnigg A et al (2002) Cognitive function and fMRI in patients with multiple sclerosis: evidence for compensatory cortical activation during an attention task. Brain 125(6):1275–1282
183. Audoin B, Ibarrola D, Ranjeva JP, Confort-Gouny S, Malikova I, Ali-Chérif A et al (2003) Compensatory cortical activation observed by fMRI during a cognitive task at the earliest stage of multiple sclerosis. Hum Brain Mapp 20(2):51–58
184. Mainero C, Caramia F, Pozzilli C, Pisani A, Pestalozza I, Borriello G et al (2004) fMRI evidence of brain reorganization during attention and memory tasks in multiple sclerosis. NeuroImage 21(3):858–867
185. Rocca MA, Valsasina P, Hulst HE, Abdel-Aziz K, Enzinger C, Gallo A et al (2014) Functional correlates of cognitive dysfunction in multiple sclerosis: a multicenter fMRI study. Hum Brain Mapp 35(12):5799–5814
186. Hulst HE, Schoonheim MM, Roosendaal SD, Popescu V, Schweren LJ, van der Werf YD et al (2012) Functional adaptive changes within the hippocampal memory system of patients with multiple sclerosis. Hum Brain Mapp 33(10):2268–2280
187. Sumowski JF, Wylie GR, DeLuca J, Chiaravalloti N (2010) Intellectual enrichment is linked to cerebral efficiency in multiple sclerosis: functional magnetic resonance imaging evidence for cognitive reserve. Brain 133(2):362–374

188. Schoonheim MM, Meijer KA, Geurts JJ (2015) Network collapse and cognitive impairment in multiple sclerosis. Front Neurol 6
189. Schoonheim MM, Geurts JJ, Barkhof F (2010) The limits of functional reorganization in multiple sclerosis. Neurology 74(16):1246–1247
190. Roosendaal SD, Schoonheim MM, Hulst HE, Sanz-Arigita EJ, Smith SM, Geurts JJ et al (2010) Resting state networks change in clinically isolated syndrome. Brain 133 (6):1612–1621
191. Rocca MA, Valsasina P, Absinta M, Riccitelli G, Rodegher ME, Misci P et al (2010) Default-mode network dysfunction and cognitive impairment in progressive MS. Neurology 74 (16):1252–1259
192. Cader S, Cifelli A, Abu-Omar Y, Palace J, Matthews PM (2006) Reduced brain functional reserve and altered functional connectivity in patients with multiple sclerosis. Brain 129 (2):527–537
193. Bonnet M, Allard M, Dilharreguy B, Deloire M, Petry K, Brochet B (2010) Cognitive compensation failure in multiple sclerosis. Neurology 75(14):1241–1248
194. Audoin B, Boulanouar K, Ibarrola D, Malikova I, Confort-Gouny S, Celsis P et al (2005) Altered functional connectivity related to white matter changes inside the working memory network at the very early stage of MS. J Cereb Blood Flow Metab 25(10):1245–1253
195. Ranjeva J-P, Audoin B, Duong MVA, Confort-Gouny S, Malikova I, Viout P et al (2006) Structural and functional surrogates of cognitive impairment at the very early stage of multiple sclerosis. J Neurol Sci 245(1):161–167
196. Damoiseaux JS, Rombouts SA, Barkhof F, Scheltens P, Stam CJ, Smith SM et al (2006) Consistent resting-state networks across healthy subjects. Proc Natl Acad Sci U S A 103 (37):13848–13853
197. Fornito A, Zalesky A, Breakspear M (2015) The connectomics of brain disorders. Nat Rev Neurosci 16(3):159–172
198. Rocca MA, Valsasina P, Martinelli V, Misci P, Falini A, Comi G et al (2012) Large-scale neuronal network dysfunction in relapsing-remitting multiple sclerosis. Neurology 79 (14):1449–1457
199. Cruz-Gómez ÁJ, Ventura-Campos N, Belenguer A, Ávila C, Forn C (2014) The link between resting-state functional connectivity and cognition in MS patients. Mult Scler J 20(3):338–348
200. Louapre C, Perlbarg V, García-Lorenzo D, Urbanski M, Benali H, Assouad R et al (2014) Brain networks disconnection in early multiple sclerosis cognitive deficits: an anatomofunctional study. Hum Brain Mapp 35(9):4706–4717
201. Bonavita S, Gallo A, Sacco R, Della Corte M, Bisecco A, Docimo R et al (2011) Distributed changes in default-mode resting-state connectivity in multiple sclerosis. Mult Scler J 17 (4):411–422
202. Leavitt VM, Paxton J, Sumowski JF (2014) Default network connectivity is linked to memory status in multiple sclerosis. J Int Neuropsychol Soc 20(09):937–944
203. Zhou F, Zhuang Y, Gong H, Wang B, Wang X, Chen Q et al (2014) Altered inter-subregion connectivity of the default mode network in relapsing remitting multiple sclerosis: a functional and structural connectivity study. PLoS One 9(7):e101198
204. de Pasquale F, Della Penna S, Snyder AZ, Lewis C, Mantini D, Marzetti L et al (2010) Temporal dynamics of spontaneous MEG activity in brain networks. Proc Natl Acad Sci 107 (13):6040–6045
205. de Pasquale F, Della Penna S, Snyder AZ, Marzetti L, Pizzella V, Romani GL et al (2012) A cortical core for dynamic integration of functional networks in the resting human brain. Neuron 74(4):753–764
206. Wojtowicz M, Mazerolle EL, Bhan V, Fisk JD (2014) Altered functional connectivity and performance variability in relapsing–remitting multiple sclerosis. Mult Scler J 1352458514524997

207. Faivre A, Rico A, Zaaraoui W, Crespy L, Reuter F, Wybrecht D et al (2012) Assessing brain connectivity at rest is clinically relevant in early multiple sclerosis. Mult Scler J 18 (9):1251–1258

208. Rocca MA, Absinta M, Amato MP, Moiola L, Ghezzi A, Veggiotti P et al (2014) Posterior brain damage and cognitive impairment in pediatric multiple sclerosis. Neurology 82 (15):1314–1321

209. Swaab D (1991) Brain aging and Alzheimer's disease, "wear and tear" versus "use it or lose it". Neurobiol Aging 12(4):317–324

210. de Haan W, Mott K, van Straaten EC, Scheltens P, Stam CJ (2012) Activity dependent degeneration explains hub vulnerability in Alzheimer's disease. PLoS Comput Biol 8(8): e1002582

211. Hämäläinen M, Hari R, Ilmoniemi RJ, Knuutila J, Lounasmaa OV (1993) Magnetoencephalography—theory, instrumentation, and applications to noninvasive studies of the working human brain. Rev Mod Phys 65(2):413

212. Supek S, Aine CJ (2014) Magnetoencephalography: from signals to dynamic cortical networks. Springer, Berlin/Heidelberg

213. Murakami S, Okada Y (2006) Contributions of principal neocortical neurons to magnetoencephalography and electroencephalography signals. J Physiol 575(3):925–936

214. da Silva FL (2013) EEG and MEG: relevance to neuroscience. Neuron 80(5):1112–1128

215. Hillebrand A, Barnes GR (2005) Beamformer analysis of MEG data. Int Rev Neurobiol 68:149–171

216. Hillebrand A, Singh KD, Holliday IE, Furlong PL, Barnes GR (2005) A new approach to neuroimaging with magnetoencephalography. Hum Brain Mapp 25(2):199–211

217. O'Neill GC, Barratt EL, Hunt BA, Tewarie PK, Brookes MJ (2015) Measuring electrophysiological connectivity by power envelope correlation: a technical review on MEG methods. Phys Med Biol 60(21):R271

218. Schoffelen JM, Gross J (2009) Source connectivity analysis with MEG and EEG. Hum Brain Mapp 30(6):1857–1865

219. Hillebrand A, Barnes GR, Bosboom JL, Berendse HW, Stam CJ (2012) Frequency-dependent functional connectivity within resting-state networks: an atlas-based MEG beamformer solution. NeuroImage 59(4):3909–3921

220. Baillet S, Mosher JC, Leahy RM (2001) Electromagnetic brain mapping. Signal Processing Magazine, IEEE 18(6):14–30

221. Van Veen BD, Buckley KM (1988) Beamforming: a versatile approach to spatial filtering. IEEE ASSP Mag 5(2):4–24

222. Robinson S, Vrba J (1999) Functional neuroimaging by synthetic aperture magnetometry (SAM). In: Recent advances in biomagnetism. Tohoku University Press, Sendai, pp 302–305

223. Pereda E, Quiroga RQ, Bhattacharya J (2005) Nonlinear multivariate analysis of neurophysiological signals. Prog Neurobiol 77(1):1–37

224. Kida T, Tanaka E, Kakigi R (2015) Multi-dimensional dynamics of human electromagnetic brain activity. Front Hum Neurosci 9:713

225. Licker MD, Geller E (2002) Dictionary of physics, XI ed. McGraw-Hill, Clarinda

226. Stam CJ, Nolte G, Daffertshofer A (2007) Phase lag index: assessment of functional connectivity from multi channel EEG and MEG with diminished bias from common sources. Hum Brain Mapp 28(11):1178–1193

227. Brookes MJ, Woolrich MW, Barnes GR (2012) Measuring functional connectivity in MEG: a multivariate approach insensitive to linear source leakage. NeuroImage 63(2):910–920

228. Stam C, Van Dijk B (2002) Synchronization likelihood: an unbiased measure of generalized synchronization in multivariate data sets. Physica D 163(3):236–251

229. Karhu J, Hari R, Mäkelä JP, Huttunen J, Knuutila J (1992) Cortical somatosensory magnetic responses in multiple sclerosis. Electroencephalogr Clin Neurophysiol 83(3):192–200

230. Kassubek J, Sörös P, Kober H, Stippich C, Vieth JB (1999) Focal slow and beta brain activity in patients with multiple sclerosis revealed by magnetoencephalography. Brain Topogr 11 (3):193–200

231. Leocani L, Locatelli T, Martinelli V, Rovaris M, Falautano M, Filippi M et al (2000) Electroencephalographic coherence analysis in multiple sclerosis: correlation with clinical, neuropsychological, and MRI findings. J Neurol Neurosurg Psychiatry 69(2):192–198

232. Cover KS, Vrenken H, Geurts JJ, van Oosten BW, Jelles B, Polman CH et al (2006) Multiple sclerosis patients show a highly significant decrease in alpha band interhemispheric synchronization measured using MEG. NeuroImage 29(3):783–788

233. Van der Meer M, Tewarie P, Schoonheim M, Douw L, Barkhof F, Polman C et al (2013) Cognition in MS correlates with resting-state oscillatory brain activity: an explorative MEG source-space study. Neuroimage Clin 2:727–734

234. Valdés-Hernández PA, Ojeda-González A, Martínez-Montes E, Lage-Castellanos A, Virués-Alba T, Valdés-Urrutia L et al (2010) White matter architecture rather than cortical surface area correlates with the EEG alpha rhythm. NeuroImage 49(3):2328–2339

235. Hindriks R, Woolrich M, Luckhoo H, Joensson M, Mohseni H, Kringelbach M et al (2015) Role of white-matter pathways in coordinating alpha oscillations in resting visual cortex. NeuroImage 106:328–339

236. Tecchio F, Zito G, Zappasodi F, Dell'Acqua ML, Landi D, Nardo D et al (2008) Intra-cortical connectivity in multiple sclerosis: a neurophysiological approach. Brain 131(7):1783–1792

237. Schoonheim MM, Geurts JJ, Landi D, Douw L, van der Meer ML, Vrenken H et al (2013) Functional connectivity changes in multiple sclerosis patients: a graph analytical study of MEG resting state data. Hum Brain Mapp 34(1):52–61

238. Tewarie P, Schoonheim MM, Stam CJ, van der Meer ML, van Dijk BW, Barkhof F et al (2013) Cognitive and clinical dysfunction, altered MEG resting-state networks and thalamic atrophy in multiple sclerosis. PLoS One 8(7):e69318

239. Tewarie P, Schoonheim MM, Schouten DI, Polman CH, Balk LJ, Uitdehaag BM et al (2015) Functional brain networks: linking thalamic atrophy to clinical disability in multiple sclerosis, a multimodal fMRI and MEG study. Hum Brain Mapp 36(2):603–618

240. Tewarie P, Steenwijk MD, Tijms BM, Daams M, Balk LJ, Stam CJ et al (2014) Disruption of structural and functional networks in long-standing multiple sclerosis. Hum Brain Mapp 35 (12):5946–5961

241. Stam CJ (2014) Modern network science of neurological disorders. Nat Rev Neurosci 15 (10):683–695

242. Cv S, Van Straaten E (2012) The organization of physiological brain networks. Clin Neurophysiol 123(6):1067–1087

243. Van Wijk BC, Stam CJ, Daffertshofer A (2010) Comparing brain networks of different size and connectivity density using graph theory. PLoS One 5(10):e13701

244. Rubinov M, Sporns O (2010) Complex network measures of brain connectivity: uses and interpretations. NeuroImage 52(3):1059–1069

245. Bullmore E, Sporns O (2012) The economy of brain network organization. Nat Rev Neurosci 13(5):336–349

246. Watts DJ, Strogatz SH (1998) Collective dynamics of 'small-world' networks. Nature 393 (6684):440–442

247. Barabási A-L, Albert R (1999) Emergence of scaling in random networks. Science 286 (5439):509–512

248. Tijms BM, Wink AM, de Haan W, van der Flier WM, Stam CJ, Scheltens P et al (2013) Alzheimer's disease: connecting findings from graph theoretical studies of brain networks. Neurobiol Aging 34(8):2023–2036

249. Diessen E, Diederen SJ, Braun KP, Jansen FE, Stam CJ (2013) Functional and structural brain networks in epilepsy: what have we learned? Epilepsia 54(11):1855–1865

250. Van Diessen E, Numan T, van Dellen E, van der Kooi A, Boersma M, Hofman D et al (2015) Opportunities and methodological challenges in EEG and MEG resting state functional brain network research. Clin Neurophysiol 126(8):1468–1481

251. Tewarie P, Van Dellen E, Hillebrand A, Stam C (2015) The minimum spanning tree: an unbiased method for brain network analysis. NeuroImage 104:177–188

252. Van Mieghem P, Magdalena SM. Phase transition in the link weight structure of networks. Phys Rev E 2005;72(5):056138

253. Meier J, Tewarie P, Van Mieghem P (2015) The Union of shortest path trees of functional brain networks. Brain Connect

254. Schoonheim MM, Hulst HE, Landi D, Ciccarelli O, Roosendaal SD, Sanz-Arigita EJ et al (2012) Gender-related differences in functional connectivity in multiple sclerosis. Mult Scler J 18(2):164–173

255. Rocca MA, Valsasina P, Meani A, Falini A, Comi G, Filippi M (2016) Impaired functional integration in multiple sclerosis: a graph theory study. Brain Struct Funct 221(1):115–131

256. Gamboa OL, Tagliazucchi E, von Wegner F, Jurcoane A, Wahl M, Laufs H et al (2014) Working memory performance of early MS patients correlates inversely with modularity increases in resting state functional connectivity networks. NeuroImage 94:385–395

257. Hardmeier M, Schoonheim MM, Geurts JJ, Hillebrand A, Polman CH, Barkhof F et al (2012) Cognitive dysfunction in early multiple sclerosis: altered centrality derived from resting-state functional connectivity using magneto-encephalography. PLoS One 7(7):e42087

258. Schoonheim M, Geurts J, Wiebenga O, De Munck J, Polman C, Stam C et al (2014) Changes in functional network centrality underlie cognitive dysfunction and physical disability in multiple sclerosis. Mult Scler J 20(8):1058–1065

259. Tewarie P, Bright M, Hillebrand A, Robson S, Gascoyne L, Morris P et al (2016) Predicting haemodynamic networks using electrophysiology: the role of non-linear and cross-frequency interactions. NeuroImage 130:273–292

260. Tewarie P, Hillebrand A, Schoonheim M, Van Dijk B, Geurts J, Barkhof F et al (2014) Functional brain network analysis using minimum spanning trees in multiple sclerosis: an MEG source-space study. NeuroImage 88:308–318

261. Sbardella E, Tona F, Petsas N, Upadhyay N, Piattella M, Filippini N et al (2015) Functional connectivity changes and their relationship with clinical disability and white matter integrity in patients with relapsing–remitting multiple sclerosis. Mult Scler J 21(13):1681–1692

262. Meier J, Tewarie P, Hillebrand A, Douw L, van Dijk B, Stufflebeam S et al (2016) A mapping between structural and functional brain networks. Brain Connect 6(4):298–311

263. Robinson P (2012) Interrelating anatomical, effective, and functional brain connectivity using propagators and neural field theory. Phys Rev E 85(1):011912

264. Deco G, Tononi G, Boly M, Kringelbach ML (2015) Rethinking segregation and integration: contributions of whole-brain modelling. Nat Rev Neurosci 16(7):430–439

265. Hindriks R, Adhikari M, Murayama Y, Ganzetti M, Mantini D, Logothetis N et al Can sliding-window correlations reveal dynamic functional connectivity in resting-state fMRI? NeuroImage 2015

266. Braun U, Schäfer A, Walter H, Erk S, Romanczuk-Seiferth N, Haddad L et al (2015) Dynamic reconfiguration of frontal brain networks during executive cognition in humans. Proc Natl Acad Sci 112(37):11678–11683

267. Baker AP, Brookes MJ, Rezek IA, Smith SM, Behrens T, Smith PJP et al (2014) Fast transient networks in spontaneous human brain activity. elife e01867:3

268. Brookes MJ, O'Neill GC, Hall EL, Woolrich MW, Baker A, Corner SP et al (2014) Measuring temporal, spectral and spatial changes in electrophysiological brain network connectivity. NeuroImage 91:282–299

269. O'Neill GC, Bauer M, Woolrich MW, Morris PG, Barnes GR, Brookes MJ (2015) Dynamic recruitment of resting state sub-networks. NeuroImage 115:85–95

270. Roebroeck A, Formisano E, Goebel R (2005) Mapping directed influence over the brain using granger causality and fMRI. NeuroImage 25(1):230–242

271. Hillebrand A, Tewarie P, van Dellen E, Yu M, Carbo EWS, Douw L et al (2016) Direction of information flow in large-scale resting-state networks is frequency dependent. Proc Natl Acad Sci
272. Lobier M, Siebenhühner F, Palva S, Palva JM (2014) Phase transfer entropy: a novel phase-based measure for directed connectivity in networks coupled by oscillatory interactions. NeuroImage 85:853–872
273. Brookes MJ, Tewarie PK, Hunt BA, Robson SE, Gascoyne LE, Liddle EB et al (2016) A multi-layer network approach to MEG Connectivity Analysis. NeuroImage 132:425–438

Chapter 9
Neuroimaging in Ataxias

C. C. Piccinin and A. D'Abreu

Introduction

The ability to visually analyze plain structural MRI sequences was a remarkable advance in understanding neurological diseases. The presence of signal abnormalities in determined sequences, structural anomalies, and measures obtained using MRI-based analyses generated clues about disease physiopathology, progression, and differential diagnosis. The parameters obtained by conventional imaging, however, have low sensitivity in detecting subtle microstructural alterations with current magnet strength limitations, and there is considerable overlap between imaging findings and different syndromes [1].

New MRI sequences, coupled with several tools for imaging processing and analysis, enhanced morphological analyses by creating parameters able to measure volume, area, and thickness of tissues, as well as the degree of molecular diffusion through them. More recently, some techniques have overcome the barrier of structural analysis, enabling the qualification and quantification of neurochemical compounds in the brain and the assessment of functional connectivity (FC) between distinct brain regions.

Voxel-based morphometry (VBM) quantifies gray matter (GM), white matter (WM), and cerebrospinal fluid [2]. Diffusion tensor imaging (DTI) can be used to estimate the integrity of connections [3]. FreeSurfer is a surface-based analysis able

C. C. Piccinin (✉)
Neuroimaging Laboratory, School of Medical Sciences, University of Campinas, Campinas, SP, Brazil

A. D'Abreu
Neuroimaging Laboratory, School of Medical Sciences, State University of Campinas, Campinas, Brazil

Neurology Department, School of Medical Sciences, State University of Campinas, Campinas, Brazil

© Springer International Publishing AG, part of Springer Nature 2018
C. Habas (ed.), *The Neuroimaging of Brain Diseases*, Contemporary Clinical Neuroscience, https://doi.org/10.1007/978-3-319-78926-2_9

to provide measures of cortical thickness and can be used in preprocessing of functional MRI (fMRI) and tractography [4]. fMRI estimates activation maps, determines functional connectivity (FC) between noncontiguous cerebral areas, associates physiological functions to specific brain regions, and detects intra- and intersubject variability and abnormal activation or deactivation in pathological processes [5]. Spectroscopy allows the in vivo qualification and quantification of brain metabolites, providing clues about neuronal and membrane integrity, the presence of ischemic and gliotic processes, as well as energy changes [6].

These advanced neuroimaging techniques have demonstrated greater ability in detecting earlier and more subtle irregularities [1, 7–10]. Compared with neuropathological studies, neuroimaging methods usually have larger samples, are performed in vivo so they are free from histological artifacts secondary to *postmortem* changes, allow for whole-brain analyses and prospective studies, are less expensive, and have much greater feasibility [11].

Ataxic disorders arise from many different etiologies but often have overlapping clinical features, and, conversely, single gene disorders often display different phenotypes within the same family [6]. Hence, clinical manifestations are usually not enough to determine the diagnosis.

Neuroimaging has allowed the detection of structural, neurochemical, and functional alterations associated with determined types of ataxia. Questions regarding sensitivity, specificity, validity, and reproducibility still need to be answered with larger, longitudinal studies. Such methods may potentially provide biomarkers predicting prognosis. Moreover, finding surrogate markers for clinical trials will greatly enhance efficiency.

Imaging Techniques

VBM

VBM compares groups by measuring the volume or density of a tissue voxel by voxel, in whole-brain or region of interest (ROI) analyses. It is an automated method with low operator dependence. The most studied diseases are those with the highest prevalence such as Friedreich's ataxia and the spinocerebellar ataxias (SCA) mostly types 1 (SCA1), 3 (SCA3), and 6 (SCA6).

Cerebellar atrophy is universally found in SCA1; however, the atrophic voxels described may vary in location [12, 13]. VBM demonstrated extensive GM decrease in the whole vermis and all cerebellar lobules except I, II, and X in a 10-SCA1-subjects sample compared with age-matched healthy controls (HC) [13]. Ataxia severity strongly correlated with the degree of GM loss in the cerebellar hemispheres [12]. Throughout the brain, GM reduction was found in small clusters, mostly sensorimotor areas [13, 14]. Decreased WM was found mostly in the cerebellar peduncles sparing the cerebral hemispheres [13, 14]. In a longitudinal study involving 37 subjects with SCA1, there was marked GM decrease in the mesencephalon as well as in the caudate and putamen bilaterally in the end of the 2-year follow-up. The rate of volume loss was higher in SCA1 compared with SCA3 and SCA6 in the brainstem, left cerebellar hemisphere, and putamen. The changes found by VBM were more sensitive than clinical scales in detecting progressive genotype-related neurodegeneration, and the authors suggested that these measurements could be used as surrogate markers [15].

For SCA2, there was GM volume loss in the whole cerebellum (sparing lobules I, II, and X), the right thalamus, the left inferior frontal operculum, the inferior parietal, the pre- and postcentral gyri [13], pons, and mesencephalon [16]. WM involvement was similar to SCA1; however, the pontine and vermal loss were more pronounced than in other SCAs [12], and cerebral areas such as the left precentral gyrus and the right middle cingulate gyrus were affected [13].

SCA3, or Machado-Joseph Disease (MJD), is the most prevalent autosomal dominant ataxia in the world [17]. All VBM studies in SCA3 detected bilateral GM atrophy in the cerebellum [12, 15, 18–21] and all but one in the brainstem [12, 15, 18–20]. The SUIT tool, a specific tool for cerebellar evaluation, reported GM loss in the posterior lobe, vermis, tonsil, inferior semilunar lobule, declive, uvula, fastigium, and tuber. The GM atrophy is widespread in the brain, involving the frontal, parietal, temporal, and occipital lobes. Areas such as the thalamus, cingulum, putamen, and caudate are more controversial, since atrophy is not consistently observed [18, 20, 21]. SCA3 showed a more pronounced vermal atrophy than SCA1 and SCA2 [12]. When compared to SCA1 and SCA6, there was a greater GM loss in the lentiform nucleus after a 2-year follow-up [15]. CAG repeat length, age, and disease duration were predictors of the GM changes. WM reduction seems to be restricted to the cerebellum [18] or symmetric and closely related to the GM changes affecting the brainstem and thalamus [20].

SCA6 is usually considered a "pure cerebellar" ataxia. Earlier studies had found GM loss in the cerebellar hemispheres and vermis with no WM changes [22]. Later studies, however, demonstrated a reduction in the brainstem volume ($P < 0.027$) less severe than in SCA3 ($P < 0.001$) [23] and volume loss in the thalamus, putamen, and pallidum; those three regions, along with the cerebellum, showed the greatest degree of deterioration after a 2-year follow-up [15].

In SCA7, ataxia is associated with a pyramidal syndrome and progressive macular degeneration that culminates in blindness. The pontine volume loss was significantly greater in SCA7 than in SCA3 and SCA6, whereas the cerebellar reduction was not different between the two latter conditions. Unlike cerebellar atrophy, pontine atrophy is detectable in early stages [24]. Bilateral GM reductions were found in the pre- and the postcentral gyri, inferior and medial frontal, parahippocampal, parietal inferior, and occipital cortices. The widespread results are compatible with the variety of clinical symptoms and previous neuropathological reports [25].

SCA17 is characterized by cerebellar, pyramidal signs, movement disorders, and neuropsychiatric symptoms. SCA17 presents with atrophy of the cerebellum, putamen, thalamus, and occipitoparietal regions [26–28]. CAG repeats negatively correlated with cerebellar GM volume [27]. In an 18-month follow-up, SCA17 patients demonstrated a greater reduction in the cerebellum, limbic system, and the precuneus than HC [28]. Negative correlations with motor clinical scores were found in the GM of the cerebellum and other motor areas, particularly the basal ganglia, while positive correlations with psychiatric scores were found in the GM of the frontotemporal regions, cuneus, and cingulum [26].

Friedreich ataxia (FRDA) is the most prevalent recessive ataxia in the world, and it is characterized by the involvement of spinal cord and dorsal root ganglia. Cerebellar atrophy was detectable in basically all VBM studies [11, 29, 30], mostly in the inferomedial portions of the cerebellar hemispheres as well as in the dorsal medulla [29, 30]. Loss of WM was present in the peridentate region [11], posterior cingulate, paracentral lobe, and middle frontal gyrus, supporting a more widespread pathology than previously hypothesized [30]. GM increase in the pulvinar correlated with improvement in ataxia scores after 1-year follow-up of FRDA patients who were started on recombinant human erythropoietin. These results support the use of VBM as an indicator of microstructural changes in both progression and treatment response in FRDA [31].

Fragile X-associated tremor-ataxia syndrome (FXTAS) is an adult-onset neuro-degenerative disorder usually affecting older male premutation carriers. To the best of our knowledge, only one study evaluated this rare condition using VBM, showing GM reduction in the cerebellum, mostly the anterior and superior posterior lobes including the correspondent vermal areas, the prefrontal cortex, anterior cingulate, precuneus, orbitofrontal cortex, amygdale, and insula. Asymptomatic mutation carriers showed reduction of GM in the anterior cerebellar areas in a ROI analysis when compared to HC, supporting the idea that microstructural abnormalities may precede the clinical manifestation of the disease [32].

DTI

DTI is a powerful tool able to estimate the molecular mobility and fiber integrity in tissues by measuring and combining variables as anisotropy and magnitude of diffusion [3]. DTI studies try to characterize specific WM alterations in certain types of ataxia, differentiate between ataxia syndromes, investigate progression of WM deterioration in longitudinal designs, and clarify possible common pathological mechanisms. Since VBM is an accurate tool for GM analysis but is limited to voxel-wise comparisons of WM, and DTI is conversely a great tool for tracing WM fiber tracts and detecting pathways' disruptions, some authors have therefore combined VBM and DTI analysis in order to increase the detail and power of the study [11, 19, 25].

DTI Using FMRIB's Diffusion Toolbox from FSL

Mean diffusivity (MD), also called apparent diffusion coefficient (ADC), corresponds to the rate of molecular diffusion and is sensitive to cellularity, edema, and necrotic processes. It was increased in the middle cerebellar peduncles (MCPs), medulla, peridentate WM in SCA1 [33, 34] and in the superior cerebellar peduncle (SCP), MCPs, transverse pontine fibers, medulla, and corticospinal tract in SCA2. There were no supratentorial abnormalities as well as no statistically significant difference between SCA1 and SCA2 in the whole-brain analysis. The measures were significantly higher in SCA2 than SCA1 [34]. A correlation between MD values and clinical severity was found in SCA1. Since the clinical deterioration is faster in SCA1 and DTI suggests a faster rate of neurodegeneration compared to SCA2, diffusion techniques may possibly be used in longitudinal studies and as surrogate or additional markers for clinical trials [33].

Fractional anisotropy (FA) reflects the range of directions of the molecular displacement by diffusion and was lower in all regions of interest for both groups, but in the transverse pontine fibers and in the cerebral peduncle at the level of the corticospinal tract, the FA decrease was more evident in SCA2 [34]. Increased MD in the brainstem and cerebellum was more pronounced in SCA2 and associated with a mild MD increase in cerebral hemispheres [13]. In contrast to the findings of Guerrini et al., SCA1 and SCA2 differ in diffusion properties, although the DTI analysis solely is limited to differentiate these two groups and controls [34].

Tract-Based Spatial Statistics (TBSS)

TBSS improved the sensitivity and the interpretation of findings in multi-subject diffusion analyses. Benefits of TBSS compared with other voxel-wise methods rely on better alignment of the subjects by matching with a common space and on the exemption from the arbitrariness in choosing the extent of spatial smoothing process [35]. DTI-TBSS analysis provides accurate multi-subject measures of diffusion properties in a whole-brain framework.

In SCA1 and SCA2, for example, the ICP (inferior cerebellar peduncle), MCP and SCP, pontine transverse fibers, medial and lateral lemnisci, spinothalamic and corticospinal tracts, and the corpus callosum showed decreased FA. These findings were more pronounced in SCA2, which also showed lower FA in the short intracerebellar tracts [13], bilateral posterior limb of the internal capsule, and superior corona radiate [36]. Increased MD was more extensive than reported by non-TBSS MD measures and was found in cerebellar WM, MCP, medial lemniscus, anterior corona radiate, posterior limb of internal capsule, and right corticospinal tract at the pons [36] (Fig. 9.1).

In SCA3, the cerebellum and brainstem showed lower FA values, while the thalamus, frontal, and occipital lobes showed increased radial diffusivity (RD) –

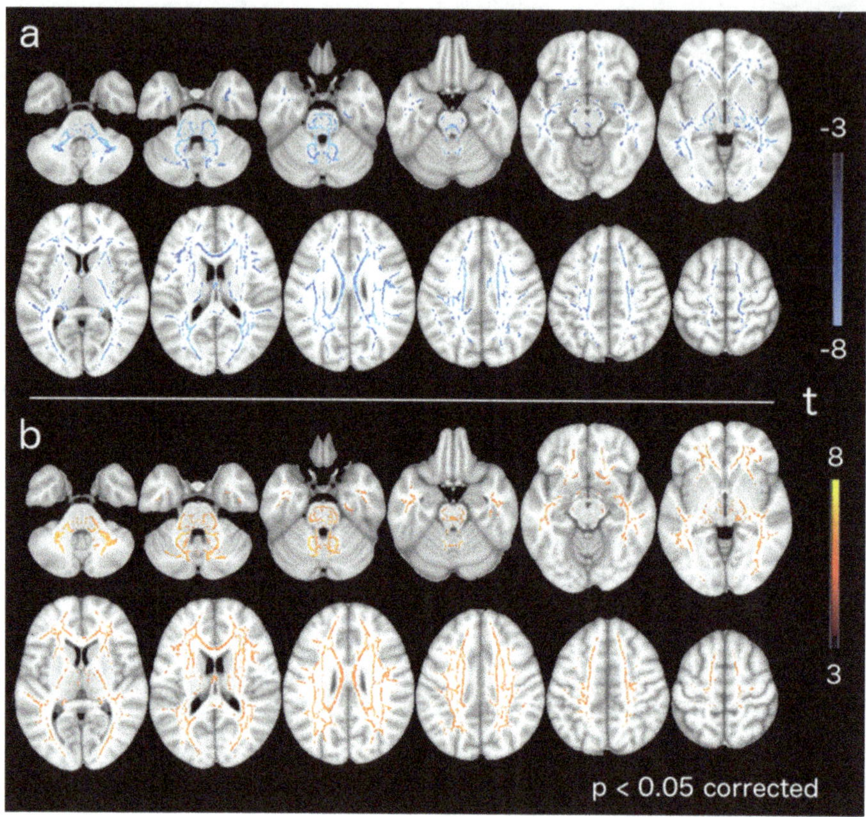

Fig. 9.1 Significant differences in diffusion properties between SCA2 patients and HC using TBSS. (**a**) Decreased fractional anisotropy in patients compared with HC (cold color). (**b**) Increased mean diffusivity in patients compared with HC (warm color). https://doi.org/10.1371/journal.pone.0135449 [36]

the rate of diffusion in the transverse direction [19]. Decreased FA in the brainstem correlated with cognitive dysfunction [21], while ataxia severity correlated negatively with FA values in a network comprising frontal, thalamic, brainstem, and left cerebellar WM [20]. Yet, WM abnormalities seem to be more widespread, involving not just the previously described structures but the bilateral frontal, parietal, temporal, and occipital lobes as well [20]. The greater number of diffusion parameters measured in the later study rendered more robust and a more detailed analysis [20].

TBSS has successfully identified affected areas with decreased FA not previously detected in SCA7, such as the deep cerebral WM and corpus callosum [25]. In FRDA, symmetrical FA decrement was found in the ICP bilaterally and corticospinal tracts in the ventral medulla (pyramids). Increased MD and decreased FA demonstrated a correlation with clinical severity in the left SCP supporting its important role in FRDA pathophysiology [29, 37].

Despite the few number of available neuroimaging studies on autosomal recessive spastic ataxia of Charlevoix-Saguenay (ARSACS), patients with this disease had extensive alterations in supra- and infratentorial WM, compatible with the wide range of its clinical manifestations [38].

Tractography

Tractography is used to segment WM based on the preferred diffusion direction in order to determine specific tracts. This is called deterministic tractography (DT). A matrix containing the relationship between every pair of voxels can be built from the tracts to illustrate homogeneous connectivity – a process called probabilistic tractography (PT) [39].

DT isolated the ICP, MCP, and SCP in subjects with SCA1, MSA-C, and other ataxias. For all patients, PT revealed decreased FA in the three tracts and correlated with severity of ataxia and decreased asymmetrically in patients with marked ataxia asymmetry. Differences in FA values in the analyzed tracts enabled the differentiation between SCA1 and MSA-C patients as well as other types of ataxia [9].

In ARSACS, tractography showed thickening of the transverse pontine fibers, which displaced and compressed the corticospinal tract with relative thinning of the midbrain and medulla. The corticospinal tract showed lower FA and increased RD and MD, consistent with the TBSS findings and supporting demyelination, while transverse pontine fibers showed higher FA, lower RD, and higher AD which might be attributed to hypertrophy or hypermyelination processes [38].

In ataxia-telangiectasia (AT), a complex autosomal recessive syndrome, cortico-ponto-cerebellar, cerebellar-thalamo-cortical, somatosensory, and lateral corticospinal tracts were extracted and analyzed using tractography. FA in AT patients was lower than in controls in all the tracts observed. Anterior tracts showed higher MD, while in cerebellar peduncles it was reduced. The authors suggested that areas such as the thalamus and pre- and postcentral gyri are likely to be affected in young AT patients [40].

FreeSurfer

FreeSurfer performs an automated segmentation of images providing information about volume and area of cortical and subcortical structures. Its major advantage is the accurate measure of the cerebral cortex thickness that is obtained from native space making it less susceptible to errors [4].VBM results measured as a function of volume are actually obtained from the interaction between surface area and cortical thickness, which are genetically and phenotypically independent measures. Thus, cortical thickness determined by FreeSurfer is a more direct and preferred parameter for genetically determined diseases [41].

SCA3 subjects displayed reduced cortical thickness in the paracentral cortex, middle frontal gyrus, superior and transverse temporal gyri, left lateral occipital cortex, and right supramarginal gyrus. The major novelty was the bilateral hippocampal atrophy, never previously reported ([42]). Three-dimensional fractal dimension thought to identify altered patterns in cortical regions demonstrated an important GM cortical atrophy mainly in the posterior lobe of cerebellum but also in other 32 areas distributed in subcortical areas and the frontal, parietal, and limbic lobes [43].

Further studies using surface analysis techniques as FreeSurfer should be performed for a better characterization of cortical abnormalities in ataxias including disruptions in the brain surface and folds.

fMRI

Resting-state functional MRI (rs-fMRI)

rs-fMRI detects blood-oxygen level dependence fluctuations that correspond to neuronal activity and translates them into patterns. Simultaneous and related activation in anatomically separated brain areas reflects a measure of functional connectivity (FC) [44]. While structural studies are important to detect neurodegeneration, functional studies are valuable to characterize the impact of such degeneration on functional integration of the remaining tissue by the analysis of the pattern of activity in FC [36].

In SCA2 a disruption in the cerebellar FC is a common finding [36, 45]. Nonetheless, there is no consensus about the pattern of FC changes between studies, mostly regarding the default mode network (DMN) [36, 45]. SCA2 can assume different phenotypes, including a parkinsonism-predominant form. Within this variety, both asymptomatic carriers and parkinsonian ones demonstrated reduced FC in the basal ganglia-thalamus-cortical loop compared to controls. While the asymptomatic group showed an increased connectivity in networks involving the cerebral cortex and motor areas, the symptomatic group showed decreased FC. The increased connectivity in those motor areas in the asymptomatic group may be a compensatory mechanism due to the disruption in the basal ganglia-thalamus-cortical loop. In parallel, the decreased connectivity in the same areas observed in the symptomatic group reveals a failure in maintaining these compensatory mechanisms, culminating in parkinsonism [46].

In a seed-based fMRI analysis, seeds were anchored in atrophic GM areas previously determined by VBM, thereafter the FC coming from these specific areas were estimated. Twenty-six genetically determined SCA7 patients compared with HC showed reduced FC between the cerebellum and the middle and superior frontal gyri and between the visual and motor cortices. An increased functional interaction was detected between the cerebellum and a range of visual cortical areas [47]. A whole-brain investigation of intrinsic networks in SCA7 showed 19 abnormal functional connections when compared to controls. There was decreased connectivity between the right cerebellar lobule crus II and the left middle frontal gyrus and increased connectivity between the left superior temporal pole and the right inferior frontal gyrus. Interestingly, the authors used measurements of FC in determined brain areas to distinguish patients from HC reaching an accuracy of 92.3%, supporting fMRI as a powerful tool as a biomarker in SCA7 [48].

A similar seed-based design was used to study SCA17. Seeds were allocated in altered areas found in previous VBM studies: the anterior lobe bilaterally and the left posterior lobe. The anterior seed revealed stronger connectivity to lobules V, VI, and to some extent I–IV, while the posterior seed connected to the posterior lobules VI–IX. Moreover, the anterior seed reported stronger FC with motor, premotor, and the primary somatosensory cortex compared with the posterior seed. These results are in line with previous structural and functional evaluations of the cerebellum that had described a dichotomization of this structure in which the anterior lobe is linked to motor functions, while the posterior lobe has predominantly cognitive and emotional roles [49].

Task-Based fMRI

Task-based fMRI or just fMRI measures brain activity during different tasks and stimuli in order to better understand how the brain is organized and to identify conditions in which this organization is disrupted.

Hand tasks were the most tested in SCAs patients. In an alternated pronation and supination task, for example, while the HC group showed activation in a network comprised by the sensorimotor cortex, supplementary motor area, cingulate motor area, putamen, and cerebellum, SCA1 patients demonstrated a large absence of activation in the cerebellum with an additional activity in contralateral cortices and thalami supporting a disruption of the cortico-cerebellar pathways in SCA1 [50].

Cerebellar activation was altered in SCA3 patients while performing hand movements. Yet, the results were no longer significant when the analysis was restricted to the cerebellar cortex or to the cerebellar nuclei. Conversely, in SCA6 and FRDA, the reduced cerebellar activation was still significant when the analysis was restricted to both cerebellar cortex and nuclei [51].

Subjects with SCA6 performed a visual task based on smooth-pursuit movement. A shift in activation was found in the vermis of presymptomatic individuals extending to the lateral cerebellum in moderate-to-severe subjects. Cerebro-cerebellar connectivity demonstrated highest effectiveness in moderate cases, while it disappeared in severe cases, supporting, respectively, the compensatory and failure of compensatory mechanisms of FC in the progression of neurodegenerative disorders [52].

In SCA17, task-based analyses were performed using the same seeds, anchored in the RS analyses described. While the anterior lobe had showed FC with motor and sensorimotor areas during RS, the task-based evaluation revealed connectivity between the posterior seed and fronto-temporo-parietal areas and the insula and the thalamus, corroborating the hypothesis of a cerebellar dichotomy in anterior-motor and posterior-cognitive roles of the cerebellum [49].

FRDA is the most commonly fMRI studied ataxia. Non-visually cued hand tapping and writing of "8" figure tasks revealed hypoactivation of the left primary sensorimotor cortex and the right cerebellum compared with HC. Reduced activation was detected in the left thalamus and right dorsolateral prefrontal cortex during hand tapping only. The writing of "8" led to a hyperactivation of the right parietal and precentral cortex and in the lentiform nucleus. The activation increased with disease severity in the right parietal, anterior cingulum and lentiform nucleus [53].

FDRA patients demonstrated some degree of higher activation in mostly the parietal and sensorimotor areas during visually cued regular and irregular single finger tapping, self-paced regular finger tapping, and multi-finger tapping tasks when compared to controls. The highest discrepancy was found during the self-paced regular finger tapping task in which higher activation was found in the cerebellar crus I/II lobules of FRDA patients, while a widespread hyperactivation was found across cerebellar and cerebral regions in the HC group, suggesting remarkable disruptions in the cortico-striatal and cortico-cerebellar loops in FRDA patients [54]. Finally, Simon effect task showed higher Simon effect, which stands for the tendency to be faster and more accurate when the stimulus matches some response feature and is calculated by incongruent minus congruent stimuli, in FRDA patients compared with HC and a reduced functional activation throughout a range of cortical, subcortical, and cerebellar regions supporting that cortico-cerebellar loops are probably disrupted in FRDA as well as attention areas are less responsive [55].

AT patients that received a 10-day course of betamethasone showed FC changes in motor areas following improvement in performance in a pronation-supination task. The longitudinal comparison showed a higher number of activated voxels within motor cortex after the end of the steroid cycle compared with baseline in the two AT patients who completed the study protocol [56].

Task-based fMRI is able to evaluate language, executive function, and memory. An interesting study recruited 15 symptomatic FXTAS subjects, 15 asymptomatic FXTAS carriers, and 12 HC and performed a verbal working memory task to identify whether areas in the prefrontal cortex were impaired in FXTAS. Reduced activation in the right ventral inferior frontal cortex and left premotor/dorsal inferior frontal cortex was found in premutation carriers compared with HC. Hypoactivation in the right premotor/dorsal inferior frontal cortex was detected in the symptomatic FXTAS group only. Activity in the right ventral inferior frontal cortex was negatively correlated with the levels of fragile X mental retardation 1 gene [57].

Magnetic Resonance Spectroscopy (MRS)

MRS is an MRI-combined technique that provides semiquantitative (ratios) and quantitative (concentrations) measurements of brain metabolites. Many nuclei can be explored for quantification, but protons nuclei (H-MRS) are easier, less expensive and provide a higher signal-to-noise ratio compared with other nuclei. The most used proton metabolites are the N-acetylaspartate (NAA, a marker of neuronal density and function), creatine/phosphocreatine (Cr, energy metabolism marker), choline compounds (Cho, marker of synthesis and degradation of cell membranes), and myoinositol (mI, found in glial tissue) [6].

In SCA1, NAA/Cr and Cho/Cr ratios reductions were detected in the basis pontis in symptomatic subjects and in the cerebellar hemisphere of SCA1 carriers [58, 59]. A quantitative study observed lower concentrations of total NAA (N-acetylaspartate + N-acetylaspartylglutamate, tNAA) and glutamate but higher concentrations of glutamine, mI, and total creatine (creatine + phosphocreatine, tCr) in the SCA1 group compared to HC. These findings are in line with neuronal dysfunction, gliosis, and alterations in glutamate-glutamine cycling and energy metabolism. Interestingly, patients could be differentiated from controls by their NAA/mI ratio of their individual spectra (100% specificity and sensitivity) [60].

In a study comparing the metabolic profiles in SCA1 and SCA2, both showed lower concentrations of NAA in the pons and cerebellum but preserved Cr and Cho concentrations. Only SCA1 showed a lactate peak in the pons and a correlation between NAA concentration and the Inherited Ataxia Clinical Rating Scale (IACRS) [33]. Conversely, more recent and larger studies described a lactate peak in the pons of SCA2 patients as well [61, 62]. NAA/Cr and Cho/Cr ratios were decreased [58, 61–64], while mI was detected in higher concentrations in the pons and cerebellum of SCA2 patients [65].

In SCA3, there was decreased NAA/Cr and Cho/Cr ratios but to a milder extent compared with SCA2 and MSA-C patients [58, 63]. Vermal NAA/Cr ratio correlated with SARA which increased progressively with age and disease duration; hence, authors support the use of MRS to predict disease onset [64]. Different neuronal markers ratios as NAA + N-acetyl-aspartate-glutamate to total creatine (NAA + NAAG/Cr + PCr) and glutamate + glutamine to total creatine (Glx/Cr + PCr) were decreased in the cerebellum of SCA3 patients using a large spectra of metabolites. These metabolites showed correlation with cognitive tests, suggesting that cerebellar dysfunction associates with executive function [21]. NAA/Cr ratio reduction in the deep white matter suggested a far more extensive neuronal dysfunction than thought at the time [66].

NAA/Cr reduction was present in SCA6 patients but less prominently than in SCA2 and MSA-C [58, 61, 63]. In contrast to SCA2 and SCA3 patients, the cerebellar hemispheric NAA/Cr ratio correlated better with SARA than vermal NAA/Cr [64]. Despite the discrete changes, SCA6 could be distinguished from other SCA by a higher lactate level and lack of pontine involvement [67].

A patterned study protocol for pontine and vermal MRS evaluation was applied in SCA1, SCA2, SCA3, and SCA7. It found lower concentrations of tNAA and glutamate but higher mI and tCr in all groups [67, 68]. Interestingly, tNAA and glutamate were inversely correlated with mI and tCr, suggesting an attempt by the glial cells to repair and to compensate for the neuronal injuries. Moreover, neurochemical profile plots with specific compounds allowed the distinction between SCA2 and SCA3 from HC [68]. Likewise, vermal mI, hemispheric mI, tNAA, glutamate and glutamine, and pontine mI, tNAA, and glutamate, when systematically combined, were capable of differentiating SCA1, SCA2, SCA6, and MSA-C. Therefore MRS may eventually be used as a biomarker for ataxias [67].

Longitudinal and retrospective studies have also assessed SCAs in order to investigate the use of MRS measurements and markers of disease progression [58, 63, 64]. The NAA/Cr ratio in the vermis and cerebellar hemispheres decreased progressively with increasing age in SCA2, SCA3, and SCA6. The association between the vermal ratio and the age of the patients was successfully used to retrospectively predict the age of clinical onset in SCA2 and SCA3 as reported by the subjects. Moreover, the vermal ratio correlated better than the hemispheric with clinical severity score (SARA) in SCA2 and SCA3, while the hemispheric ratio correlated better in SCA6 patients [64]. SCA2 and MSA-C patients showed a progressive reduction in NAA/Cr ratio during spectra assessments culminating in a lower ratio when compared with SCA1, SCA3, SCA6 patients [58, 63]. Lower vermal NAA/Cr ratio at initial assessments and lower hemispheric NAA/Cr at late assessments were found in MSA-C patients compared with SCA2 [63] Fig. 9.2.

In autosomal recessive disorders, FRDA showed decreased NAA/Cr ratio with normal Cho/Cr ratio in the pons, deep cerebellar hemispheres, and cerebral WM [30, 69]. The concentration of Cho itself, however, was decreased compared with controls [65]. Higher levels of mI and glutamine were detected in the vermis, while tCr was elevated in the cerebellar hemispheres [70].

Two studies evaluated AT patients in both cerebral and cerebellar areas of interest. Metabolic dysfunction was not found in basal ganglia or the parieto-occipital WM [71, 72]. In the cerebellum, dentate nucleus showed decreased NAA/Cho but increased Cho/Cr, suggesting an increase in Cho signal [72]. Vermal quantitative analysis, however, demonstrated profound reduction of NAA and Cho [71].

Ataxia with oculomotor apraxia type 2 (AOA2), a rare autosomal recessive form of ataxia, showed reduction of NAA in the vermis and cerebellar hemispheres and of glutamate in the vermis compared with HC. Vermal and pontine mI were found elevated indicating gliosis. Gluten ataxia (GA), the most common neurological manifestation of gluten sensitivity, showed decreased NAA/Cr ratio, NAA/Cho ratio, and NAA concentration with elevated Cho/Cr ratio in the right dentate nucleus at long time echo (TE = 135 ms) compared with HC. For short time echo (TE = 20 ms) only NAA demonstrated significant reduction. There was no correlation between atrophy and cerebellar metabolic status suggesting that NAA abnormalities are independent of structural changes [10].Two studies, involving a total of four patients, evaluated FXTAS [73, 74]. Thus larger samples are needed to better characterize its neurochemical profile.

31P-MRS measures phosphorus-derived metabolites such as phosphocreatine, adenosine triphosphate, and inorganic phosphate, but it requires larger voxels than H-MRS. Episodic ataxia type 2 (EA2), characterized by paroxysmal attacks of ataxia, vertigo, and nauseas [75], demonstrated reduction in high-energy phosphate ratios. A higher pH was found in the cerebellum, which extended to the occipital lobes. Lactate peaks were found in half of the patients' group [76]. Decreased Cr concentration was detected in the vermis and deep cerebellar hemispheres [77]. All these neurometabolic abnormalities are consistent with the pathology involved in EA2 [76, 77]. An altered energy state was also detected in childhood ataxia with

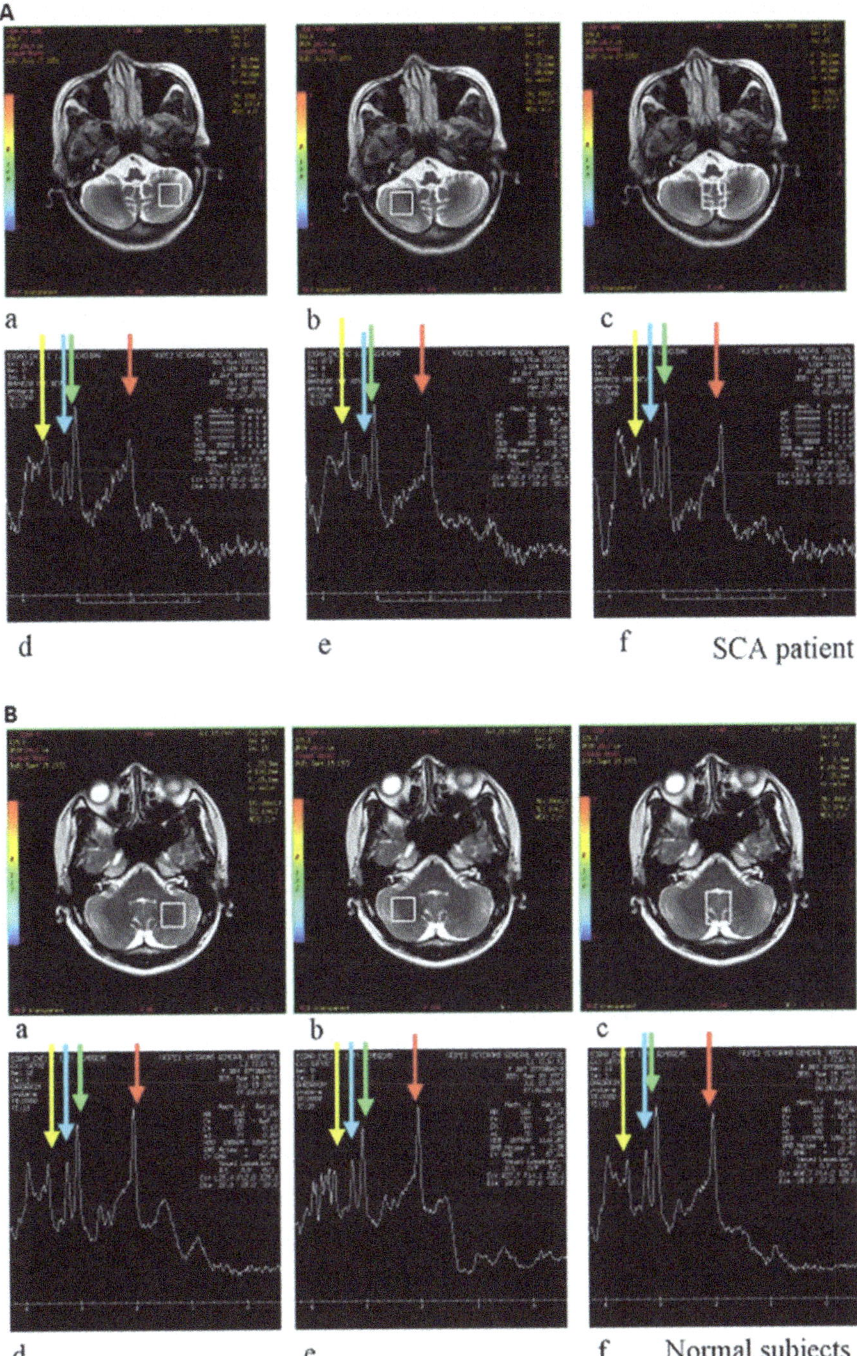

Fig. 9.2 Representative cerebellar proton magnetic resonance spectra (H-MRS). (**a**) (Upper panels) planes of the cerebellar hemispheres and vermis in a normal subject at which the MRS signal was acquired; (lower panels) corresponding MR spectra with green arrow (creatine [Cr]), red arrow (N-acetyl aspartate [NAA]), blue arrow (choline [Cho]), and yellow arrow (myoinositol [MI]). (**b**) (upper panels) planes of the cerebellar hemispheres and vermis in a patient with spinocerebellar ataxia (SCA) at which the MRS signal was acquired; (lower panels) corresponding MR spectra with green arrow (Cr), red arrow (NAA), blue arrow (Cho), and yellow arrow (MI). https://doi.org/10.1371/journal.pone.0118828 [43]

hypomyelination evaluation [78]. NAA, Cho, and Cr signals were diminished along the WM of all patients, and lactate excess was detected in half of them [79].

Conclusion

Modern MRI imaging acquisition and analysis have become widespread in the evaluation of ataxias, and despite the low prevalence of certain subtypes, some studies have analyzed a substantial number of subjects. The results have brought to light the advantages of such methods. These new techniques clarified many aspects of the pathophysiology of these disorders, generating valuable insights on new regions of interest to be investigated, new methods to be applied, and hopefully on the development of treatments. fMRI and MRS findings were successful in distinguishing patients from controls with accuracies of 92.3% and 100%, respectively [48, 60]. The correlation between imaging parameters and clinical data such as severity scores, cognitive tests, motor clinical scores, or trinucleotide repeat length demonstrates the potential use of these techniques as biomarkers and surrogate markers. The combination of different measurements within a certain methodology, as FA and MD in DTI or different chemical compounds in MRS, was used to clarify different degenerative processes, and multimodal studies have contributed to the better characterization of ataxias syndromes.

However, there still are many limitations on the application of those MRI techniques to determine the diagnosis of ataxias due to the lack of specificity of the findings in most of them. Moreover, some studies have reported conflicting results, making it difficult to establish a specific pattern of cerebral involvement for an ataxia subtype. Further studies should address the issues remaining so that imaging analysis becomes a more effective tool in diagnosis, clinical progression evaluation, and prognosis as well as monitoring individual treatment response in a variety of ataxia subtypes.

References

1. Baldarçara L, Currie S, Hadjivassiliou M et al (2015) Consensus paper: radiological biomarkers of cerebellar diseases. Cerebellum 14:175–196. https://doi.org/10.1007/s12311-014-0610-3
2. Ashburner JFK (2000) Voxel-based morphometry--the methods. NeuroImage 11:805–821
3. Alexander AL, Lee JE, Lazar M, Field AS (2007) Diffusion tensor imaging of the brain. Neurotherapeutics 4:316–329
4. Fischl B (2012) FreeSurfer. NeuroImage 62:774–781. https://doi.org/10.1016/j.neuroimage.2012.01.021.FreeSurfer
5. Lee MH, Smyser CD, Shimony JS (2014) Resting-state fMRI: a review of methods and clinical applications. AJNR Am J Neuroradiol 34:1866–1872. https://doi.org/10.3174/ajnr.A3263.Resting

6. Viau M, Boulanger Y (2004) Characterization of ataxias with magnetic resonance imaging and spectroscopy. Parkinsonism Relat Disord 10:335–351. https://doi.org/10.1016/j.parkreldis.2004.02.006

7. Konaka K, Kaido M, Okuda Y et al (2000) Proton magnetic resonance spectroscopy of a patient with Gerstmann-Straussler-Scheinker disease. Neuroradiology 42:662–665

8. Mascalchi M, Vella A (2011) Magnetic resonance and nuclear medicine imaging in ataxias. Handb Clin Neurol 103:85–110. https://doi.org/10.1016/B978-0-444-51892-7.00004-8

9. Prakash N, Hageman N, Hua X et al (2009) Patterns of fractional anisotropy changes in white matter of cerebellar peduncles distinguish spinocerebellar ataxia-1 from multiple system atrophy and other ataxia syndromes. NeuroImage 47:T72–T81. https://doi.org/10.1016/j.neuroimage.2009.05.013

10. Wilkinson ID, Hadjivassiliou M, Dickson JM et al (2005) Cerebellar abnormalities on proton MR spectroscopy in gluten ataxia. J Neurol Neurosurg Psychiatry 76:1011–1013. https://doi.org/10.1136/jnnp.2004.049809

11. Della Nave R, Ginestroni A, Tessa C et al (2008b) Brain white matter tracts degeneration in Friedreich ataxia. An in vivo MRI study using tract-based spatial statistics and voxel-based morphometry. Neuroimage 40:19–25. https://doi.org/10.1016/j.neuroimage.2007.11.050

12. Goel G, Pal PK, Ravishankar S et al (2011) Gray matter volume deficits in spinocerebellar ataxia: an optimized voxel based morphometric study. Parkinsonism Relat Disord 17:521–527. https://doi.org/10.1016/j.parkreldis.2011.04.008

13. Della NR, Ginestroni A, Tessa C et al (2008) Brain white matter damage in SCA1 and SCA2. An in vivo study using voxel-based morphometry, histogram analysis of mean diffusivity and tract-based spatial statistics. NeuroImage 43:10–19. https://doi.org/10.1016/j.neuroimage.2008.06.036

14. Ginestroni A, Della Nave R, Tessa C et al (2008) Brain structural damage in spinocerebellar ataxia type 1 : a VBM study. J Neurol 255:1153–1158. https://doi.org/10.1007/s00415-008-0860-4

15. Reetz K, Costa AS, Mirzazade S et al (2013) Genotype-specific patterns of atrophy progression are more sensitive than clinical decline in SCA1, SCA3 and SCA6. Brain 136:905–917. https://doi.org/10.1093/brain/aws369

16. Brenneis C, Bösch S, Schocke M et al (2003) Atrophy pattern in SCA2 determined by voxel-based morphometry. Neuroreport 14:1799–1802

17. Bird T (2016) Hereditary Ataxia overview. In: Pagon RA, Adam MP, Ardinger HH et al (eds) GeneReviews®. University of Washington, Seattle

18. D'Abreu A, França MC, Yasuda CL et al (2012) Neocortical atrophy in Machado-Joseph disease: a longitudinal neuroimaging study. J Neuroimaging 22:285–291. https://doi.org/10.1111/j.1552-6569.2011.00614.x

19. Guimarães RP, D'Abreu A, Yasuda CL et al (2013) A multimodal evaluation of microstructural white matter damage in spinocerebellar ataxia type 3. Mov Disord 28:1125–1132. https://doi.org/10.1002/mds.25451

20. Kang JS, Klein JC, Baudrexel S et al (2014) White matter damage is related to ataxia severity in SCA3. J Neurol 261:291–299. https://doi.org/10.1007/s00415-013-7186-6

21. Lopes TM, D'Abreu A, França MC et al (2013) Widespread neuronal damage and cognitive dysfunction in spinocerebellar ataxia type 3. J Neurol 260:2370–2379. https://doi.org/10.1007/s00415-013-6998-8

22. Lukas C, Schöls L, Bellenberg B et al (2006) Dissociation of grey and white matter reduction in spinocerebellar ataxia type 3 and 6: a voxel-based morphometry study. Neurosci Lett 408:230–235. https://doi.org/10.1016/j.neulet.2006.09.007

23. Eichler L, Bellenberg B, Hahn HK et al (2011) Quantitative assessment of brain stem and cerebellar atrophy in spinocerebellar ataxia types 3 and 6: impact on clinical status. Am J Neuroradiol 32:890–897. https://doi.org/10.3174/ajnr.A2387

24. Bang OY, Lee PH, Kim SY et al (2004) Pontine atrophy precedes cerebellar degeneration in spinocerebellar ataxia 7: MRI-based volumetric analysis. J Neurol Neurosurg Psychiatry 75:1452–1456. https://doi.org/10.1136/jnnp.2003.029819
25. Alcauter S, Barrios FA, Díaz R, Fernández-Ruiz J (2011) Gray and white matter alterations in spinocerebellar ataxia type 7: an in vivo DTI and VBM study. NeuroImage 55:1–7. https://doi.org/10.1016/j.neuroimage.2010.12.014
26. Lasek K, Lencer R, Gaser C et al (2006) Morphological basis for the spectrum of clinical deficits in spinocerebellar ataxia 17 (SCA17). Brain 129:2341–2352. https://doi.org/10.1093/brain/awl148
27. Reetz K, Kleinman A, Klein C et al (2011) CAG repeats determine brain atrophy in spinocerebellar ataxia 17: a VBM study. PLoS One. https://doi.org/10.1371/journal.pone.0015125
28. Reetz K, Lencer R, Hagenah JM et al (2010) Structural changes associated with progression of motor deficits in spinocerebellar ataxia 17. Cerebellum 9:210–217. https://doi.org/10.1007/s12311-009-0150-4
29. Della Nave R, Ginestroni A, Giannelli M et al (2008a) Brain structural damage in Friedreich's ataxia. J Neurol Neurosurg Psychiatry 79:82–85. https://doi.org/10.1136/jnnp.2007.124297
30. França MC, D'Abreu A, Yasuda CL et al (2009) A combined voxel-based morphometry and 1H-MRS study in patients with Friedreich's ataxia. J Neurol 256:1114–1120. https://doi.org/10.1007/s00415-009-5079-5
31. Santner W, Schocke M, Boesch S et al (2014) A longitudinal VBM study monitoring treatment with erythropoietin in patients with Friedreich ataxia. Acta Radiol short Rep 3:2047981614531573. https://doi.org/10.1177/2047981614531573
32. Hashimoto RI, Javan AK, Tassone F et al (2011b) A voxel-based morphometry study of grey matter loss in fragile X-associated tremor/ataxia syndrome. Brain 134:863–878. https://doi.org/10.1093/brain/awq368
33. Guerrini L, Lolli F, Ginestroni A et al (2004) Brainstem neurodegeneration correlates with clinical dysfunction in SCA1 but not in SCA2. A quantitative volumetric, diffusion and proton spectroscopy MR study. Brain 127:1785–1795. https://doi.org/10.1093/brain/awh201
34. Mandelli ML, De Simone T, Minati L et al (2007) Diffusion tensor imaging of spinocerebellar ataxias types 1 and 2. Am J Neuroradiol 28:1996–2000. https://doi.org/10.3174/ajnr.A0716
35. Smith SM, Jenkinson M, Johansen-Berg H et al (2006) Tract-based spatial statistics: Voxelwise analysis of multi-subject diffusion data. NeuroImage 31:1487–1505. https://doi.org/10.1016/j.neuroimage.2006.02.024
36. Hernandez-Castillo CR, Galvez V, Mercadillo R et al (2015) Extensive white matter alterations and its correlations with ataxia severity in SCA 2 patients. PLoS One 10:1–10. https://doi.org/10.1371/journal.pone.0135449
37. Karuta S, Raskin S, de Carvalho NA et al (2015) Diffusion tensor imaging and tract-based spatial statistics analysis in Friedreich's ataxia patients. Parkinsonism Relat Disord 21:504–508. https://doi.org/10.1016/j.parkreldis.2015.02.021
38. Oguz KK, Haliloglu G, Temucin C et al (2013) Assessment of whole-brain white matter by DTI in autosomal recessive spastic ataxia of Charlevoix-Saguenay. Am J Neuroradiol 34:1952–1957. https://doi.org/10.3174/ajnr.A3488
39. Nucifora PGP, Verma R, Lee S, Melhem ER (2007) Diffusion-tensor MR imaging and tractography: exploring brain microstructure and connectivity. Radiology 245:367–384
40. Sahama I, Sinclair K, Fiori S et al (2015) Motor pathway degeneration in young ataxia telangiectasia patients: a diffusion tractography study. Neuroimage Clin 9:206–215. https://doi.org/10.1016/j.nicl.2015.08.007
41. Winkler (2011) Cortical thickness or Grey matter. Neuroimage 53:1135–1146. https://doi.org/10.1016/j.neuroimage.2009.12.028.Cortical
42. de Rezende TJR, D'Abreu A, Guimarães RP et al (2015) Cerebral cortex involvement in Machado-Joseph disease. Eur J Neurol 22:277–283. https://doi.org/10.1111/ene.12559

43. Wang TY, Jao CW, Soong BW et al (2015) Change in the cortical complexity of spinocerebellar ataxia type 3 appears earlier than clinical symptoms. PLoS One 10:1–18. https://doi.org/10.1371/journal.pone.0118828

44. van den Heuvel MP, Hulshoff Pol HE (2010) Exploring the brain network: a review on resting-state fMRI functional connectivity. Eur Neuropsychopharmacol 20:519–534. https://doi.org/10.1016/j.euroneuro.2010.03.008

45. Cocozza S, Saccà F, Cervo A et al (2015) Modifications of resting state networks in spinocerebellar ataxia type 2. Mov Disord 00:1–9. https://doi.org/10.1002/mds.26284

46. Wu T, Wang C, Wang J et al (2013) Preclinical and clinical neural network changes in SCA2 parkinsonism. Parkinsonism Relat Disord 19:158–164. https://doi.org/10.1016/j.parkreldis.2012.08.011

47. Hernandez-Castillo CR, Alcauter S, Galvez V et al (2013) Disruption of visual and motor connectivity in spinocerebellar Ataxia type 7. Mov Disord 28:1708–1716. https://doi.org/10.1002/mds.25618

48. Hernandez-Castillo CR, Galvez V, Morgado-Valle C, Fernandez-Ruiz J (2014) Whole-brain connectivity analysis and classification of spinocerebellar ataxia type 7 by functional MRI. Cerebellum Ataxias 1:2. https://doi.org/10.1186/2053-8871-1-2

49. Reetz K, Dogan I, Rolfs A et al (2012) Investigating function and connectivity of morphometric findings - exemplified on cerebellar atrophy in spinocerebellar ataxia 17 (SCA17). NeuroImage 62:1354–1366. https://doi.org/10.1016/j.neuroimage.2012.05.058

50. Jayakumar PN, Desai S, Pal PK et al (2008) Functional correlates of incoordination in patients with spinocerebellar ataxia 1: a preliminary fMRI study. J Clin Neurosci 15:269–277. https://doi.org/10.1016/j.jocn.2007.06.021

51. Stefanescu MR, Dohnalek M, Maderwald S et al (2015) Structural and functional MRI abnormalities of cerebellar cortex and nuclei in SCA3, SCA6 and Friedreich's ataxia. Brain 138:1182–1197. https://doi.org/10.1093/brain/awv064

52. Falcon M, Gomez C, Chen E et al (2015) Early cerebellar network shifting in spinocerebellar Ataxia type 6. Cereb Cortex. https://doi.org/10.1093/cercor/bhv154

53. Ginestroni A, Diciotti S, Cecchi P et al (2012) Neurodegeneration in Friedreich's ataxia is associated with a mixed activation pattern of the brain. A fMRI study. Hum Brain Mapp 33:1780–1791. https://doi.org/10.1002/hbm.21319

54. Akhlaghi H, Corben L, Georgiou-Karistianis N et al (2012) A functional MRI study of motor dysfunction in Friedreich's ataxia. Brain Res 1471:138–154. https://doi.org/10.1016/j.brainres.2012.06.035

55. Georgiou-Karistianis N, Akhlaghi H, Corben LA et al (2012) Decreased functional brain activation in Friedreich ataxia using the Simon effect task. Brain Cogn 79:200–208. https://doi.org/10.1016/j.bandc.2012.02.011

56. Quarantelli M, Giardino G, Prinster A et al (2013) Steroid treatment in Ataxia-telangiectasia induces alterations of functional magnetic resonance imaging during prono-supination task. Eur J Paediatr Neurol 17:135–140. https://doi.org/10.1016/j.ejpn.2012.06.002

57. Hashimoto R, Backer K, Tassone F et al (2011a) An fMRI study of the prefrontal activity during the performance of a working memory task in premutation carriers of the fragile X mental retardation 1 gene with and without fragile X-associated tremor/ataxia syndrome (FXTAS). J Psychiatr Res 45:36–43. https://doi.org/10.1016/j.jpsychires.2010.04.030

58. Lirng JF, Wang PS, Chen HC et al (2012) Differences between spinocerebellar ataxias and multiple system atrophy-cerebellar type on proton magnetic resonance spectroscopy. PLoS One 7:1–7. https://doi.org/10.1371/journal.pone.0047925

59. Mascalchi M, Tosetti M, Plasmati R et al (1998) Proton magnetic resonance spectroscopy in an Italian family with spinocerebellar ataxia type 1. Ann Neurol 43:244–252

60. Oz G, Hutter D, Tkac I, Clark H (2010) Neurochemical alterations in spinocerebellar ataxia type 1 and their correlations with clinical status. Mov Disord 25:1253–1261. doi: https://doi.org/10.1002/mds.23067.NEUROCHEMICAL

61. Boesch S, Schocke M, Bürk K et al (2001) Proton magnetic resonance spectroscopic imaging reveals differences in spinocerebellar ataxia types 2 and 6. J Magn Reson Imaging 13:553–559
62. Boesch S, Wolf C, Seppi K et al (2007) Differentiation of SCA2 from MSA-C using proton magnetic resonance spectroscopic imaging. J Magn Reson Imaging 25:564–569. https://doi.org/10.1002/jmri.20846
63. Chen HC, Lirng JF, Soong BW et al (2014) The merit of proton magnetic resonance spectroscopy in the longitudinal assessment of spinocerebellar ataxias and multiple system atrophy-cerebellar type. Cerebellum Ataxias 1:17. https://doi.org/10.1186/s40673-014-0017-4
64. Wang P-S, Chen H-C, Wu H-M et al (2012) Association between proton magnetic resonance spectroscopy measurements and CAG repeat number in patients with spinocerebellar ataxias 2, 3, or 6. PLoS One 7:e47479. https://doi.org/10.1371/journal.pone.0047479
65. Viau M, Marchand L, Bard C, Boulanger Y (2005) (1)H magnetic resonance spectroscopy of autosomal ataxias. Brain Res 1049:191–202. https://doi.org/10.1016/j.brainres.2005.05.015
66. D'Abreu A, França M, Appenzeller S et al (2009) Axonal dysfunction in the deep white matter in Machado-Joseph disease. J Neuroimaging 19:9–12. https://doi.org/10.1111/j.1552-6569.2008.00260.x
67. Oz G, Iltis I, Hutter D et al (2011) Distinct neurochemical profiles of spinocerebellar ataxias 1, 2, 6, and cerebellar multiple system atrophy. Cerebellum 10:208–217. https://doi.org/10.1007/s12311-010-0213-6
68. Adanyeguh I, Henry P, Nguyen T et al (2015) In vivo neurometabolic profiling in patients with spinocerebellar ataxia types 1, 2, 3, and 7. Mov Disord 30:662–670. https://doi.org/10.1002/mds.26181
69. Mascalchi M, Cosottini M, Lolli F et al (2002) Proton MR spectroscopy of the cerebellum and pons in patients with degenerative ataxia. Radiology 223:371–378
70. Iltis I, Hutter D, Bushara K et al (2010) (1)H MR spectroscopy in Friedreich's ataxia and ataxia with oculomotor apraxia type 2. Brain Res 1358:200–210. https://doi.org/10.1016/j.brainres.2010.08.030
71. Lin D, Crawford T, Lederman H, Barker P (2006) Proton MR spectroscopic imaging in ataxia-telangiectasia. Neuropediatrics 37:241–246. https://doi.org/10.1055/s-2006-924722
72. Wallis LI, Griffiths PD, Romanowski CA et al (2007) Proton spectroscopy and imaging at 3T in ataxia telangiectasia. AJNR Am J Neuroradiol 28:79–83
73. Ginestroni A, Guerrini L, Della Nave R et al (2007) Morphometry and 1H-MR spectroscopy of the brain stem and cerebellum in three patients with fragile X-associated tremor/ataxia syndrome. AJNR Am J Neuroradiol 28:486–488
74. Sarac H, Henigsberg N, Markeljević J et al (2011) Fragile X-premutation tremor/ataxia syndrome (FXTAS) in a young woman: clinical, genetics, MRI and 1H-MR spectroscopy correlates. Coll Antropol 35:327–332
75. Spacey S (2015) Episodic Ataxia type. In: Pagon R, Adam M, Ardinger H et al (eds) GeneReviews [Internet]. University of Washington, Seattle, p 2
76. Sappey-Marinier D, Vighetto A, Peyron R et al (1999) Phosphorus and proton magnetic resonance spectroscopy in episodic ataxia type 2. Ann Neurol 46:256–259
77. Harno H, Heikkinen S, Kaunisto M et al (2005) Decreased cerebellar total creatine in episodic ataxia type 2: a 1H MRS study. Neurology 64:542–544. https://doi.org/10.1212/01.WNL.0000150589.26350.3D
78. Blüml S, Philippart M, Schiffmann R et al (2003) Membrane phospholipids and high-energy metabolites in childhood ataxia with CNS hypomyelination. Neurology 61:648–654
79. Tedeschi G, Schiffmann R, Barton N et al (1995) Proton magnetic resonance spectroscopic imaging in childhood ataxia with diffuse central nervous system hypomyelination. Neurology 45:1526–1532

Chapter 10
Recent Insights from fMRI Studies into the Neural Basis of Reciprocal Imitation in Autism Spectrum Disorders

Yuko Okamoto and Hirotaka Kosaka

Background

Autism spectrum disorder (ASD) is a neurodevelopmental disorder characterized by the *Diagnostic and Statistical Manual of Mental Disorders*, fifth edition (DSM-5) as (1) the presence of deficits in social communication and social interaction and (2) restricted, repetitive patterns of behavior, interests, or activities [1]. ASD was first reported in the early 1940s [2, 3], yet the exact pathophysiology of ASD remains elusive. As a result, the diagnosis of ASD is still based on behavioral features rather than on physiological clinical findings, almost 70 years after the first reported case.

To elucidate the pathophysiology of ASD, many researchers have tried to associate the characteristic behavioral symptoms of ASD with brain activity patterns using functional magnetic resonance imaging (fMRI). Indeed, the number of published research papers using MRI (Fig. 10.1a) and published research papers using MRI of individuals with ASD (Fig. 10.1b) has gradually increased over the past two decades (i.e., from 1995 to 2015). The ratio of MRI-based publications on ASD to all MRI-based publications has also increased during this period (Fig. 10.1c). These trends express a growing interest in the role of altered brain function in ASD.

Individuals with ASD have various social dysfunctions related to both verbal and nonverbal communication. With regard to nonverbal communication, abnormal imitative exchanges are one remarkable social deficit found in ASD patients [4, 5]. Because preverbal children are considered to acquire various social skills such as theory of mind and turn taking through imitative interaction with their caregivers [4, 6], it is reasonable to hypothesize that altered brain activation during reciprocal imitation (i.e., imitation of others' actions and being imitated by another

Y. Okamoto (✉) · H. Kosaka
Research Center for Child Mental Development, University of Fukui, Fukui, Japan
e-mail: yokamoto@u-fukui.ac.jp

© Springer International Publishing AG, part of Springer Nature 2018
C. Habas (ed.), *The Neuroimaging of Brain Diseases*, Contemporary Clinical Neuroscience, https://doi.org/10.1007/978-3-319-78926-2_10

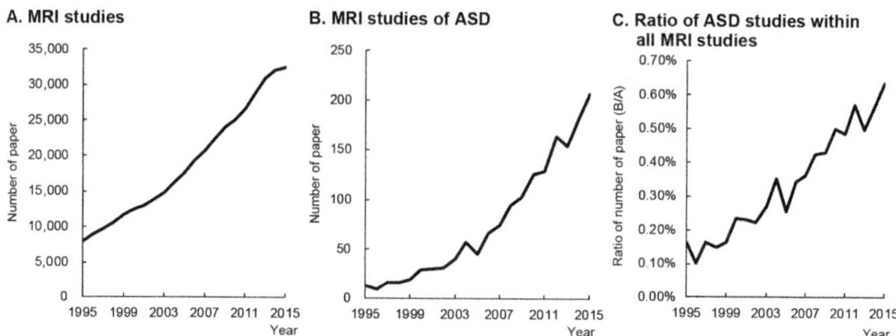

Fig. 10.1 Numbers of magnetic resonance imaging (MRI) papers published on autism spectrum disorder (ASD), by year. (**a**) Number of papers found by searching on "MRI." (**b**) Number of papers found by searching on "MRI and autism." (**c**) Ratio of number of MRI papers associated with autism to all MRI papers, calculated as B/A. (These searches were conducted in PubMed on 2016.8.26)

person) is a basic pathophysiology of ASD. Several fMRI studies have examined the neural mechanisms underlying impairments in reciprocal imitation and, more specifically, imitation deficits in ASD [7–10], as well as atypical social responses to being imitated [11, 12]. However, these fMRI studies were conducted under simple and well-controlled conditions that did not well represent imitative difficulties observed in daily life, which can be complex and vary hour-to-hour. For example, imitative tendencies are significantly affected by social context ([13]; see also the review by [14]); however, no study to date has considered the effects of social modulation on imitative interaction. Furthermore, no studies have examined brain activation when the participants communicate with a real person using reciprocal imitation. Therefore, the complete picture of abnormal imitative interactions in individuals with ASD has yet to be revealed.

In this chapter, we will review fMRI studies examining the neural basis of the characteristic reciprocal imitation deficits in individuals with ASD. We will further describe the limitations of present studies and suggest further approaches to a better understanding of the neural substrates of reciprocal imitation in ASD.

The Neural Basis of Imitation Deficits in ASD

In typically developing (TD) children, imitation plays an important role in the development of social skills as well as in the acquisition of new skills such as language [4, 6]. Under the assumption that the observation and execution of actions are innately coupled, Meltzoff [6] proposed that newborn children learn to understand the thoughts and intentions of others through imitation. For example, he observed that neonates could imitate facial movements such as tongue protrusion. After birth, infants learn to associate their own bodily states with specific mental states; when an infant observes the actions of another individual, the infant projects the associated mental state onto the other individual [6]. Imitation is also important for interpersonal communication in adulthood; during conversation, adults frequently imitate the actions or facial expressions of others in an unconscious manner. These implicit imitative interactions lead to the development of friendship and rapport among interacting people [15].

In contrast, individuals with ASD exhibit difficulties in imitating others [5, 16]. Williams and co-workers [5] systematically reviewed 21 well-controlled studies examining imitation deficits in individuals with ASD and concluded that children with ASD perform poorly in imitative tasks, especially tasks involving the imitation of non-meaningful gestures. Based on these facts, they proposed that the imitation difficulties are potentially associated with deficits in self-other mapping for actions, which relies on integration of visual and motor modalities.

The neural basis of self-other mapping during imitation of another's actions was in part deciphered with the discovery of mirror neurons. Mirror neurons respond to both the performance and the observation of a given action (e.g., grasping) and were originally identified in the monkey ventral premotor [17–19] and parietal brain regions [20, 21]. Thereafter, several fMRI studies confirmed the presence of similar activations in the frontal and parietal regions of the human brain [22–26]. From a functional perspective, mirror neurons automatically activate the corresponding motor representation of a given action to facilitate imitation in the absence of specific motor information. Therefore, the discovery of mirror neurons resolved the correspondence problem [27], which refers to the fact that, when observing the actions of another person, we cannot see the patterns of muscle activation required for successful imitation.

The frontal and parietal regions have reciprocal connections with the lateral occipitotemporal cortex (LOTC) [28]. This area includes the extrastriate body area (EBA), which is sensitive to viewing body parts [29], and the posterior superior temporal sulcus, which is sensitive to viewing biological motion [30]. The frontal region, the parietal region, and the LOTC comprise the human mirror neuron system (MNS). A meta-analysis of TD individuals demonstrated the involvement of the MNS in imitation [31]. From a computational perspective, the MNS comprises an internal model with forward and inverse components [28, 32]. Of note, although the internal model was originally proposed as a computational model for motor control, the same model can also be used to explain social interactions. For application of the internal model to imitation, an inverse model that describes the representation of visual information about an observed action is available in the LOTC, and the transmission of this information to the frontal MNS enables the planning of self-actions (Fig. 10.2a). By contrast, a forward model available in the frontal MNS generates a copy of motor commands (i.e., the efference copy) and sends this copy to the LOTC to (1) compare executed and observed actions and (2) to calculate the similarity or divergence of these actions (Fig. 10.2b). Both mechanisms are relevant for imitation and lead to the hypothesis that imitation deficits in individuals with ASD are associated with abnormal function of the internal model of the MNS.

A. Inverse model

B. Forward model

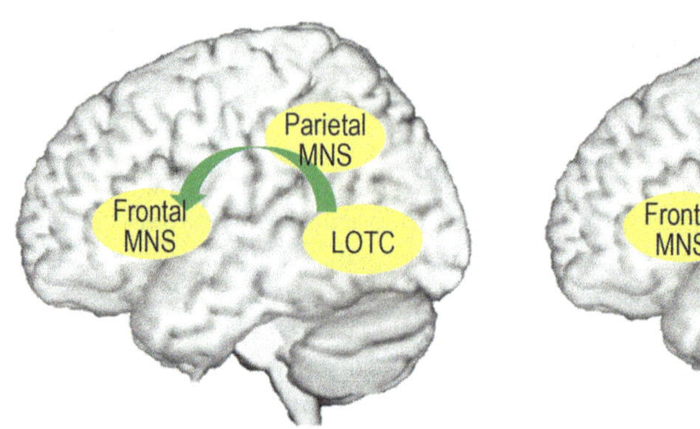

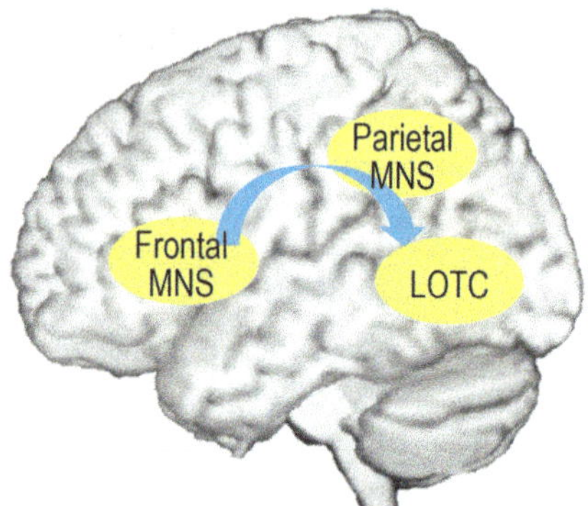

Fig. 10.2 Internal models within the mirror neuron system (MNS). (**a**) Pathway of control by inverse model. (**b**) Pathway of control by forward model; LOTC, lateral occipitotemporal cortex

The first report of altered brain activation during imitation in children with ASD was published by Dapretto and colleagues [7]. In this study, nine children with ASD and nine TD children were instructed to imitate photographs of emotional facial expressions during MRI scanning. Dapretto and colleagues found that activation in the inferior frontal gyrus (IFG) was weaker in children with ASD compared with TD children. Furthermore, in children with ASD, activation in the interior frontal gyrus was significantly correlated with the social subscales of the Autism Diagnostic Interview-Revised (ADI-R) and the Autism Diagnostic Observation Schedule-Generic (ADOS-G) [7]. The results suggested that MNS dysfunction might underlie imitation impairments and other social deficits in children with ASD. Following this study, several other fMRI experiments identified ASD-related abnormalities in brain activation during the imitation of others' actions [8–10]. Williams and co-workers [10] measured brain activation during the imitative finger-tapping task developed by Iacoboni and co-workers [33] in 16 adolescents with ASD and in 15 age- and IQ-matched TD control participants. No atypical activation of the inferior frontal gyrus was observed; instead, the authors reported decreased activation in the right inferior parietal lobule and right LOTC during the imitation condition relative to the rest condition in ASD participants versus TD participants. Taken together, the above findings suggested that the observation of abnormalities in brain activation in ASD depends on the type of action imitated. Poulin-Lord and co-workers [9] performed fMRI on 23 adolescent ASD patients and 22 TD adolescents during an imitation task involving finger gestures, to evaluate differences in spatial (topological) variability in brain activation. The authors found that spatial variability in the superior parietal cortex and associated visual regions (Brodmann areas 18 and 19) was higher in the ASD group than in the TD group; that is, individuals with ASD showed greater individual variations in the location of task-evoked activation than did TD individuals. Based on these results, it was speculated that individuals with ASD might use more variable and unique strategies to perform imitation than do TD individuals

[9]. Jack and Morris [8] used fMRI to examine alterations in functional connectivity during an imitative finger-tapping task in 15 adolescents with ASD and 15 TD adolescents. During imitation, the inferior frontal gyrus and LOTC were more strongly activated in TD individuals than in ASD individuals. Furthermore, a psychophysiological interaction analysis revealed that stronger connectivity between the posterior superior temporal sulcus (a region in the LOTC) and Crus I of the neocerebellum was associated with greater mentalizing ability among adolescents with ASD. Recent resting-state fMRI studies have confirmed the presence of connectivity alterations between brain regions associated with imitation [34, 35]. Using resting-state fMRI, Nebel and co-workers [35] evaluated relationships between resting-state brain activity and imitation as well as social skills in 50 children with ASD and in 50 TD children. An independent components analysis was used to extract temporally coherent brain networks from the fMRI data. Functional connectivity between the LOTC and the lateral upper limb areas within the precentral and postcentral gyri was found to be impaired in children with ASD relative to TD children. In the ASD group, functional connectivity between the above areas was correlated with social skill but not with imitation skill [35]. This suggested that children with ASD might acquire imitation skills by recruiting compensatory networks; however, it remains unclear whether reduced resting-state network synchrony contributes to imitation deficits in ASD. In summary, although different studies have highlighted different regions of interest as neurological bases for imitation deficits in ASD, reduced activation of the inferior frontal gyrus, inferior parietal lobule, and LOTC were most commonly identified. These findings partially support the hypothesis that imitation deficits in ASD are related to dysfunction of the internal model within the MNS.

However, in addition to impaired imitation ability, some individuals with ASD demonstrate *excessive* imitation behaviors such as echolalia. If the observed action never automatically maps onto the frontoparietal MNS in ASD, excessive imitation such as echolalia should not occur; the "broken mirror theory" above cannot account for this. Recently, a theoretical account of the observations that provides an alternative to the "broken mirror theory" has been proposed [14, 36]; the existence of a top-down neural modulation of imitation was posited, which would determine whether a given observed action is imitated or not. Furthermore, this modulation effect was proposed to be dependent on social context. For example, automatic mimicry is more common when TD individuals interact with a familiar person than with an unfamiliar person [15]. Cook and co-workers [14] conducted a behavioral study to examine the effects of social context on imitative tendencies in individuals with ASD. ASD and TD participants were asked to press a key based on a visually presented number cue. At the same time, the researchers presented a moving finger indicating a key that was compatible or incompatible with the participant's correct action. Generally, participants' finger movements tend to be lured toward the observed action; reaction time is therefore expected to be longer in the incompatible condition than in the compatible condition. Thus, tendency to imitate was determined by subtracting the reaction time in the incompatible condition from that in the compatible condition. Social context was further manipulated using a priming task before the imitation task, wherein participants rearranged a series of words to form a grammatically correct sentence. For half of the participants in each group, the sentence included pro-social words such as "friend" and "sociable" (i.e., the social condition), whereas for the other half, the sentence included nonsocial words such as "selfish" and "alone" (i.e., the nonsocial condition). TD individuals in the social condition showed enhanced imitative tendency relative to those in the nonsocial condition. In contrast, there were no differences between the social and nonsocial conditions for individuals with ASD [13]. These findings suggested that TD individuals were better able to alter their imitative tendencies according to social context, whereas individuals with ASD did not use social context to modulate imitative tendency. Cook and co-workers [14] proposed a role for the dorsal prefrontal cortex in modulating connectivity within the MNS and suggested that individuals with ASD might exhibit abnormalities in this area. A recent meta-analysis of fMRI studies on action observation and imitation in ASD, which focused on activation likelihood estimation, has provided evidence in support of this perspective [37]. Yang and colleagues identified altered activation in the anterior inferior parietal lobule, LOTC, and dorsolateral prefrontal cortex as common observations in various imitation and observation of action fMRI studies of ASD. Altered brain activation in the anterior inferior parietal lobule and LOTC could result from dysfunction of the internal model of the MNS; however, the dorsolateral prefrontal cortex is thought to play a role in the top-down control of imitation [38]. Therefore, altered activation in the dorsolateral prefrontal cortex suggests that behavioral abnormalities in imitation in individuals with ASD can also result from abnormal top-down control of the internal model.

In summary, although recent fMRI studies provide important evidence informing our thinking on the neural substrates of imitation deficits in individuals with ASD, the complete pathological picture underlying these abnormalities remains unclear. An important issue is whether imitation deficits result from altered top-down control of the MNS. If the hypotheses of Cook and co-workers [14] and Hamilton [36] are true, connectivity within the MNS should be modulated by the dorsal prefrontal cortex according to social context in TD individuals, and therefore the dorsal prefrontal cortex should be an area of dysfunction in ASD. In addition, a better understanding of imitation deficits in different task contexts is required to determine whether abnormal top-down modulation of the MNS depends in patients with ASD on the action to be imitated.

The Neural Basis of Atypical Responses to Being Imitated in ASD

Children with ASD also exhibit abnormal behaviors when they are imitated by other people [4, 39, 40]. For example, when imitated, TD children frequently change their actions and look at the person with which they are interacting to test the intentions of the imitator. In contrast, most children with ASD do not display these behaviors [4], and social responses to being imitated in children with ASD were less frequent compared with TD children [39]. These findings suggest that children with ASD may be less sensitive to or less aware of imitators than TD children.

To provide a theoretical basis for imitation recognition, Gergely and Watson [41] proposed the existence of a contingency detection module (CDM) that establishes the primary representation of the bodily self as well as later orientations toward reactive social objects. In CDM theory, this module is innately set to explore preferentially perfect response-contingent stimulations such as visual feedback of one's own actions. At around 3 months of age, the CDM is thought switch toward a preference for less-than-perfectly contingent actions, such as the actions of others engaged in imitation [41, 42]. Children with ASD are thought to fail to switch their preference from perfect to less-than-perfect contingency during this developmental period. As a result, ASD children become less sensitive to less-than-perfectly contingent situations and spend more time in repetitive motor activity to satisfy a preference for self-related, perfect contingency [40].

Based on the above theory, we formulated the following hypotheses: (1) in normal individuals, the neural substrates of CDM are differently activated when observed and executed actions are congruent versus incongruent; and (2) individuals with ASD show atypical brain activation during congruent versus incongruent conditions. To test these hypotheses, we performed fMRI on ASD and TD individuals during reciprocal imitation of finger gestures [12]. In TD individuals, EBA, which is a component of the LOTC, was strongly activated during reciprocal imitation compared with a non-imitation condition [12]. An important question stemming from this observation is whether the EBA is a neural substrate of the CDM. The EBA was originally identified as a brain region activated in response to viewing non-face body parts such as the hand or foot [43]. Subsequently, Astafiev and co-workers [44] found that the EBA also responded to motor action planning and execution in the absence of visual feedback, indicating that an efference copy of the executed action was sent to the EBA. Based on these properties, the EBA was proposed to play a role in detecting perfect contingency [45], and this role was confirmed by subsequent fMRI and transcranial magnetic stimulation studies [46, 47]. Because our study showed that the EBA was also involved in detecting less-than-perfect contingency [12], it is reasonable to hypothesize that the EBA subserves the CDM in TD individuals.

In the second part of our study, we explored whether EBA activation was different in ASD versus TD individuals. We observed attenuation of the contingency effect in the EBA of ASD individuals during the congruent condition [12]. Therefore, it is reasonable to propose dysfunction of the EBA (i.e., the CDM) as a pathological substrate of decreased imitation sensitivity in ASD. Areas of the LOTC including the EBA are thought to receive efference copy from the frontal MNS, communicating the forward component of the internal model [28]. Therefore, reduced activation in the EBA might also relate to abnormalities in the internal model.

Research has only begun to address the question of why children with ASD have difficulty detecting imitation. At present, it is unclear whether atypical responses to being imitated are due to dysfunction of the forward model itself. Cook and co-workers [14] argued that top-down modulation is relevant for both action imitation and social perception and that both of these aspects are dysfunctional in ASD patients. Similarly, in this case, reduced activation in the EBA might result from abnormal control of the internal model. If this theory is true, the attenuated congruency effect in the EBA should depend on social context. This hypothesis should be carefully addressed in future fMRI studies.

Brain Regions Mediating Social Responses to Imitation in ASD

Berger and Ingersoll [39] found that social responses to being imitated were correlated with imitation skill and with symptom severity in ASD. This suggests that imitation might have utility as an intervention tool to improve social deficits in individuals with ASD. To this end, some studies have indicated that children with ASD show more social behavior when they are repeatedly imitated by another person ([48, 49]; see review by [50]). Escalona and co-workers [48] explored the social effects of being imitated in persons with ASD, by examining differences in the reactions of children with ASD to adults who were either imitating them or interacting with them normally. They found that children with ASD spent more time touching (i.e., attempting to initiate an interaction with) adults when the adult was imitating them [48]. Accordingly, imitation appears to facilitate affiliative behavior in children with ASD.

The neural mechanisms of the abovementioned behavioral phenomenon were recently explored by Delaveau et al. [11], who investigated neural responses to being imitated using fMRI to study six high-functioning male adults with ASD. This experiment consists with the following three steps. At first, the participants conducted an fMRI task where participant hand movements were imitated by an experimenter. After that, they participated in behavioral task, where participant action was imitated by experimenter outside MR scanner. Again, they performed fMRI task same as initially performed. Thus, examining difference of brain activation between second and first MR session arrow to elucidate changes of awareness of being imitated through the experience that one's own action was imitated by another person, which is exposed in behavioral task. As a result, the authors identified enhanced activation in the insula and reduced activation in the precuneus and inferior parietal lobe (part of the default mode network) in second MR session relative to first MR session. The insula is part of the salience network, which detects and integrates behaviorally relevant stimuli; thus, these findings suggest that being imitated may increase the salience of social signals through modulation of strategic brain regions involved in self- and other-processing [11]. Although the above finding is interesting, the study suffered from several methodological weaknesses. For example, the sample size was limited, and the statistical threshold was not sufficiently strict. Furthermore, the study did not have a control group of individuals who were not imitated, in contrast to other behavioral studies examining the social effects of being imitated [48, 49]. Therefore, it is possible that the observed changes in activity were related to MRI session repetition (i.e., learning or adaptation). Even so, the authors' paradigm should be improved and replicated in future studies using a larger sample size and more strictly controlled conditions, to connect neuroimaging and behavioral intervention studies.

Summary and Future Perspectives of fMRI Studies Examining the Neural Substrates of Reciprocal Imitation in ASD

In this chapter, we introduced recent fMRI studies examining the neural substrates of imitative exchanges in individuals with ASD, which includes both imitating another's action [7–10] and recognizing that one is being imitated by another person [12]. Overall, the deficit in imitative exchanges seen in individuals with ASD is likely explainable by an abnormality in an internal model represented in the MNS. However, as we described above, it is unclear whether reduced activation in the MNS is due to dysfunction of the internal model itself or to aberrant top-down modulation of the internal model induced by different activation levels within the MNS. To test these possibilities, we should ask whether reduced activation in the MNS is consistently observed in ASD in various social situations. If aberrant activation in the MNS depends on the social context, we should provide supporting evidence that aberrant top-down modulation of the internal model is the core pathophysiology of abnormal imitative interaction in individuals with ASD.

Recently, a new fMRI technique has been developed, namely, "hyper-scanning," which measures the brain activation of two interacting persons. For instance, Tanabe and co-workers [51] have used this method to examine the neural basis of eye contact and joint attention in individuals with ASD. In this study, the brain activation of two people was measured when they interacted. The researchers compared brain activations when TD participants interacted with TD participants and when TD participants interacted with ASD participants. They found relatively reduced occipital pole activation for TD/ASD interactions, which should be associated with abnormal detection of another's gaze. In addition, enhanced activation was found in bilateral occipital cortex and in the right prefrontal area for TD/ASD interactions. The authors proposed that this reflects compensatory workload. Furthermore, intra-brain functional connectivity in the right IFG and superior temporal sulcus for TD/ASD pairing was reduced compared with TD/TD pairing. In this way, hyper-scanning allows us to address how the characteristics of two people (e.g., individual characteristics like ASD or social ability or the nature of the relationship, such as friend or family) affect brain activation and how the brain activations of two people are related. Thus, research is needed to provide further insight into the neural substrates of imitative exchanges in ASD under more naturalistic interaction conditions.

References

1. American Psychiatric Association (2013) Diagnostic and statistical manual of mental disorders, 5th edn. American Psychiatric Association, Washington, DC
2. Asperger H (1944) Die"Autistischen Psychopathen"im Kindesalter. Archiv fur Psychiatrie und Nervenkrankheiten 117:76–136
3. Kanner L (1943) Autistic disturbances of affective contact. Nervous Child 2:217–250
4. Nadel J (2002) Imitation and imitation recognition: functional use in preverbal infants and nonverbal children with autism. In: Meltzoff AN, Prinz W (eds) The imitative mind: development evolution and brain basis. Cambridge Univer-sity Press, Cambridge, pp 42–62
5. Williams JH, Whiten A, Singh T (2004) A systematic review of action imitation in autistic spectrum disorder. J Autism Dev Disord 34(3):285–299
6. Meltzoff AN (2005) Imitation and other minds: the "like me" hypothesis. In: Hurley S, Chater N (eds) Perspectives on imitation: from neuroscience to social science, vol 2. MIT Press, Cambridge, pp 55–77
7. Dapretto M, Davies MS, Pfeifer JH, Scott AA, Sigman M, Bookheimer SY, Iacoboni M (2006) Understanding emotions in others: mirror neuron dysfunction in children with autism spectrum disorders. Nat Neurosci 9(1):28–30
8. Jack A, Morris JP (2014) Neocerebellar contributions to social perception in adolescents with autism spectrum disorder. Dev Cogn Neurosci 10:77–92
9. Poulin-Lord MP, Barbeau EB, Soulieres I, Monchi O, Doyon J, Benali H, Mottron L (2014) Increased topographical variability of task-related activation in perceptive and motor associative regions in adult autistics. Neuroimage Clin 4:444–453
10. Williams JH, Waiter GD, Gilchrist A, Perrett DI, Murray AD, Whiten A (2006) Neural mechanisms of imitation and 'mirror neuron' functioning in autistic spectrum disorder. Neuropsychologia 44(4):610–621
11. Delaveau P, Arzounian D, Rotge JY, Nadel J, Fossati P (2015) Does imitation act as an oxytocin nebulizer in autism spectrum disorder? Brain 138:e360
12. Okamoto Y, Kitada R, Tanabe HC, Hayashi MJ, Kochiyama T, Munesue T, Ishitobi M, Saito DN, Yanaka HT, Omori M, Wada Y, Okazawa H, Sasaki AT, Morita T, Itakura S, Kosaka H, Sadato N (2014) Attenuation of the contingency detection effect in the extrastriate body area in autism spectrum disorder. Neurosci Res 87:66–76
13. Cook JL, Bird G (2012) Atypical social modulation of imitation in autism spectrum conditions. J Autism Dev Disord 42(6):1045–1051
14. Cook J, Barbalat G, Blakemore SJ (2012) Top-down modulation of the perception of other people in schizophrenia and autism. Front Hum Neurosci 6:175
15. Chartrand TL, van Baaren R (2009) Chapter 5 human mimicry. Adv Exp Soc Psychol 41:219–274
16. Smith IM, Bryson SE (1994) Imitation and action in autism: a critical review. Psychol Bull 116 (2):259–273
17. di Pellegrino G, Fadiga L, Fogassi L, Gallese V, Rizzolatti G (1992) Understanding motor events: a neurophysiological study. Exp Brain Res 91(1):176–180
18. Gallese V, Fadiga L, Fogassi L, Rizzolatti G (1996) Action recognition in the premotor cortex. Brain 119:593–609
19. Rizzolatti G, Fadiga L, Gallese V, Fogassi L (1996) Premotor cortex and the recognition of motor actions. Brain Res Cogn Brain Res 3(2):131–141

20. Fogassi L, Ferrari PF, Gesierich B, Rozzi S, Chersi F, Rizzolatti G (2005) Parietal lobe: from action organization to intention understanding. Science 308(5722):662–667
21. Rozzi S, Ferrari PF, Bonini L, Rizzolatti G, Fogassi L (2008) Functional organization of inferior parietal lobule convexity in the macaque monkey: electrophysiological characterization of motor, sensory and mirror responses and their correlation with cytoarchitectonic areas. Eur J Neurosci 28(8):1569–1588
22. Chong TT, Cunnington R, Williams MA, Kanwisher N, Mattingley JB (2008) fMRI adaptation reveals mirror neurons in human inferior parietal cortex. Curr Biol 18(20):1576–1580
23. de la Rosa S, Schillinger FL, Bulthoff HH, Schultz J, Uludag K (2016) fMRI adaptation between action observation and action execution reveals cortical areas with mirror neuron properties in human BA 44/45. Front Hum Neurosci 10:78
24. Gazzola V, Keysers C (2009) The observation and execution of actions share motor and somatosensory voxels in all tested subjects: single-subject analyses of unsmoothed fMRI data. Cereb Cortex 19(6):1239–1255
25. Molenberghs P, Cunnington R, Mattingley JB (2009) Is the mirror neuron system involved in imitation? A short review and meta-analysis. Neurosci Biobehav Rev 33(7):975–980
26. Oosterhof NN, Wiggett AJ, Diedrichsen J, Tipper SP, Downing PE (2010) Surface-based information mapping reveals crossmodal vision-action representations in human parietal and occipitotemporal cortex. J Neurophysiol 104(2):1077–1089
27. Brass M, Heyes C (2005) Imitation: is cognitive neuroscience solving the correspondence problem? Trends Cogn Sci 9(10):489–495
28. Sasaki AT, Kochiyama T, Sugiura M, Tanabe HC, Sadato N (2012) Neural networks for action representation: a functional magnetic-resonance imaging and dynamic causal modeling study. Front Hum Neurosci 6:236
29. Peelen MV, Downing PE (2007) The neural basis of visual body perception. Nat Rev Neurosci 8(8):636–648
30. Puce A, Perrett D (2003) Electrophysiology and brain imaging of biological motion. Philos Trans R Soc Lond Ser B Biol Sci 358(1431):435–445
31. Caspers S, Zilles K, Laird AR, Eickhoff SB (2010) ALE meta-analysis of action observation and imitation in the human brain. NeuroImage 50(3):1148–1167
32. Kilner JM, Friston KJ, Frith CD (2007) Predictive coding: an account of the mirror neuron system. Cogn Process 8(3):159–166
33. Iacoboni M, Woods RP, Brass M, Bekkering H, Mazziotta JC, Rizzolatti G (1999) Cortical mechanisms of human imitation. Science 286(5449):2526–2528
34. Fishman I, Datko M, Cabrera Y, Carper RA, Muller RA (2015) Reduced integration and differentiation of the imitation network in autism: a combined functional connectivity magnetic resonance imaging and diffusion-weighted imaging study. Ann Neurol 78(6):958–969
35. Nebel MB, Eloyan A, Nettles CA, Sweeney KL, Ament K, Ward RE, Choe AS, Barber AD, Pekar JJ, Mostofsky SH (2016) Intrinsic visual-motor synchrony correlates with social deficits in autism. Biol Psychiatry 79(8):633–641
36. Hamilton AF (2013) Reflecting on the mirror neuron system in autism: a systematic review of current theories. Dev Cogn Neurosci 3:91–105
37. Yang J, Hofmann J (2016) Action observation and imitation in autism spectrum disorders: an ALE meta-analysis of fMRI studies. Brain Imaging Behav 10(4):960–969
38. Ubaldi S, Barchiesi G, Cattaneo L (2015) Bottom-up and top-down visuomotor responses to action observation. Cereb Cortex 25(4):1032–1041
39. Berger NI, Ingersoll B (2015) An evaluation of imitation recognition abilities in typically developing children and young children with autism Spectrum disorder. Autism Res 8 (4):442–453
40. Gergely G (2001) The obscure object of desire: 'nearly, but clearly not, like me': contingency preference in normal children versus children with autism. Bull Menn Clin 65(3):411–426

41. Gergely G, Watson JS (1999) Early socio-emotional development: contingency perception and the social-biofeedback model. In: Mahwah NJ (ed) Social cognition: understanding others in the first months of life. Lawrence Erlbaum Associates, p 101–36 NJ, United States.
42. Bahrick LE, Watson JS (1985) Detection of intermodal proprioceptive-visual contingency as a potential basis of self-perception in infancy. Dev Psychol 21(6):963–973
43. Downing PE, Jiang Y, Shuman M, Kanwisher N (2001) A cortical area selective for visual processing of the human body. Science 293(5539):2470–2473
44. Astafiev SV, Stanley CM, Shulman GL, Corbetta M (2004) Extrastriate body area in human occipital cortex responds to the performance of motor actions. Nat Neurosci 7(5):542–548
45. Jeannerod M (2004) Visual and action cues contribute to the self-other distinction. Nat Neurosci 7(5):422–423
46. David N, Cohen MX, Newen A, Bewernick BH, Shah NJ, Fink GR, Vogeley K (2007) The extrastriate cortex distinguishes between the consequences of one's own and others' behavior. NeuroImage 36(3):1004–1014
47. David N, Jansen M, Cohen MX, Osswald K, Molnar-Szakacs I, Newen A, Vogeley K, Paus T (2009) Disturbances of self-other distinction after stimulation of the extrastriate body area in the human brain. Soc Neurosci 4(1):40–48
48. Escalona A, Field T, Nadel J, Lundy B (2002) Brief report: imitation effects on children with autism. J Autism Dev Disord 32(2):141–144
49. Field T, Field T, Sanders C, Nadel J (2001) Children with autism display more social behaviors after repeated imitation sessions. Autism 5(3):317–323
50. Contaldo A, Colombi C, Narzisi A, Muratori F (2016) The social effect of "being imitated" in children with autism Spectrum disorder. Front Psychol 7:726
51. Tanabe HC, Kosaka H, Saito DN, Koike T, Hayashi MJ, Izuma K, Komeda H, Ishitobi M, Omori M, Munesue T, Okazawa H, Wada Y, Sadato N (2012) Hard to "tune in": neural mechanisms of live face-to-face interaction with high-functioning autistic spectrum disorder. Front Hum Neurosci 6:268

Chapter 11
Functional Connectivity in Dementia

Hugo Botha and David T. Jones

Introduction

The human brain is organized on many different but interdependent levels (Fig. 11.1). Novel structural and functional techniques have allowed in vivo analysis of large, distributed brain networks. Probing the systems or network level has been of particular interest in the field of neurodegenerative disease as it has become increasingly clear that degenerative diseases target large-scale brain networks, with most dementias having distinct network-level signatures [1]. In this regard, large-scale networks or systems are typically identified and their integrity assessed by evaluating the functional connectivity between regions or subsystems. Functional connectivity generally refers to measures of temporal correlation of a signal among two or more spatially distinct regions. The signal of interest should be a surrogate for neural activity and may be obtained via electrophysiologic means, as in EEG or MEG, or through neuroimaging, such as the BOLD signal in fMRI. Task-free functional MRI (TF-fMRI) is a particularly appealing technique since it is safe, requires minimal patient cooperation (important in diseased populations), can be performed on most commercial MRI machines, and results in data that can be shared between centers easily [2]. It is little surprise that it has emerged as the modality of choice to study functional connectivity in dementia. Because of this, it will be the primary focus of this review of functional connectivity as it pertains to dementia. In the first section, we will provide a brief introduction to TF-fMRI, including the principles underlying the BOLD signal, image preprocessing, and data analysis. Next, we review age-related connectivity changes before moving on to Alzheimer's disease, which will be our primary focus within the dementias. We then briefly review alterations in function connectivity in non-Alzheimer's dementias. Finally,

H. Botha (✉) · D. T. Jones
Department of Neurology, Mayo Clinic, Rochester, MN, USA
e-mail: Botha.Hugo@mayo.edu; Jones.David@mayo.edu

© Springer International Publishing AG, part of Springer Nature 2018

C. Habas (ed.), *The Neuroimaging of Brain Diseases*, Contemporary Clinical Neuroscience, https://doi.org/10.1007/978-3-319-78926-2_11

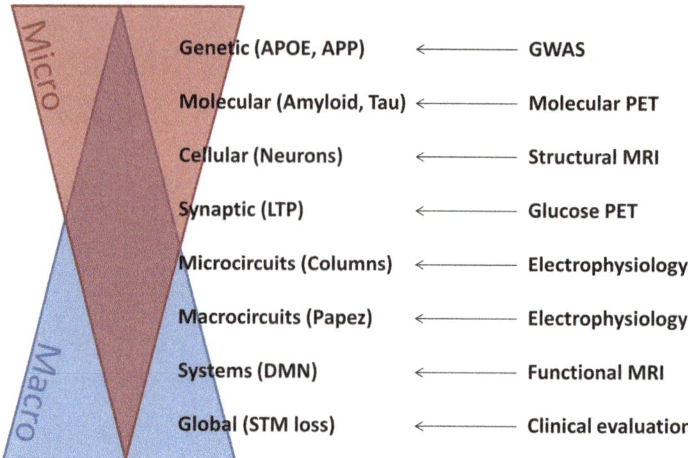

Fig. 11.1 Bidirectional hierarchical multiscale organization of the brain. Micro level phenomena give rise to macro level manifestations in a bottom-up manner (e.g., risk factor genes resulting in abnormal proteins, neuronal dysfunction, and eventual memory loss), but at the same time, macro level phenomena alter the micro-environment (e.g., stress resulting in system-level dysfunction and eventually altered genetic expression). Abbreviations: APOE apolipoprotein E, APP amyloid precursor protein, DMN default mode network, GWAS genome-wide association study, LTP long-term potentiation, MRI magnetic resonance imaging, PET positron emission tomography, STM short-term memory

we discuss the future of functional connectivity in neurodegenerative research and clinical practice.

Overview of Task-Free fMRI

A detailed discussion of task-free functional fMRI (TF-fMRI) data acquisition, preprocessing, and analysis is beyond the scope of this chapter. However, some familiarity with the process is necessary to appreciate the connectivity literature, and as such we have provided a brief overview of the process and point out aspects that may influence measures of connectivity, particularly in older or impaired subjects.

The BOLD Signal and Task-Free fMRI

The principle underlying fMRI is the relationship between neuronal activity and the blood-oxygen-level dependent (BOLD) signal, a direct result of neurovascular coupling. In essence, the sequence used in fMRI is sensitive to changes in the decay of transverse magnetization, which results from molecular interactions and local inhomogeneities in the magnetic field (together represented by the time constant T2*). Deoxygenated blood is paramagnetic, and a change in its ratio to oxygenated blood, which is diamagnetic, will result in a measurable change in the T2*. By virtue of neurovascular coupling, increased neural activity results in a rise in oxygenated blood that exceeds the increase in deoxygenated blood from the change in activity. The reduction in paramagnetic deoxyhemoglobin results in less signal loss or an increase in the measured MRI signal, which can then be localized spatially. Unsurprisingly many other factors also result in changes in this signal, and a complex set of preprocessing steps (see below) are required to extract the signal changes thought to result from local neuronal activity.

Early fMRI work tended to use task-based designs, where the differences in BOLD signal between volumes during the task and those at rest represented the task-related neural activity. Several spatially distributed networks were identified using such designs. However, in the mid-1990s, Bhrat Biswal and colleagues observed that changes in the BOLD signal occurring at rest, at low frequency (0.01–0.1 Hz), were correlated between functionally related areas [3]. Subsequent work showed that this signal can be used to recapitulate the networks found in task-related designs [4]. This important realization – that correlated activity in specialized networks was present at all times – leads to the rise of "resting state" or "task-free" fMRI designs (TF-fMRI), from which "intrinsic connectivity networks" (ICNs) were derived [5]. Among its many benefits, the fact that TF-fMRI work could be done without the need for MRI-compatible task-related equipment and required minimal cooperation from subjects were especially helpful, particularly as it allowed for multicenter collaborative studies on cognitively impaired subjects. With the emergence of open-source software packages for preprocessing and analysis of fMRI data, TF-fMRI rose to the most widely implemented imaging modality for evaluating functional connectivity in humans.

Preprocessing

Preprocessing refers to the series of steps that are undertaken prior to image analysis, usually aimed at reducing contamination of the BOLD signal by non-neuronal sources. For example, subject motion, cardiac and respiratory cycle-related changes, changes in the magnetic field, and non-neuronal intracranial sources of signal such as white matter and CSF need to be accounted for.

There is no single, universally accepted preprocessing pipeline, and a discussion of the trade-offs and controversies regarding particular approaches are beyond the scope of this article. Furthermore, although some preprocessing steps are nearly always performed it is important to realize that the order of steps affect the final results too. With that in mind, we will only touch on the most common aspects of preprocessing and what these steps try to address.

The first few volumes of a series are usually removed, since it takes a while to reach steady-state magnetization, and the initial volumes are more contaminated.

Although not universally applied, despiking of the time series of each voxel can improve realignment and reduce motion artifacts [6]. The sequential and interleaved nature of MRI acquisition results in the later slices in volumes being acquired significantly later than the first slices and significant discrepancies in acquisition time between adjacent slices. Slice time correction addresses this by temporal interpolation, by which the signal at a specific time point (usually the middle slice in interleaved acquisition) is estimated from nearby time points at each voxel. Of note, performing either of these steps prior to estimating motion will tend to reduce motion estimates [7].

Motion correction or realignment tools usually assume that only head positional changes (as opposed to size and shape) need to be accounted for, so called "rigged body transformation." As such these methods cannot account for disruption of image intensity that may occur if motion happens during the acquisition phases. Although motion can be measured directly, it is usually estimated from the fMRI signal. Motion correction usually involves selecting a target image (e.g., the middle image) and then estimating the relative displacement (e.g., right-left, anterior-posterior, superior-inferior, pitch, roll, and yaw) between a given time point and the reference image.

There are several options for addressing non-neuronal physiologic variables, such as artifacts from the respiratory and cardiac cycles. Some involve additional monitoring, such as pulse oximetry, in a manner that is time locked with the MR acquisition. Others involve creating a "noise ROI" consisting of white matter and CSF and regressing out the signal from this ROI [8]. Increasingly, exploratory data analysis tools such as independent component analysis (ICA) are used to detect artifacts which can then be removed from the data [9]. For example, a component that changes significantly between slices may represent motion, given the interleaved nature of most acquisitions, as may a component time course with large spikes. Nuisance variables can also be identified using these methods and then regressed out, for example, physiological sources of motion or alterations in the magnetic field induced by the cardiac and respiratory cycles.

Spatial normalization allows for individual scans to be aligned (subject space) and to be transformed to a reference image (template space) to allow for comparison between subjects. This is crucial for group analyses, since it reduces individual variability and also allows for inter-study comparisons when standard coordinate systems are used (standard space). Spatial smoothing can increase the signal-to-noise ratio and reduce between subject mismatches. Because TF-fMRI primarily involves the analysis of low-frequency fluctuations, band-pass filtering is usually applied (e.g., 0.009–0.08 Hz).

There are several other possible steps and alternative techniques for addressing the abovementioned issues that we have not discussed, including distortion correction, silencing of volumes that contain motion beyond a specified degree, regression of the "global signal," and so on and so forth (Fig. 11.2).

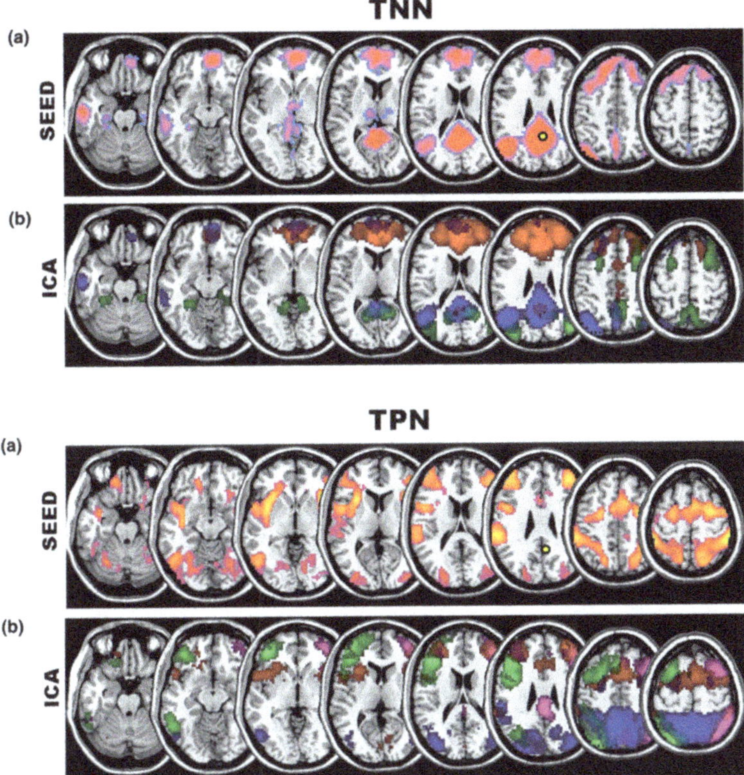

Fig. 11.2 The task-negative network (TNN), shown in the top panel, and the task-positive network, shown in the bottom panel, identified with both seed-based and low-dimensional ICA in a group analysis of 341 elderly healthy control subjects. (Adapted from [10])

Data Analysis

In order to assess the system-level changes in aging and disease, the four-dimensional preprocessed fMRI sequences have to be analyzed within a particular framework, which usually result in temporal information (e.g., a particular time course), spatial information (e.g. maps of a particular network), and/or summary metrics of network topology (as in graph theory) that can be correlated with clinical metrics or compared between groups. As will become clear later, it is important to note that the control group used in such analyses should be age matched as aging is associated with several network-level changes.

Seed-Based/Model-Based

The simplest and most widely used analysis methods involve placing a seed, or multiple seeds, at a particular location and then extracting the time series from all the voxels in the seed. The average time series can then be compared to the time series at a different seed (seed-to-seed), with some measure of similarity (usually correlation) taken as surrogate for functional connectivity. Alternatively, the time series from the seed can be regressed against the signal from the rest of the voxels in the brain (seed-to-brain), yielding a spatial map of voxels that have similar time series. These spatial maps can be compared between groups, or used in a model with clinical variables. Seed based analyses usually depend on a model, in that the regions have to be selected at the outset. This may be based on activation loci from task-based fMRI studies or atrophy patterns from volumetric imaging for example.

Data-Driven/Model-Independent

In contrast, several data-driven methods do not require the a priori selection of regions or seeds. The most widely used method, which will be our focus here, is independent component analysis (ICA). ICA is a modality independent method to separate a multivariate signal into statistically independent subcomponents. Given the multivariate nature of the BOLD signal, where even the portion that is the result of neural activity is the summation of activity in several networks, ICA is ideally suited for TF-fMRI analysis. Typically, the ICA is run on group level, either in the population being studied or an independent group, which results in spatial maps for independent networks. These are then used in a spatial-temporal dual regression to back-construct subject level spatial maps for each network and to extract the mean time series for the network. The spatial maps can be compared between groups, or correlated with clinical measures, and often these are converted to z-scores which allow for easy analysis of the clinical correlates of group differences. The time series can be used to assess the between-network connectivity. There are ICA toolboxes available for most imaging analysis platforms (e.g., the Group ICA Toolbox from MIALAB available at https://www.nitrc.org/projects/gift/), and there are population-based connectivity atlases containing the spatial maps from ICA in large cohorts of cognitively normal subjects (e.g., the Mayo Clinic Study of Aging Functional Connectivity Atlas [11]).

Graph Theory

Graph theory is a branch of mathematics that is central to much of the modern "network neuroscience." It is premised on representing a system or network as a collection of nodes, with the interaction between these represented by edges. From its origin in the eighteenth century with Euler's solution to the Bridges of

Königsberg problem, it has been applied to many systems, across numerous disciplines. For instance, the so-called 6 degrees of separation observed by Travers and Milgram [12] during their famous experiment in the 1960 was formalized mathematically using graph theory by Watts and Strogatz [13]. The resulting network revealed the now famous small-world phenomenon – that is, a high clustering coefficient (local regularity) but short average path length as a result of hubs connecting the clusters, resulting in efficient information transfer. Brain networks have been shown to exhibit these properties, and based on the network degeneration that is central to neurodegenerative disease, graph theory metrics are increasingly being used to study these disorders.

A particularly appealing aspect of graph theory is the relative ease with which summary metrics of network organization and function can be derived. This simplifies the task of comparing the complex, multidimensional activity of the human brain among subjects. With regard to network construction, brain regions are typically used as nodes or vertices and the Pearson correlation coefficient of the time series as the edges between them. A limitation of correlation-based metrics not discussed above is "transitive closure," whereby temporal correlation as a result of indirect connections is typically represented as direct edges between nodes, resulting in networks that are denser, more modular, and with a higher clustering coefficient than expected based on the underlying anatomy.

A network can be represented as an adjacency matrix, where the rows and columns represent nodes and the values in the cells represent the edges between them. Further analyses may involve thresholding the network, maintaining only edges above a certain cut point, which may then be binarized or left in a weighted state. It is typically recommended to explore the graph properties of interest across a range of thresholds. Alternatively, graph metrics can be calculated on the fully connected network, which may offer advantages. Commonly, global metrics are calculated, such as characteristic path length (mean shortest path between all pairs of nodes), global efficiency (inverse of characteristic path length), or measures of modularity, although there are indices calculated within a module or even at the nodal level, such as betweenness centrality (proportion of shortest paths that pass through the node). The resulting graph properties should be compared to appropriate null models. These measures can also be compared between groups or correlated with clinical data. There are also statistical frameworks for comparing networks in their entirety in a t-test or ANOVA design (e.g., the Network Based Statistics Toolbox from Andrew Zalesky available at https://www.nitrc.org/projects/nbs/).

Despite the impressive growth of "network neuroscience," its application to degenerative disease is still in its infancy, and many unresolved theoretical and methodological issues remain. These are beyond the scope of this chapter, but the interested reader is directed to the recently published *Fundamentals of Brain Network Analysis* [14].

Age-Related Connectivity Changes

There have been hundreds of studies using TF-fMRI to study "normal aging," using seed-based, data-driven, and graph theoretical analyses (for recent reviews, see [15] and [16]). One ongoing issue is the definition of "normal aging" and separating this from preclinical degenerative disorders or untreated psychiatric disease. Most studies have focused on adults across a range of ages, with no evidence of cognitive impairment and no history of psychiatric diagnoses, often reporting a diffuse increase in predominantly inter-network correlation with increasing age [17, 18]. In keeping with this, others have shown that networks that are functionally independent in younger adults have more inter-network or inter-module connections [19, 20], and may even merge, in older subjects [18]. Work in this area also suggests a decrease in complexity with older age [21]. Some have proposed a "dedifferentiation" process that occurs in the aging brain [22, 23]. In addition to increases in positive correlations, there are reductions in negative correlations (anticorrelations) between networks in the aging brain, most notably between the DMN and the attentional network(s) [17, 24, 25]. Interestingly, this loss of anticorrelation between the task-negative DMN and task-positive networks is a reversal of an important neurodevelopmental stage occurring in late adolescence or early adulthood [26].

In contrast to the abovementioned between-network changes, aging is typically associated with decreases in positive correlations *within* networks, especially the DMN, salience, and executive control networks [27, 17]. Functional integration (represented by positive correlations) and segregation (represented by anticorrelations) are both required for optimal information processing [28–30] and are both dependent on structural integrity. As such, it may be that the changes discussed above result from age-related brain "disconnectivity" [24], although it should be noted that the changes remain significant after correcting for atrophy (e.g., [17] and several others). Other factors often implicated include neurotransmitter deficits, especially dopamine, loss of white matter integrity, and amyloid deposition or preclinical Alzheimer's more generally. This latter possibility is important to consider, since a limitation in many of the studies on aging is the fact that subjects are usually not screened with Alzheimer's disease biomarkers such as amyloid PET. Some have argued that a large proportion of the connectivity changes involving the DMN and dorsal attention networks observed as part of "normal" aging is in fact driven by preclinical Alzheimer's pathology [31], but an alternative interpretation would be that preclinical and early Alzheimer's, marked by amyloid accumulation, results in accelerated changes seen in aging [32] (discussed below). It has also been suggested that changes in some networks, such as the executive control and sensorimotor networks, and the interaction between the DMN and the dorsal attention network are more independent of AD-related changes [31]. A further area of controversy relates to the cause of the age-related changes – whether or not is due to network reorganization or the result of, for instance, injury or vascular insults. The former seems to be favored, but numerous models have been proposed to account for these changes, such as hemispheric asymmetry reduction in older adults (HAROLD)

[33], compensation-related utilization of neural circuits (CRUNCH) [34], and posterior-anterior shift with aging (PASA) [35]. With age being the single greatest risk factor for neurodegenerative disease, and a major driver of phenotype within a given degenerative disease, it is clear that a better understanding of age-related changes is needed.

Alzheimer's Disease

First described by Alois Alzheimer in 1906, Alzheimer's disease is a neurodegenerative disease characterized by progressive cognitive decline leading to dementia and is associated with accumulation of amyloid plaques and neurofibrillary tangles consisting of tau. Terminology in this area can be confusing, as Alzheimer's disease is at times used to refer to cases with autopsy-confirmed accumulation of amyloid and tau, while it is also used to refer to a clinical presentation of amnestic cognitive impairment leading to dementia. We will use Alzheimer's disease dementia to refer to the typical clinical presentation characterized by early amnesia, medial temporal atrophy, and posterior cingulate and lateral temporoparietal hypometabolism, which progresses over time to involve other cognitive faculties with associated temporoparietal and frontal atrophy and hypometabolism. Atypical Alzheimer's disease dementia will refer to the group of phenotypes that do not follow this pattern, such as the language variant (logopenic progressive aphasia), dysexecutive variant (usually early onset), and visual variant (posterior cortical atrophy). Alzheimer's disease will refer to the disease process, necessitating the presence of amyloid and tau, confirmed either through molecular imaging or post mortem examination. These distinctions are far from trivial, since the phenotype is closely linked to the spatial distribution of connectivity changes and systems affected, and the clinical diagnosis of Alzheimer's disease dementia is associated with Alzheimer's disease pathology or positive AD biomarkers in only 75–80% of cases. Mild cognitive impairment refers to objective evidence of cognitive impairment but not meeting criteria for dementia owing to minimal impairment in activities of daily living. Amnestic MCI refers to the subset with memory impairment, either in isolation or with other domains, and represents an early stage of cognitive impairment through which patients with Alzheimer's disease dementia pass. While Alzheimer's disease is by no means the only cause of amnestic MCI among older subjects, for our purposes, we will consider it as a prodromal stage of Alzheimer's disease dementia.

Amnestic MCI and Alzheimer's Disease Dementia

Since the seminal observation that the DMN is affected in Alzheimer's disease dementia [36], it has been the primary focus for connectivity work in the field, with earlier studies analyzing the network as a single system and later studies breaking it into sub-components. Viewing the DMN as made up of smaller networks makes sense, given the different cognitive functions attributed to the different components, but is also crucial to capture the changes that occur over the course of aging and Alzheimer's disease dementia [37]. Changes in DMN connectivity in amnestic MCI and early Alzheimer's disease closely resemble those seen in the aging brain [32]. However, the decrease in functional connectivity between anterior and posterior portions of the DMN is more severe, as is the decreased within network connectivity in the posterior DMN. The early connectivity literature was discordant regarding primarily frontal lobe and anterior DMN connectivity, with some reporting increased connectivity and some reporting decreased connectivity.

As it turns out, these results are not incompatible. Subsequent work has shown that networks might undergo phases of increased and decreased connectivity over the course of normal aging or disease. Studies that split amnestic MCI into mild-moderate or early and moderate-severe or late groups have shown that the milder stage is associated with increased DMN connectivity, which declines in the more severe phase [38, 39]. This is consistent with longitudinal work, showing an initial increase in hippocampal connectivity to the posterior DMN, declining later, and an ongoing decline in posterior connectivity in MCI with increased anterior connectivity over a 3-year follow-up [40, 41]. The Alzheimer's disease dementia literature is consistent with this sequence of events. There is ongoing posterior and anterior disconnection, with longitudinal decline in posterior DMN integrity and a biphasic connectivity pattern in the anterior DMN, where increased connectivity is seen earlier and decreased connectivity later [42]. The subsequent decline in frontal connectivity may indicate a failure in compensation, or failure of a region overloaded by processing demands from the original failing network, signaling impending clinical deterioration [43] (see "Connectivity based models" below). Several studies have shown convincingly that these DMN connectivity changes can differentiate patients from controls and that connectivity metrics correlate with clinical symptoms [44–46]. In addition, a decline in angular gyrus connectivity was shown to predict conversion from aMCI to dementia [47], as did the severity of decrease DMN connectivity [48], arguing that connectivity measures may serve as biomarkers (see "Biomarkers" below).

Atypical Alzheimer's Disease Dementia

While this focus on the DMN in typical, late-onset, amnestic Alzheimer's disease dementia seems justified, there are several atypical phenotypes. As expected, the network disruptions and changes over time differ based on phenotype. The posterior DMN remains vulnerable and is involved in these variants too, but other specific networks, depending on the phenotype, are involved early [49, 50]. Early-onset Alzheimer's disease dementia, often characterized clinically by executive and working memory dysfunction, is associated with disruptions in the salience and executive control networks [51, 52]. Logopenic progressive aphasia is associated with disrupted language network connectivity, although younger subjects also have left working memory involvement, which correlates with clinical measures [53]. Posterior cortical atrophy results in early primary and higher order visual network disruption [49, 54]. More longitudinal studies are needed in atypical phenotypes to determine if there are phases of increased connectivity involving some of these non-DMN networks and to determine candidate biomarkers.

Preclinical and At-Risk Populations

In anticipation of the fact that disease-modifying therapies are likely to be most effective if instituted early in the disease course, and ideally before cognitive impairment, much of the biomarker literature has focused on early identification. There are several options for studying preclinical Alzheimer's. Although rare, autosomal dominant Alzheimer's disease can result from mutations in presenilin (*PS*) or the amyloid precursor protein (*APP*) genes. Mutation carriers can be identified decades before symptoms. There are also risk-modifying polymorphisms, the most common of which is the APOE *E4* allele, which confers an increased risk for Alzheimer's disease. Even in the absence of any of these known genetic markers, families with a large burden of late-life Alzheimer's disease can highlight potentially undiscovered polymorphisms or lifestyle and environmental factors. Finally, biomarkers such as PET imaging of amyloid, and more recently tau, allow for an in vivo diagnosis of Alzheimer's disease long before symptom onset. Connectivity has been assessed in all of these populations.

A recent cross-sectional study showed that posterior DMN connectivity changes were present in asymptomatic APOE *E4* allele carriers before any molecular (e.g., amyloid PET) abnormalities were detected [55]. Mirroring the Alzheimer's dementia results discussed above, both increased and decreased connectivity between the hippocampus and the precuneus/posterior cingulate (posterior DMN) and medial prefrontal cortex/anterior cingulate has been documented in asymptomatic E4 allele and PS/APP mutation carriers [56–58]. Presumably the same pattern of increased connectivity followed by a decline will be found in longitudinal studies, since connectivity appears to decline in APP/PS mutation carriers as the time of expected disease onset approaches [59]. Cognitively normal subjects with a family history of Alzheimer's disease dementia, and were APOE E4 negative, have been shown to have reduced posterior cingulate to medial temporal connectivity compared to age-matched controls [60].

Finally, with regard to biomarker-based preclinical populations, a longitudinal study found that incident amyloid PET-positive subjects, meaning subjects that became amyloid positive during the study, had increased posterior DMN connectivity at baseline [61]. More recently, researchers have used amyloid and tau PET in cognitively normal subjects to show that amyloidosis in the absence of significant tau is associated with increased DMN connectivity, whereas tau is associated with

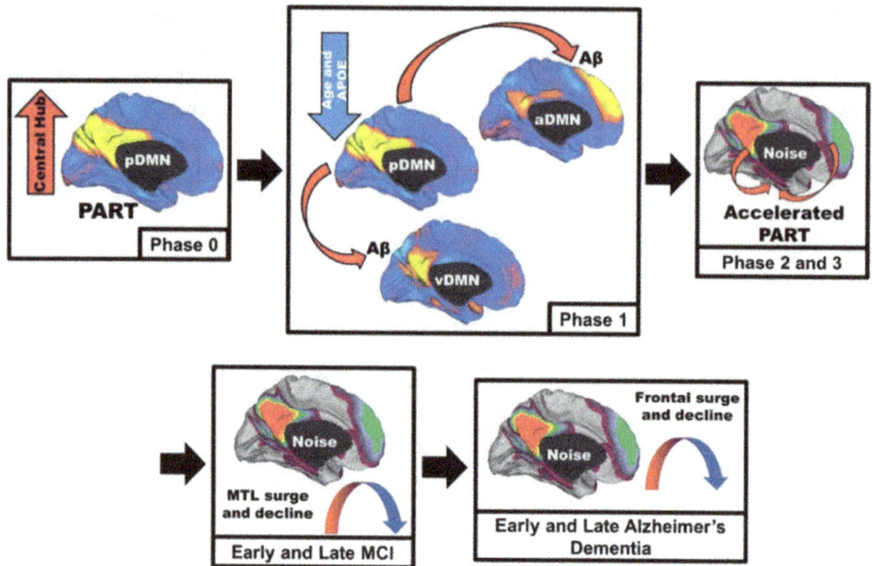

Fig. 11.3 The cascading network failure model of Alzheimer's disease. (Permission pending from [43])

decreased connectivity, supporting the sequence of hyper- and hypo-connectivity discussed earlier [62–64].

Connectivity-Based Models of Alzheimer's Disease

Recently, Jones et al. [43] proposed a model for Alzheimer's disease at the systems level based on a cross-sectional study of DMN connectivity patterns across the various stages of the disease. The authors suggest that connectivity changes follow a predictable pattern, starting with the highly connected pDMN and progressing, through phases of increased connectivity, along a posterior-anterior gradient. Furthermore, the network changes mirror the clinical, structural imaging and glucose PET imaging trajectories mentioned above (See Fig. 11.3). This framework, termed the cascading network failure model, can incorporate much of the abovementioned literature on network changes in aging, at risk individuals, aMCI and ADD, and the transient phases of increased connectivity followed by connectivity decline spreading through various systems may explain some of the discordant results mentioned previously. For example, prior longitudinal work has confirmed that the decline in pDMN connectivity is initially accompanied by increased connectivity in the anterior and ventral DMN, both of which then decline during follow-up as these subsystems fail under increased load. Cross-sectional studies of the early phase would report increased connectivity, while those in the latter stage will report decline. The model also relates network changes to the molecular and cellular pathophysiology of AD in a way that can account for the spatial and temporal discrepancies between amyloid and tau. Several prior studies have shown that amyloid deposition as measured by amyloid PET occurs primarily in hubs of high connectivity, most notably the pDMN. In fact, increased connectivity in the pDMN predicted incident amyloidosis in a large cohort [61]. The authors of the cascading network model argued that tau deposition patterns are associated with distinct connectivity profiles in different phases of the disease, with the relationship being mediated by amyloid especially early on [62, 65]. Advances in our understanding of the connectivity changes in AD have led to connectivity-based biomarkers discussed further below.

Non-Alzheimer's Dementias

A detailed discussion of all the non-Alzheimer's dementias is beyond the scope of this chapter. We will briefly review significant findings for three groups of disorders: the typical and atypical parkinsonian disorders with associated dementia, the progressive aphasias and apraxia of speech, and finally the behavioral variant frontotemporal dementia. It is worth noting, however, that connectivity changes have been documented in many diseases not covered here, such as clinical and preclinical Huntington's disease, vascular dementia, et cetera.

The Parkinsonian Dementia Syndromes

Parkinson's disease and dementia with Lewy bodies (DLB) are both characterized pathologically by intra-neuronal inclusions of alpha-synuclein. While DLB subjects have dementia by definition, only a subset of Parkinson's disease subjects is diagnosed with dementia at some point. However, some degree of cognitive impairment is seen in most subjects at some point over the course of the disease. Parkinson's disease subjects have been shown to have reduced posterior DMN connectivity in the absence of cognitive impairment, whereas those with impairment had reduced connectivity in the frontoparietal control network [66]. DLB subjects had reduced connectivity of dorsal and ventral attention networks [67], with left frontoparietal connectivity associated with the frequency and severity of fluctuation [68]. Parkinson's disease subjects with dementia had similar but less severe functional connectivity disruptions as DLB subjects, except for the motor system where connectivity was more disrupted in Parkinson's disease [69]. Some studies have reported increased connectivity in DLB compared to controls, involving the posterior cingulate [70].

As for atypical parkinsonian disorders, progressive supranuclear palsy (PSP), characterized by parkinsonism, vertical gaze paresis, and early falls, often with associate neuropsychological or behavioral deficits resembling frontotemporal dementia, has been evaluated using TF-fMRI in a few studies. Reduced connectivity involving the prefrontal cortex, supplementary motor area, insula, deep gray nuclei, midbrain, and cerebellum has been reported [71–73], with some of these findings correlating with cognitive and motor impairment. Increased connectivity has also been reported between the midbrain and thalamus [71] and the dentate and subcortical structures [74].

Progressive Aphasia and Apraxia of Speech

Using the area of peak atrophy in patients as a seed for a seed-to-brain functional connectivity analysis has been popular in PPA. There is significant overlap between the overall atrophy pattern and the spatial map in controls for the semantic and agrammatic variants [75, 76]. In addition, functional connectivity differences between subjects with agrammatic PPA and controls may be evident before changes in gray matter volume [77]. Connectivity measures have also been linked to clinical severity, with the core deficits in semantic dementia being associated with reduced degree centrality of the left anterior hippocampus and reduced connectivity between it and the left temporal pole, insula and left middle frontal gyrus [78] and connectivity within the language network negatively correlated with semantic task accuracy [79]. Progressive apraxia of speech, a disorder of motor planning sparing language function for several years, was linked to reduced connectivity between the supplementary motor area and the rest of the speech and language network [80].

Behavioral Variant Frontotemporal Dementia

The earliest network affected in behavioral variant frontotemporal dementia (bvFTD) appears to be the salience network [1, 81, 82]. Researchers have shown that connectivity in the salience network correlate with psychologic and behavioral measures in controls, and several of the core neuropsychological and neuropsychiatric deficits in bvFTD have been linked to connectivity changes [83]. For example, baseline salience network connectivity predicted subsequent behavioral changes [84], global network properties were shown to correlate with executive dysfunction [85], and prefrontal hyperconnectivity has been associated with apathy [86]. In fact, it has been argued that distinct subtypes of bvFTD results from different patterns of network disruption [87]. Unsurprisingly, some studies have found increased connectivity too, most notably in parts of the DMN, the dorsal attention network and regions within the prefrontal lobe [88, 86, 89].

The three most common causes of genetic bvFTD are mutations in progranulin (GRN) and microtubule-associated protein tau (MAPT) and hexanucleotide repeat expansions in C9orf72. Connectivity changes mimicking those in the symptomatic phase have been found in preclinical mutation carriers, much like was the case for

Alzheimer's disease [90]. Asymptomatic MAPT mutation carriers were found to have decreased connectivity in the temporal and anterior DMN regions and increased connectivity in posterior DMN regions [82]. Symptomatic C9orf72 carriers had disruption of salience and sensorimotor networks similar to those seen in non-carriers [91], with similar but less severe changes found in the presymptomatic phases [92]. Preclinical GRN mutation carriers had frontal and parietal connectivity changes decades before expected symptom onset [93].

Future of TF-fMRI and Functional Connectivity

Technical Advances

The field of connectomics has grown exponentially over the past decade. Software and hardware advances continue to expand the use of MRI in general and fMRI in particular. Multiband imaging, where several slices are excited and acquired simultaneously, is increasingly being used, which allows for a much higher temporal and spatial resolution. Higher-strength (7 T) MR scanners have advanced the spatial resolution at which we can probe the brain. Graph metrics that were originally calculated on static, thresholded networks can now be computed on unthresholded, multilayer, and multimodel networks. Advances in cortical parcellation and imaging analyses pipelines that resulted from the human connectome project will be applied to large cohorts of Alzheimer's and frontotemporal dementia subjects in the near future.

Connectivity-Based Biomarkers

Given the association between clinical measures and connectivity metrics, discussed above, and the ability for TF-fMRI to differentiate patients from controls before structural MR abnormalities are evident, connectivity-based biomarkers are an active area of research. In the case of Alzheimer's disease, the "network failure quotient" was recently proposed [94]. This is defined as the ratio between posterior to anterior dorsal and posterior to ventral DMN (numerator) divided by the intrinsic connectivity in the posterior and ventral DMN (denominator). It performed well when Alzheimer's disease patients were compared to controls and other disorders and correlated well with numerous clinical measures. Although more work is needed in this area, there is a real possibility that connectivity-based biomarkers will form part of subsequent trial designs.

Challenges

Despite the advances, several challenges remain. Firstly, most connectivity metrics have been defined at group level, whereas any clinical application will need single-subject resolution. Secondly, the lack of standardized and generalizable connectivity analyses, both due to hardware and software constraints, has made multicenter studies harder to complete. Thirdly, most current connectivity studies in disease start with clinically phenotyped subjects, which may limit the information that can be garnered from the data. It may be helpful to start with the imaging data and use data-driven approaches to group subjects. Lastly, there are several statistical pitfalls that plague fMRI research.

Conclusions

In this chapter, we briefly reviewed the principles behind functional connectivity analysis and discussed some of the important findings in normal aging, Alzheimer's disease dementia and non-Alzheimer's dementias. Although the application of TF-fMRI to neurodegenerative disease is still relatively new, several important studies have established that connectivity measures have the potential to aid the early diagnosis, longitudinal progression tracking, and phenotyping of these disorders. Future work will have to identify robust biomarkers for the various disorders and phenotypes, standardize the analysis techniques to allow for multicenter studies, and bring connectivity measures to the single subject level.

References

1. Seeley WW, Crawford RK, Zhou J, Miller BL, Greicius MD (2009) Neurodegenerative diseases target large-scale human brain networks. Neuron 62(1):42–52. https://doi.org/10.1016/j.neuron.2009.03.024
2. Fleisher AS, Sherzai A, Taylor C, Langbaum JB, Chen K, Buxton RB (2009) Resting-state BOLD networks versus task-associated functional MRI for distinguishing Alzheimer's disease risk groups. Neuroimage 47(4):1678–1690. https://doi.org/10.1016/j.neuroimage.2009.06.021
3. Biswal B, Yetkin FZ, Haughton VM, Hyde JS (1995) Functional connectivity in the motor cortex of resting human brain using echo-planar MRI. Magn Reson Med 34(4):537–541
4. Smith SM, Fox PT, Miller KL, Glahn DC, Fox PM, Mackay CE, Filippini N, Watkins KE, Toro R, Laird AR, Beckmann CF (2009) Correspondence of the brain's functional architecture during activation and rest. Proc Natl Acad Sci U S A 106(31):13040–13045. https://doi.org/10.1073/pnas.0905267106
5. Damoiseaux JS, Rombouts SA, Barkhof F, Scheltens P, Stam CJ, Smith SM, Beckmann CF (2006) Consistent resting-state networks across healthy subjects. Proc Natl Acad Sci U S A 103 (37):13848–13853. https://doi.org/10.1073/pnas.0601417103

6. Jo HJ, Gotts SJ, Reynolds RC, Bandettini PA, Martin A, Cox RW, Saad ZS (2013) Effective preprocessing procedures virtually eliminate distance-dependent motion artifacts in resting state FMRI. J Appl Math 2013. https://doi.org/10.1155/2013/935154

7. Power JD, Plitt M, Kundu P, Bandettini PA, Martin A (2017) Temporal interpolation alters motion in fMRI scans: magnitudes and consequences for artifact detection. PLoS One 12(9): e0182939. https://doi.org/10.1371/journal.pone.0182939

8. Behzadi Y, Restom K, Liau J, Liu TT (2007) A component based noise correction method (CompCor) for BOLD and perfusion based fMRI. Neuroimage 37(1):90–101. https://doi.org/10.1016/j.neuroimage.2007.04.042

9. Pruim RH, Mennes M, van Rooij D, Llera A, Buitelaar JK, Beckmann CF (2015) ICA-AROMA: a robust ICA-based strategy for removing motion artifacts from fMRI data. Neuroimage 112:267–277. https://doi.org/10.1016/j.neuroimage.2015.02.064

10. Vemuri P, Jones DT, Jack CR Jr (2012) Resting state functional MRI in Alzheimer's Disease. Alzheimer's Res Ther 4(1):2. https://doi.org/10.1186/alzrt100

11. Jones DT, Vemuri P, Murphy MC, Gunter JL, Senjem ML, Machulda MM, Przybelski SA, Gregg BE, Kantarci K, Knopman DS, Boeve BF, Petersen RC, Jack CR Jr (2012) Non-stationarity in the "resting brain's" modular architecture. PLoS One 7(6):e39731. https://doi.org/10.1371/journal.pone.0039731

12. Travers J, Milgram S (1969) An experimental study of the small world problem. Sociometry 32(4):425–443. https://doi.org/10.2307/2786545

13. Watts DJ, Strogatz SH (1998) Collective dynamics of "small-world" networks. Nature 393 (6684):3

14. Fornito A, Zalesky A, Edward T (2016) Bullmore. Academic Press, San Diego. ISBN 9780124079083

15. Sala-Llonch R, Bartres-Faz D, Junque C (2015) Reorganization of brain networks in aging: a review of functional connectivity studies. Front Psych 6:663. https://doi.org/10.3389/fpsyg.2015.00663

16. Damoiseaux JS (2017) Effects of aging on functional and structural brain connectivity. Neuroimage. https://doi.org/10.1016/j.neuroimage.2017.01.077

17. Ferreira LK, Regina AC, Kovacevic N, Martin Mda G, Santos PP, Carneiro Cde G, Kerr DS, Amaro E Jr, McIntosh AR, Busatto GF (2016) Aging effects on whole-brain functional connectivity in adults free of cognitive and psychiatric disorders. Cereb Cortex 26 (9):3851–3865. https://doi.org/10.1093/cercor/bhv190

18. Geerligs L, Renken RJ, Saliasi E, Maurits NM, Lorist MM (2015) A brain-wide study of age-related changes in functional connectivity. Cereb Cortex 25(7):1987–1999. https://doi.org/10.1093/cercor/bhu012

19. Chan MY, Park DC, Savalia NK, Petersen SE, Wig GS (2014) Decreased segregation of brain systems across the healthy adult lifespan. Proc Natl Acad Sci U S A 111(46):E4997–E5006. https://doi.org/10.1073/pnas.1415122111

20. Song J, Birn RM, Boly M, Meier TB, Nair VA, Meyerand ME, Prabhakaran V (2014) Age-related reorganizational changes in modularity and functional connectivity of human brain networks. Brain Connect 4(9):662–676. https://doi.org/10.1089/brain.2014.0286

21. Grady CL, Garrett DD (2014) Understanding variability in the BOLD signal and why it matters for aging. Brain Imaging Behav 8(2):274–283. https://doi.org/10.1007/s11682-013-9253-0

22. Park DC, Reuter-Lorenz P (2009) The adaptive brain: aging and neurocognitive scaffolding. Annu Rev Psychol 60:173–196. https://doi.org/10.1146/annurev.psych.59.103006.093656

23. Goh JO (2011) Functional dedifferentiation and altered connectivity in older adults: neural accounts of cognitive aging. Aging Dis 2(1):30–48

24. Wu JT, Wu HZ, Yan CG, Chen WX, Zhang HY, He Y, Yang HS (2011) Aging-related changes in the default mode network and its anti-correlated networks: a resting-state fMRI study. Neurosci Lett 504(1):62–67. https://doi.org/10.1016/j.neulet.2011.08.059

25. Meier TB, Desphande AS, Vergun S, Nair VA, Song J, Biswal BB, Meyerand ME, Birn RM, Prabhakaran V (2012) Support vector machine classification and characterization of age-related

reorganization of functional brain networks. Neuroimage 60(1):601–613. https://doi.org/10.1016/j.neuroimage.2011.12.052

26. Chai XJ, Ofen N, Gabrieli JD, Whitfield-Gabrieli S (2014) Selective development of anticorrelated networks in the intrinsic functional organization of the human brain. J Cogn Neurosci 26(3):501–513. https://doi.org/10.1162/jocn_a_00517

27. Ferreira LK, Busatto GF (2013) Resting-state functional connectivity in normal brain aging. Neurosci Biobehav Rev 37(3):384–400. https://doi.org/10.1016/j.neubiorev.2013.01.017

28. Stam CJ (2010) Characterization of anatomical and functional connectivity in the brain: a complex networks perspective. Int J Psychophysiol 77(3):186–194. https://doi.org/10.1016/j.ijpsycho.2010.06.024

29. Hawellek DJ, Hipp JF, Lewis CM, Corbetta M, Engel AK (2011) Increased functional connectivity indicates the severity of cognitive impairment in multiple sclerosis. Proc Natl Acad Sci U S A 108(47):19066–19071. https://doi.org/10.1073/pnas.1110024108

30. Bonnelle V, Ham TE, Leech R, Kinnunen KM, Mehta MA, Greenwood RJ, Sharp DJ (2012) Salience network integrity predicts default mode network function after traumatic brain injury. Proc Natl Acad Sci U S A 109(12):4690–4695. https://doi.org/10.1073/pnas.1113455109

31. Brier MR, Thomas JB, Snyder AZ, Wang L, Fagan AM, Benzinger T, Morris JC, Ances BM (2014) Unrecognized preclinical Alzheimer disease confounds rs-fcMRI studies of normal aging. Neurology 83(18):1613–1619. https://doi.org/10.1212/WNL.0000000000000939

32. Jones DT, Machulda MM, Vemuri P, McDade EM, Zeng G, Senjem ML, Gunter JL, Przybelski SA, Avula RT, Knopman DS, Boeve BF, Petersen RC, Jack CR Jr (2011) Age-related changes in the default mode network are more advanced in Alzheimer's disease. Neurology 77 (16):1524–1531. https://doi.org/10.1212/WNL.0b013e318233b33d

33. Cabeza R (2002) Hemispheric asymmetry reduction in older adults: the HAROLD model. Psychol Aging 17(1):85–100

34. Reuter-Lorenz PA, Cappell KA (2008) Neurocognitive aging and the compensation hypothesis. Curr Dir Psychol Sci 17(3):177–182

35. Davis SW, Dennis NA, Daselaar SM, Fleck MS, Cabeza R (2008) Que. PASA? The posterior-anterior shift in aging. Cereb Cortex 18(5):1201–1209. https://doi.org/10.1093/cercor/bhm155

36. Greicius MD, Srivastava G, Reiss AL, Menon V (2004) Default-mode network activity distinguishes Alzheimer's disease from healthy aging: evidence from functional MRI. Proc Natl Acad Sci U S A 101(13):4637–4642. https://doi.org/10.1073/pnas.0308627101

37. Raichle ME (2015) The brain's default mode network. Annu Rev Neurosci 38:433–447. https://doi.org/10.1146/annurev-neuro-071013-014030

38. Lee ES, Yoo K, Lee YB, Chung J, Lim JE, Yoon B, Jeong Y, Alzheimer's Disease Neuroimaging I (2016) Default mode network functional connectivity in early and late mild cognitive impairment: results from the Alzheimer's disease neuroimaging initiative. Alzheimer Dis Assoc Disord 30(4):289–296. https://doi.org/10.1097/WAD.0000000000000143

39. Tao W, Li X, Zhang J, Chen Y, Ma C, Liu Z, Yang C, Wang W, Chen K, Wang J, Zhang Z (2017) Inflection point in course of mild cognitive impairment: increased functional connectivity of default mode network. J Alzheimers Dis 60(2):679–690. https://doi.org/10.3233/jad-170,252

40. Wang Z, Liang P, Jia X, Jin G, Song H, Han Y, Lu J, Li K (2012) The baseline and longitudinal changes of PCC connectivity in mild cognitive impairment: a combined structure and resting-state fMRI study. PLoS One 7(5):e36838. https://doi.org/10.1371/journal.pone.0036838

41. Wang Z, Liang P, Jia X, Qi Z, Yu L, Yang Y, Zhou W, Lu J, Li K (2011) Baseline and longitudinal patterns of hippocampal connectivity in mild cognitive impairment: evidence from resting state fMRI. J Neurol Sci 309(1–2):79–85. https://doi.org/10.1016/j.jns.2011.07.017

42. Damoiseaux JS, Prater KE, Miller BL, Greicius MD (2012) Functional connectivity tracks clinical deterioration in Alzheimer's disease. Neurobiol Aging 33(4):828.e819–828.e830. https://doi.org/10.1016/j.neurobiolaging.2011.06.024

43. Jones DT, Knopman DS, Gunter JL, Graff-Radford J, Vemuri P, Boeve BF, Petersen RC, Weiner MW, Jack CR Jr, Alzheimer's Disease Neuroimaging I (2016) Cascading network

failure across the Alzheimer's disease spectrum. Brain 139(Pt 2):547–562. https://doi.org/10.1093/brain/awv338

44. Li M, Zheng G, Zheng Y, Xiong Z, Xia R, Zhou W, Wang Q, Liang S, Tao J, Chen L (2017) Alterations in resting-state functional connectivity of the default mode network in amnestic mild cognitive impairment: an fMRI study. BMC Med Imaging 17(1):48. https://doi.org/10.1186/s12880-017-0221-9

45. Chen G, Ward BD, Xie C, Li W, Wu Z, Jones JL, Franczak M, Antuono P, Li SJ (2011) Classification of Alzheimer disease, mild cognitive impairment, and normal cognitive status with large-scale network analysis based on resting-state functional MR imaging. Radiology 259 (1):213–221. https://doi.org/10.1148/radiol.10100734

46. Zhou B, Yao H, Wang P, Zhang Z, Zhan Y, Ma J, Xu K, Wang L, An N, Liu Y, Zhang X (2015) Aberrant functional connectivity architecture in Alzheimer's disease and mild cognitive impairment: a whole-brain, data-driven analysis. Biomed Res Int 2015:495375. https://doi.org/10.1155/2015/495375

47. Li Y, Wang X, Li Y, Sun Y, Sheng C, Li H, Li X, Yu Y, Chen G, Hu X, Jing B, Wang D, Li K, Jessen F, Xia M, Han Y (2016) Abnormal resting-state functional connectivity strength in mild cognitive impairment and its conversion to Alzheimer's disease. Neural Plast 2016:4680972. https://doi.org/10.1155/2016/4680972

48. Binnewijzend MA, Schoonheim MM, Sanz-Arigita E, Wink AM, van der Flier WM, Tolboom N, Adriaanse SM, Damoiseaux JS, Scheltens P, van Berckel BN, Barkhof F (2012) Resting-state fMRI changes in Alzheimer's disease and mild cognitive impairment. Neurobiol Aging 33(9):2018–2028. https://doi.org/10.1016/j.neurobiolaging.2011.07.003

49. Lehmann M, Madison CM, Ghosh PM, Seeley WW, Mormino E, Greicius MD, Gorno-Tempini ML, Kramer JH, Miller BL, Jagust WJ, Rabinovici GD (2013) Intrinsic connectivity networks in healthy subjects explain clinical variability in Alzheimer's disease. Proc Natl Acad Sci U S A 110(28):11606–11,611. https://doi.org/10.1073/pnas.1221536110

50. Ranasinghe KG, Hinkley LB, Beagle AJ, Mizuiri D, Dowling AF, Honma SM, Finucane MM, Scherling C, Miller BL, Nagarajan SS, Vossel KA (2014) Regional functional connectivity predicts distinct cognitive impairments in Alzheimer's disease spectrum. Neuroimage Clin 5:385–395. https://doi.org/10.1016/j.nicl.2014.07.006

51. Lehmann M, Madison C, Ghosh PM, Miller ZA, Greicius MD, Kramer JH, Coppola G, Miller BL, Jagust WJ, Gorno-Tempini ML, Seeley WW, Rabinovici GD (2015) Loss of functional connectivity is greater outside the default mode network in nonfamilial early-onset Alzheimer's disease variants. Neurobiol Aging 36(10):2678–2686. https://doi.org/10.1016/j.neurobiolaging.2015.06.029

52. Dickerson BC, Brickhouse M, McGinnis S, Wolk DA (2017) Alzheimer's disease: The influence of age on clinical heterogeneity through the human brain connectome. Alzheimers Dement 6:122–135. https://doi.org/10.1016/j.dadm.2016.12.007

53. Whitwell JL, Jones DT, Duffy JR, Strand EA, Machulda MM, Przybelski SA, Vemuri P, Gregg BE, Gunter JL, Senjem ML, Petersen RC, Jack CR Jr, Josephs KA (2015) Working memory and language network dysfunctions in logopenic aphasia: a task-free fMRI comparison with Alzheimer's dementia. Neurobiol Aging 36(3):1245–1252. https://doi.org/10.1016/j.neurobiolaging.2014.12.013

54. Migliaccio R, Gallea C, Kas A, Perlbarg V, Samri D, Trotta L, Michon A, Lacomblez L, Dubois B, Lehericy S, Bartolomeo P (2016) Functional connectivity of ventral and dorsal visual streams in posterior cortical atrophy. J Alzheimers Dis 51(4):1119–1130. https://doi.org/10.3233/jad-150934

55. Sheline YI, Morris JC, Snyder AZ, Price JL, Yan Z, D'Angelo G, Liu C, Dixit S, Benzinger T, Fagan A, Goate A, Mintun MA (2010) APOE4 allele disrupts resting state fMRI connectivity in the absence of amyloid plaques or decreased CSF Abeta42. J Neurosci 30(50):17035–17040. https://doi.org/10.1523/jneurosci.3987-10.2010

56. Zheng LJ, Su YY, Wang YF, Schoepf UJ, Varga-Szemes A, Pannell J, Liang X, Zheng G, Lu GM, Yang GF, Zhang LJ (2017) Different Hippocampus functional connectivity patterns in

healthy young adults with mutations of APP/Presenilin-1/2 and APOEepsilon4. Mol Neurobiol. https://doi.org/10.1007/s12035-017-0540-4

57. Shen J, Qin W, Xu Q, Xu L, Xu J, Zhang P, Liu H, Liu B, Jiang T, Yu C (2017) Modulation of APOE and SORL1 genes on hippocampal functional connectivity in healthy young adults. Brain Struct Funct 222(6):2877–2889. https://doi.org/10.1007/s00429-017-1377-3

58. Machulda MM, Jones DT, Vemuri P, McDade E, Avula R, Przybelski S, Boeve BF, Knopman DS, Petersen RC, Jack CR Jr (2011) Effect of APOE epsilon4 status on intrinsic network connectivity in cognitively normal elderly subjects. Arch Neurol 68(9):1131–1136. https://doi.org/10.1001/archneurol.2011.108

59. Thomas JB, Brier MR, Bateman RJ, Snyder AZ, Benzinger TL, Xiong C, Raichle M, Holtzman DM, Sperling RA, Mayeux R, Ghetti B, Ringman JM, Salloway S, McDade E, Rossor MN, Ourselin S, Schofield PR, Masters CL, Martins RN, Weiner MW, Thompson PM, Fox NC, Koeppe RA, Jack CR Jr, Mathis CA, Oliver A, Blazey TM, Moulder K, Buckles V, Hornbeck R, Chhatwal J, Schultz AP, Goate AM, Fagan AM, Cairns NJ, Marcus DS, Morris JC, Ances BM (2014) Functional connectivity in autosomal dominant and late-onset Alzheimer disease. JAMA Neurol 71(9):1111–1122. https://doi.org/10.1001/jamaneurol.2014.1654

60. Wang L, Roe CM, Snyder AZ, Brier MR, Thomas JB, Xiong C, Benzinger TL, Morris JC, Ances BM (2012) Alzheimer disease family history impacts resting state functional connectivity. Ann Neurol 72(4):571–577. https://doi.org/10.1002/ana.23643

61. Jack CR Jr, Wiste HJ, Weigand SD, Knopman DS, Lowe V, Vemuri P, Mielke MM, Jones DT, Senjem ML, Gunter JL, Gregg BE, Pankratz VS, Petersen RC (2013) Amyloid-first and neurodegeneration-first profiles characterize incident amyloid PET positivity. Neurology 81 (20):1732–1740. https://doi.org/10.1212/01.wnl.0000435556.21319.e4

62. Jones DT, Graff-Radford J, Lowe VJ, Wiste HJ, Gunter JL, Senjem ML, Botha H, Kantarci K, Boeve BF, Knopman DS, Petersen RC, Jack CR Jr (2017) Tau, amyloid, and cascading network failure across the Alzheimer's disease spectrum. Cortex. https://doi.org/10.1016/j.cortex.2017.09.018

63. Schultz AP, Chhatwal JP, Hedden T, Mormino EC, Hanseeuw BJ, Sepulcre J, Huijbers W, LaPoint M, Buckley RF, Johnson KA, Sperling RA (2017) Phases of hyperconnectivity and hypoconnectivity in the default mode and salience networks track with Amyloid and Tau in clinically normal individuals. J Neurosci 37(16):4323–4331. https://doi.org/10.1523/jneurosci.3263-16.2017

64. Sepulcre J, Sabuncu MR, Li Q, El Fakhri G, Sperling R, Johnson KA (2017) Tau and amyloid-beta proteins distinctively associate to functional network changes in the aging brain. Alzheimers Dement. https://doi.org/10.1016/j.jalz.2017.02.011

65. Jones DT, Lowe VJ, Wiste HJ, Senjem ML, Radford JG, Boeve BF, Knopman DS, Petersen RC, Jack CR Jr (2016) NETWORK-BASED TAU DEPOSITION PATTERNS ARE RELATED TO FUNCTIONAL NETWORK FAILURE LARGELY VIA BETA-AMYLOID ACROSS THE ALZHEIMER’S SPECTRUM. Alzheimers Dement 12(7):P11–P12. https://doi.org/10.1016/j.jalz.2016.06.074

66. Amboni M, Tessitore A, Esposito F, Santangelo G, Picillo M, Vitale C, Giordano A, Erro R, de Micco R, Corbo D, Tedeschi G, Barone P (2015) Resting-state functional connectivity associated with mild cognitive impairment in Parkinson's disease. J Neurol 262(2):425–434. https://doi.org/10.1007/s00415-014-7591-5

67. Kobeleva X, Firbank M, Peraza L, Gallagher P, Thomas A, Burn DJ, O'Brien J, Taylor JP (2017) Divergent functional connectivity during attentional processing in Lewy body dementia and Alzheimer's disease. Cortex 92:8–18. https://doi.org/10.1016/j.cortex.2017.02.016

68. Peraza LR, Kaiser M, Firbank M, Graziadio S, Bonanni L, Onofrj M, Colloby SJ, Blamire A, O'Brien J, Taylor JP (2014) fMRI resting state networks and their association with cognitive fluctuations in dementia with Lewy bodies. Neuroimage Clin 4:558–565. https://doi.org/10.1016/j.nicl.2014.03.013

69. Peraza LR, Colloby SJ, Firbank MJ, Greasy GS, McKeith IG, Kaiser M, O'Brien J, Taylor JP (2015) Resting state in Parkinson's disease dementia and dementia with Lewy bodies:

commonalities and differences. Int J Geriatr Psychiatry 30(11):1135–1146. https://doi.org/10.1002/gps.4342

70. Kenny ER, Blamire AM, Firbank MJ, O'Brien JT (2012) Functional connectivity in cortical regions in dementia with Lewy bodies and Alzheimer's disease. Brain 135(Pt 2):569–581. https://doi.org/10.1093/brain/awr327

71. Rosskopf J, Gorges M, Muller HP, Lule D, Uttner I, Ludolph AC, Pinkhardt E, Juengling FD, Kassubek J (2017) Intrinsic functional connectivity alterations in progressive supranuclear palsy: differential effects in frontal cortex, motor, and midbrain networks. Mov Disord 32 (7):1006–1015. https://doi.org/10.1002/mds.27039

72. Whitwell JL, Avula R, Master A, Vemuri P, Senjem ML, Jones DT, Jack CR Jr, Josephs KA (2011) Disrupted thalamocortical connectivity in PSP: a resting-state fMRI, DTI, and VBM study. Parkinsonism Relat Disord 17(8):599–605. https://doi.org/10.1016/j.parkreldis.2011.05.013

73. Bharti K, Bologna M, Upadhyay N, Piattella MC, Suppa A, Petsas N, Gianni C, Tona F, Berardelli A, Pantano P (2017) Abnormal resting-state functional connectivity in progressive supranuclear palsy and corticobasal syndrome. Front Neurol 8:248. https://doi.org/10.3389/fneur.2017.00248

74. Upadhyay N, Suppa A, Piattella MC, Gianni C, Bologna M, Di Stasio F, Petsas N, Tona F, Fabbrini G, Berardelli A, Pantano P (2017) Functional disconnection of thalamic and cerebellar dentate nucleus networks in progressive supranuclear palsy and corticobasal syndrome. Parkinsonism Relat Disord 39:52–57. https://doi.org/10.1016/j.parkreldis.2017.03.008

75. Collins JA, Montal V, Hochberg D, Quimby M, Mandelli ML, Makris N, Seeley WW, Gorno-Tempini ML, Dickerson BC (2017) Focal temporal pole atrophy and network degeneration in semantic variant primary progressive aphasia. Brain 140(2):457–471. https://doi.org/10.1093/brain/aww313

76. Mandelli ML, Vilaplana E, Brown JA, Hubbard HI, Binney RJ, Attygalle S, Santos-Santos MA, Miller ZA, Pakvasa M, Henry ML, Rosen HJ, Henry RG, Rabinovici GD, Miller BL, Seeley WW, Gorno-Tempini ML (2016) Healthy brain connectivity predicts atrophy progression in non-fluent variant of primary progressive aphasia. Brain 139(Pt 10):2778–2791. https://doi.org/10.1093/brain/aww195

77. Bonakdarpour B, Rogalski EJ, Wang A, Sridhar J, Mesulam MM, Hurley RS (2017) Functional connectivity is reduced in early-stage primary progressive Aphasia when atrophy is not prominent. Alzheimer Dis Assoc Disord 31(2):101–106. https://doi.org/10.1097/wad.0000000000000193

78. Chen Y, Chen K, Ding J, Zhang Y, Yang Q, Lv Y, Guo Q, Han Z (2017) Brain network for the core deficits of Semantic Dementia: a neural network connectivity-behavior mapping study. Front Hum Neurosci 11:267. https://doi.org/10.3389/fnhum.2017.00267

79. Sonty SP, Mesulam MM, Weintraub S, Johnson NA, Parrish TB, Gitelman DR (2007) Altered effective connectivity within the language network in primary progressive aphasia. J Neurosci 27(6):1334–1345. https://doi.org/10.1523/jneurosci.4127-06.2007

80. Botha H, Jones DT, Whitwell JL, Duffy JR, Strand EA, Machulda MM, Knopman DS, Petersen R, Jack CR, Josephs KA (2016) Language network dysfunction in primary progressive apraxia of speech. Paper presented at the fifth biennial conference on resting state brain connectivity, Vienna, Austria,

81. Zhou J, Greicius MD, Gennatas ED, Growdon ME, Jang JY, Rabinovici GD, Kramer JH, Weiner M, Miller BL, Seeley WW (2010) Divergent network connectivity changes in behavioural variant frontotemporal dementia and Alzheimer's disease. Brain 133 (Pt 5):1352–1367. https://doi.org/10.1093/brain/awq075

82. Whitwell JL, Josephs KA, Avula R, Tosakulwong N, Weigand SD, Senjem ML, Vemuri P, Jones DT, Gunter JL, Baker M, Wszolek ZK, Knopman DS, Rademakers R, Petersen RC, Boeve BF, Jack CR Jr (2011) Altered functional connectivity in asymptomatic MAPT subjects: a comparison to bvFTD. Neurology 77(9):866–874. https://doi.org/10.1212/WNL.0b013e31822c61f2

83. Seeley WW, Menon V, Schatzberg AF, Keller J, Glover GH, Kenna H, Reiss AL, Greicius MD (2007) Dissociable intrinsic connectivity networks for salience processing and executive control. J Neurosci 27(9):2349–2356. https://doi.org/10.1523/jneurosci.5587-06.2007

84. Day GS, Farb NA, Tang-Wai DF, Masellis M, Black SE, Freedman M, Pollock BG, Chow TW (2013) Salience network resting-state activity: prediction of frontotemporal dementia progression. JAMA Neurol 70(10):1249–1253. https://doi.org/10.1001/jamaneurol.2013.3258

85. Agosta F, Sala S, Valsasina P, Meani A, Canu E, Magnani G, Cappa SF, Scola E, Quatto P, Horsfield MA, Falini A, Comi G, Filippi M (2013) Brain network connectivity assessed using graph theory in frontotemporal dementia. Neurology 81(2):134–143. https://doi.org/10.1212/WNL.0b013e31829a33f8

86. Farb NA, Grady CL, Strother S, Tang-Wai DF, Masellis M, Black S, Freedman M, Pollock BG, Campbell KL, Hasher L, Chow TW (2013) Abnormal network connectivity in frontotemporal dementia: evidence for prefrontal isolation. Cortex 49(7):1856–1873. https://doi.org/10.1016/j.cortex.2012.09.008

87. Ranasinghe KG, Rankin KP, Pressman PS, Perry DC, Lobach IV, Seeley WW, Coppola G, Karydas AM, Grinberg LT, Shany-Ur T, Lee SE, Rabinovici GD, Rosen HJ, Gorno-Tempini ML, Boxer AL, Miller ZA, Chiong W, DeMay M, Kramer JH, Possin KL, Sturm VE, Bettcher BM, Neylan M, Zackey DD, Nguyen LA, Ketelle R, Block N, Wu TQ, Dallich A, Russek N, Caplan A, Geschwind DH, Vossel KA, Miller BL (2016) Distinct subtypes of behavioral variant Frontotemporal Dementia based on patterns of network degeneration. JAMA Neurol 73(9):1078–1088. https://doi.org/10.1001/jamaneurol.2016.2016

88. Rytty R, Nikkinen J, Paavola L, Abou Elseoud A, Moilanen V, Visuri A, Tervonen O, Renton AE, Traynor BJ, Kiviniemi V, Remes AM (2013) GroupICA dual regression analysis of resting state networks in a behavioral variant of frontotemporal dementia. Front Hum Neurosci 7:461. https://doi.org/10.3389/fnhum.2013.00461

89. Filippi M, Agosta F, Scola E, Canu E, Magnani G, Marcone A, Valsasina P, Caso F, Copetti M, Comi G, Cappa SF, Falini A (2013) Functional network connectivity in the behavioral variant of frontotemporal dementia. Cortex 49(9):2389–2401. https://doi.org/10.1016/j.cortex.2012.09.017

90. Dopper EG, Rombouts SA, Jiskoot LC, den Heijer T, de Graaf JR, de Koning I, Hammerschlag AR, Seelaar H, Seeley WW, Veer IM, van Buchem MA, Rizzu P, van Swieten JC (2014) Structural and functional brain connectivity in presymptomatic familial frontotemporal dementia. Neurology 83(2):e19–e26. https://doi.org/10.1212/wnl.0000000000000583

91. Lee SE, Khazenzon AM, Trujillo AJ, Guo CC, Yokoyama JS, Sha SJ, Takada LT, Karydas AM, Block NR, Coppola G, Pribadi M, Geschwind DH, Rademakers R, Fong JC, Weiner MW, Boxer AL, Kramer JH, Rosen HJ, Miller BL, Seeley WW (2014) Altered network connectivity in frontotemporal dementia with C9orf72 hexanucleotide repeat expansion. Brain 137 (Pt 11):3047–3060. https://doi.org/10.1093/brain/awu248

92. Lee SE, Sias AC, Mandelli ML, Brown JA, Brown AB, Khazenzon AM, Vidovszky AA, Zanto TP, Karydas AM, Pribadi M, Dokuru D, Coppola G, Geschwind DH, Rademakers R, Gorno-Tempini ML, Rosen HJ, Miller BL, Seeley WW (2017) Network degeneration and dysfunction in presymptomatic C9ORF72 expansion carriers. Neuroimage Clin 14:286–297. https://doi.org/10.1016/j.nicl.2016.12.006

93. Premi E, Cauda F, Gasparotti R, Diano M, Archetti S, Padovani A, Borroni B (2014) Multimodal FMRI resting-state functional connectivity in granulin mutations: the case of frontoparietal dementia. PLoS One 9(9):e106500. https://doi.org/10.1371/journal.pone.0106500

94. Wiepert DA, Lowe VJ, Knopman DS, Boeve BF, Graff-Radford J, Petersen RC, Jack CR Jr, Jones DT (2017) A robust biomarker of large-scale network failure in Alzheimer's disease. Alzheimers Dement 6:152–161. https://doi.org/10.1016/j.dadm.2017.01.004

Chapter 12
MRI Neuroimaging and Psychiatry

Laura Hatchondo

Introduction

Neuroimaging techniques have greatly been developed over the past 10 years, allowing access to brain anatomy, function, and metabolism in vivo. The last 20 years have seen a significant and constant increase of the number of studies using these techniques to explore psychiatric diseases. Indeed, human neuropsychology studies and experimental animal neurophysiology studies have led to a main hypothesis that there would be an anatomical and/or functional and/or metabolism substratum to psychiatric disorders.

Therefore, neuroimaging studies have helped us so far to gradually better understand pathophysiological mechanisms underlying psychiatric disorders. Some studies were able to make link between clinical symptoms and brain abnormalities or to show changes of brain features before and after treatment. But it is still a research tool in this area, whereas it is daily used in clinical routine for brain tumors, for instance.

There is still a long way to go for neuroimaging techniques to take the same place for psychiatric disorders. Nevertheless, it is worth it as it could hold some potential roles for diagnosis, therapeutic choice, and treatment follow-up, the ultimate goal being to provide the patient better care.

As it is a broad topic, the following chapter will only present a non-exhaustive review of the contribution of MRI neuroimaging in psychiatry. MRI is a very interesting imaging technique as it allows access to the anatomic, functional, and metabolic features of an in vivo brain, without exposing the body to ionizing radiation.

L. Hatchondo (✉)
University Hospital of Poitiers, Poitiers, France

DACTIM-MIS team LMA/ CNRS 7348, Poitiers University, Poitiers, France
e-mail: laura.hatchondo@univ-poitiers.fr

© Springer International Publishing AG, part of Springer Nature 2018
C. Habas (ed.), *The Neuroimaging of Brain Diseases*, Contemporary Clinical
Neuroscience, https://doi.org/10.1007/978-3-319-78926-2_12

We decided to focus our work on the three main psychiatric diseases: bipolar disorder (BD), schizophrenia, and obsessive-compulsive disorder (OCD). The last part will be dedicated to the effects of treatments (antidepressants, antipsychotics, and electroconvulsive therapy ECT). Each part will be divided in the same three axes: structural MRI techniques (VBM, DTI), functional MRI techniques (fMRI), and metabolic MRI techniques (MRS).

As a reminder:

- *Voxel-based morphometry (VBM)* is a neuroimaging analysis technique that allows investigation of focal volume differences in brain anatomy.
- *Diffusion tensor imaging (DTI)*, at the macrostructural level, makes it possible to 3D model white matter fiber tracts and explore the interconnectivity between different brain regions. At the microstructural level, this technique makes it possible to measure the rate and the direction of diffusion of the water molecules of the brain, which is constrained by the presence of the axons. The measurement of fractional anisotropy (FA) gives an indication of the integrity of the white substance for each voxel. Elevated FA values are observed in highly myelinated beams, whereas low FA values indicate neuronal loss or demyelination. Therefore, it makes it possible to evaluate the organization and coherence of white matter fiber tracts.
- *Functional magnetic resonance imaging (fMRI)* measures brain activity by detecting changes associated with blood flow using the blood oxygen level-dependent (BOLD) contrast. The ultimate goal of this technique is to detect correlations between brain activation and rest or a task performance (cognitive states, such as memory and recognition, emotional states)
- *Magnetic resonance spectroscopy (MRS)* is a noninvasive imaging technique that provides biochemical information on the composition of the tissues explored. Indeed, human biological tissues are composed of metabolites whose concentration varies between the physiological and pathological state. There are many different MRS sequences, the main ones being proton (MRS-^{1}H) and phosphorus (MRS-^{31}P).

Bipolar Disorder

Bipolar disorder (BD) is a mood disorder characterized by unusual shifts in mood, energy, and activity levels. Schematically, it leads to alternative mood changes between depressive episodes and manic episodes. Less severe manic periods are known as hypomanic episodes. There are two main types [1, 2]:

- BD type I (BD-1), defined by manic episodes that last at least 7 days or by manic symptoms that are so severe that the person needs immediate hospital care. Usually, depressive episodes occur as well, typically lasting at least 2 weeks
- BD type II (BD-2), defined by a pattern of depressive episodes and hypomanic episodes, but not the full-blown manic episodes described above

This psychiatric disease is a real public health issue as it causes an important disability among young people, leading to cognitive and functional impairment and raised mortality, particularly death by suicide.

Treatments are mainly based on mood stabilizers, atypical antipsychotics, antidepressants, psychotherapy, cognitive behavioral therapy (CBT), and electroconvulsive therapy (ECT).

Neuroimaging studies on BD patients have made it possible to advance in the knowledge of the physiopathology, in particular in the conceptualization of certain neurocircuit abnormalities mainly involving the genesis and regulation of emotions (e.g., fronto-limbic network).

Structural Imaging

VBM

Many morphometric studies have been performed so far and mainly have focused on the amygdala, considering its role in the emotional regulation. Unfortunately, the results remain contradictory and inconsistent confirmed by Hajek et al. meta-analysis which showed significant smaller left amygdala in children and bipolar adolescents, whereas it was tendentially larger in bipolar adults [3]. This lack of overall differences in amygdala volumes could be related to the heterogeneity of studies included and a potential patient subgroups factor.

Another meta-analysis found contradictory results with an association between BD and gray matter reduction in left rostral anterior cingulate cortex (ACC) and right fronto-insular cortex whereas the ACC volume tended to enlarge with the illness duration [4]. No volume modification was found in patients with their first episode of BD.

Nevertheless, these gray matter variations mainly occurred in anterior limbic regions that the authors relate to executive control and emotional-processing abnormalities in BD patients.

Ellison-Wright and Bullmore meta-analysis found gray matter reduction in insula and ACC, particularly in the pregenual region which seems to differentiate BD patients from schizophrenic ones [5]. Again, this work showed abnormalities in paralimbic regions implicated in emotional processing.

Moreover, according to Birur et al. review, decrease of cortical gray matter is more pronounced and extended in schizophrenia rather than in BD [6].

Finally, Lee et al. used region-of-interest (ROI) voxel-based morphometry analysis in patients with BD-I compared to healthy control subjects [7]. They found a decrease in gray matter volumes in the left ventromedial prefrontal cortex (VMPFC), left dorsomedial prefrontal cortex (DMPFC), and left ventrolateral prefrontal cortex (VLPFC) in patients with BD-I. Those volume abnormalities localized in the reward-processing neural circuitry could support its role in the pathophysiology of BD-I.

DTI

White matter abnormalities are known to be extended to many brain regions in BD. Throughout the studies, both types of DTI analysis, ROI or whole brain, have shown a decrease in FA in the anterior commissure [8]; ACC and posterior cingulum [9]; prefrontal, subcortical, and thalamic regions [10–14]; and corpus callosum, cortical, and thalamic associative fibers [15, 16], in BD patients compared to healthy control subjects.

Linke et al. focused on BD-I patients and their healthy first-degree relatives [17]. The results showed that both groups had lower FA in the right anterior limb of the internal capsule and right uncinate fasciculus. Only BD-I patients had reduced white matter integrity in corpus callosum. The very interesting point of this study is the significant correlations between, on one hand, lower FA values in uncinate fasciculus and higher risk-taking rank and, on the other hand, lower FA values in anterior limb of the internal capsule and higher number of errors during set shifting and increased risk-taking. Then, those results could be a first step toward defining vulnerability marker of BD-I and disease marker.

Nortje et al. meta-analysis, based on 15 DTI studies in BD, showed reduced FA in the majority of tracts compared to healthy control subjects. Despite these widespread abnormalities, they found three clusters of lower FA in BD: right posterior temporoparietal, left posterior middle cingulate gyrus, and left anterior cingulate gyrus [18].

In first episode BD patients, only one study showed decreased FA in the cingulum, internal capsule, and posterior brain regions, whereas nothing significant was found in first-episode schizophrenia [19]. Regarding those results, the authors suggested that BD and schizophrenia present different pathophysiological mechanisms, with a potential structural disorganization in fiber tract coherence or myelin alteration in BD.

Finally and very recently, a study examined longitudinal differences (2 years apart) in white matter integrity in youth at high familial risk for BD compared to healthy control subjects [20]. The results showed a significant and widespread decrease in FA in all the groups leading to no conclusion concerning potential vulnerability markers or longitudinal preclinical traits in BD.

Functional Imaging

To help physicians to better understand links between clinical symptoms and underlying unobservable pathophysiology, fMRI is one the best neuroimaging sequence.

First of all, we will focus on the studies which used the resting-state functional connectivity magnetic resonance imaging. Resting-state network is the default mode network, a large-scale brain network that is more active at rest and has been implicated in self-referential thinking. A review study found variations among the studies' results, but it appeared that all, in a way or another, implied modification of the connectivity in the medial prefrontal cortex and ACC with limbic-striatal structures, therefore supporting the cortico-limbic hypothesis [21].

A more recent review showed several interesting results on the default mode network modifications in BD: on one hand, reduced connectivity to the hippocampus, the fusiform gyrus, the medial prefrontal cortex, the ACC, and the posterior cingulate cortex and, on the other hand, increased connectivity to primary visual cortex, fronto-temporal cortex, and mesolimbic regions. The last two were specific of BD compared to healthy controls and schizophrenia [6].

Another study examined the resting-state fMRI connectivity impairment in schizophrenia and BD [22]. Their results showed lower global connectivity in schizophrenia patients compared to healthy control subjects, whereas BD patients had intermediate global connectivity strength that differed significantly from both schizophrenia and healthy controls, leading to support the hypothesis of a continuum of brain global dysconnectivity between schizophrenia and BD.

fMRI can also examine the brain connectivity during emotional or cognitive task.

More than a decade ago, fMRI studies of BD have found abnormally high amygdala activity in response to emotional stimuli [23, 24]. More recently, a meta-analysis using facial affect processing paradigms in BD patients versus healthy control subjects showed significant higher activity in parahippocampal gyrus and amygdala, bilaterally, and lower activity in the ventrolateral prefrontal cortex in BD [25].

Moreover, in their synthesis work on neuroimaging and BD, Brooks and Vizueta found interesting results first in bipolar depression state and then in bipolar mania state [26]. Thus, they found evidence suggesting an increased activity in amygdala during BD-I depression and during mania. On the contrary, they found a decreased activity in the dorsolateral and orbitofrontal regions of prefrontal cortex in BD-I and BD-II patients, regardless of mood states. As the authors tried to find potential predictive biomarker, they considered the prefrontal cortex activation abnormalities as a potential trait of BD.

Another study observed a significant bilateral decreased activity in the ventrolateral and dorsolateral prefrontal cortex, ACC, posterior cingulate cortex, and medial frontal gyrus during emotion downregulation in BD-I patients [27]. They also found decreased functional connectivity between the ventrolateral prefrontal cortex and the amygdala during the treatment of positive and negative emotions and emotional regulation tasks in BD patients.

In working memory tasks, most studies observed abnormal connectivity and activation in prefrontal networks, known to be involved in working memory [28]. Thus, McKenna et al. found a decreased activation of prefrontal cortex, caudate, thalamus, and posterior visual regions during the encoding interval [29]. BD patients also showed deficits in task accuracy compared to healthy control subject, and the level of brain activation in the prefrontal cortex was higher with greater medication load.

Another study reported an increased activation of dorsolateral prefrontal cortex in euthymic BD patients during working memory tasks, with the greater BOLD signal in BD-I patients and an intermediate pattern for BD-II patients, both compared to healthy control subjects [30]. The authors suggested that their results could be linked to working memory-processing impairments in BD patients.

Finally, other fMRI studies have focused on the reward system. Most of them observed abnormal activities in the ventral striatum, the prefrontal cortex, and the amygdala during the reward process. But the results remain heterogeneous. Indeed, some studies found a significant decreased activation of frontal regions: ACC in depressed patients with BD-I during reward expectancy and dorsolateral prefrontal cortex in adolescents with BD during both target anticipation and feedback anticipation [31, 32], while there was a significant increased activation in the left ventrolateral prefrontal cortex during anticipation phases in depressed patients with BD-I [31].

Trost et al. showed a decreased activation in the ventral striatum during reward stimulus [33], whereas a very recent study found an increased connectivity between the ventral striatum and both orbitofrontal cortex and amygdala [34]. This last study is very interesting as it proposed potential mechanisms to the elevated reward sensitivity in BD by making the difference in their analysis between reward receipt and reward omission phases.

At last, Berghorst et al. focused their work on the impact of stress on reward-related neural functioning in BD [35]. They found that stress had a real impact on the amygdala activation as it increased during the no-stress condition and decreased during stress condition, compared to healthy control subjects.

Metabolic Imaging

Numerous studies have been conducted so far to explore potential brain metabolism modifications in BD. Despite heterogenous results, some major trends seem to be emerging.

In ^{1}H-MRS, one meta-analysis and recent studies have shown a significant decrease of NAA levels in the frontal lobe (e.g., medial prefrontal cortex), basal ganglia, left Heschl's gyrus, and parietal cortex [36–38]. Those results suggest potential neuronal suffering and/or dysfunction in these regions.

Besides, other studies found increased levels of Glx (a combination of glutamate and glutamine) in frontal regions and ACC, of glutamine in ACC and left basal ganglia, and of the glutamine to glutamate ratio (Gln/Glu) in ACC [38–40]. Those results support the hypothesis of aberrant glutamate neurotransmission in the pathophysiology of BD.

But some studies did not find such results and, on the contrary, have concluded to a decrease of glutamate level in the left Heschl's gyrus [37] or to no alteration of NAA and Glx levels in the hippocampus, specifically in early stages of BD

[41]. This last point suggests a potential neuroprogression of BD with an interesting potentially early window to target for therapeutic interventions.

^{31}P-MRS studies in BD have shown some results that seem consistent with decreased mean of PME/PDE ratio over the brain [42] and decreased of glycerophosphocholine + phosphocholine (GPC + PC) in parietal cortex relative to healthy control subjects [38]. These data suggest a potential altered phospholipid membrane turn over in BD.

Moreover, Li et al. found lower phosphocreatine + creatine (PCr + Cr) levels in parietal cortex suggesting a potential alteration in cellular energy buffering and energy transport in neurons [43].

Schizophrenia

Schizophrenia (SZ) is a severe psychiatric disorder characterized by positive symptoms such as hallucinations, delusions, and thought disorder; negative symptoms such as reduced expression of emotions, difficulty beginning and sustaining activities, and reduced speaking; and cognitive symptoms such as poor executive functions, trouble on focusing or paying attention, and alteration of working memory. Those symptoms typically come on gradually, begin in young adulthood, and last a long time. It affects approximately 1% of the population [2, 44].

Treatments are based on antipsychotic medications and psychosocial treatments.

A large number of neuroimaging studies have been published on schizophrenia so far. They have supported the hypotheses of an imbalance of the brain chemicals or neurotransmitters (dopamine, glutamate, and serotonin) and abnormalities of brain connectivity, though variability in results has not allowed defining a consensus on neural alterations correlated with schizophrenia.

Structural Imaging

VBM

Neuroimaging studies have early focused their research on gray matter in psychiatric illnesses.

Several brain regions have mainly been explored in schizophrenia using a whole-brain analysis with voxel-based morphometry (VBM), a volumetric MRI technique. VBM have made possible to quantify more precisely the regional changes in volume and density at work in schizophrenia. Many authors defend the hypothesis of diffuse structural anomalies in schizophrenia, mainly involving the heteromodal associative cortex.

Gray and white matter deficits in patients with schizophrenia were reported in many brain regions. The most consistent findings were deficits in the left superior temporal gyrus and the left medial temporal lobe [45].

Recently, a meta-analysis of VBM studies in schizophrenia reports deficits of gray matter in frontal, temporal, cingulate, and insular cortex and thalamus and increased gray matter in the basal ganglia [5].

Even if most studies have found diffuse structural anomalies, deficits in the temporal gyrus seem to emerge more often. The upper left temporal gyrus, and more particularly its posterior part, is a central region for two main functions: hearing and language. This is consistent with Tim Crow's theory that schizophrenia is related to a failure in the establishment of hemispheric dominance for the language [46]. Rajarethinam et al. have also found a correlation between the volume decrease of the left superior temporal gyrus and the intensity of hallucinations and thought disorders [47].

A recent longitudinal review allows us to have a global perception of the structural brain changes in schizophrenia at different stages of illness [48]. The results showed a greater cortical gray matter loss in anterior regions in patients with schizophrenia who later made transition to psychosis or in patients with chronic schizophrenia and poor outcome.

Moreover, Sumner et al. performed a systematic review of structural neuroimaging studies exploring thought disorder (TD), a syndrome observed in many schizophrenia patients [49]. They found evidence that implicate the left superior temporal gyrus implication but also anterior and medial cerebral regions (e.g., orbitofrontal cortex, nucleus accumbens, amygdala).

DTI

Growing evidence of white matter abnormalities, related to a disturbance in connectivity between different brain regions, has been reported so far in patients with schizophrenia. This hypothesis puts forward the responsibility of an abnormal connectivity rather than abnormalities within separate regions.

The review of Wheeler et al. led to some interesting results. They showed notably an altered structural integrity of white matter in frontal and temporal brain regions and tracts such as the cingulum bundles, uncinate fasciculi, internal capsule, and corpus callosum, associated with the illness. Those results suggest that not one but multiple brain circuits are impaired in schizophrenia [50].

An activation likelihood estimation meta-analysis identified decreased FA in the white matter of the right deep frontal and left deep temporal lobes in first-episode schizophrenia compared to healthy controls [51]. They also provide evidence of the lack of connection in the fronto-limbic circuitry at the early stages of the disease.

Finally, a recent systematic review of MRI literature in schizophrenia and bipolar disorder found decreased FA in similar regions between those two diseases such as the inferior fronto-occipital fasciculus, uncinate fasciculus, corona radiata, anterior limb of the internal capsule, anterior and posterior thalamic radiation, and corpus callosum [6].

Functional Imaging

fMRI is a very interesting technique allowing the investigation of the cerebral activity and behavior relationships which can help us to better understand the neural substrates of negative and positive symptoms and thought disorder in schizophrenia.

A review on task-related fMRI studies found meaningful small to moderate associations between specific symptom dimensions and regional brain activity [52]. Nevertheless, it seems that ventrolateral prefrontal cortex and ventral striatum are related to negative symptoms, while medial prefrontal cortex, amygdala, and hippocampus are related to positive symptoms (persecutory ideation). Dorsolateral prefrontal cortex is related to disorganization symptoms.

Recently, Mwansisya et al. conducted a systematic literature review on fMRI studies, distinguishing resting state and cognitive task, in first-episode schizophrenia [53]. In resting state, they found decreased signal or functional connectivity in the prefrontal cortex and increased signal or functional connectivity in the DLPFC, whereas increased or decreased signal or functional connectivity were observed in the temporal lobe. In response to cognitive task, the decreased brain activity was found in prefrontal cortex and associated with word fluency, eye movement, color word Stroop and AX continuous performance tasks, explicit emotion discrimination task, drum beats, and working memory cognitive tasks. On the contrary, the increased brain activity was found in the VLPFC and was associated with the working memory cognitive task.

Therefore, this recent review showed very interesting results with a convergent brain dysfunction during task and resting states within the fronto-temporal pathway with convergence in the DLPFC, the orbital frontal cortex, and the left STG. Those results support the fronto-temporal hypothesis of schizophrenia and proposed the disruption in prefrontal and STG as the pathophysiology of schizophrenia disorder at a relatively early stage.

Birur et al. showed in their comparative review between schizophrenia and BD an activation of the ventral striatum during reward anticipation in schizophrenia but not in BD patients in a manic state [6]. This result is interesting since it leads to the striatal dopamine dysfunction hypothesis, which could be clinically expressed as anhedonia.

Finally, Poels et al. suggest that imaging genetic studies are an effective and productive methodology for fMRI [54–56]. Indeed, a translational approach is more than likely to provide consistent results to better understand and cure psychiatric diseases.

Metabolic Imaging

Spontaneously, the hypothesis of metabolic modifications underlying the fMRI abnormalities in schizophrenia has been proposed. Thereby, magnetic resonance spectroscopy (MRS) has become widely used.

Many studies using proton magnetic resonance spectroscopy (^{1}H-MRS) have been conducted so far to compare schizophrenia metabolism to the one of healthy control. The main results show reduced NAA in frontal, medial temporal, and basal ganglia regions. No changes in Cho or Cr concentrations have been found [36, 57, 58].

Other ^{1}H-MRS studies have examined glutamatergic and GABA abnormalities.

The glutamate hypothesis in schizophrenia is based on impairment of *N*-methyl-d-aspartate (NMDA) subtypes of glutamate receptors that can cause schizophrenia-like symptoms in healthy individuals and exacerbate symptoms in individuals with schizophrenia.

The review of Poels et al. suggests elevated levels of glutamatergic concentration in medial prefrontal cortex, striatum, and hippocampus in medication-naïve or medication-free patients. Clinical and neuropsychological correlations did not lead to consistent results, except for in hippocampus/medial temporal lobe in which elevated glutamate levels are related to worse executive functioning and global clinical state [54].

On the other hand, Taylor et al. focused their review on GABA abnormalities in schizophrenia [59]. Unfortunately, results remain inconsistent probably due to medication status as a confounding factor. With that in mind, two studies have found increased GABA in the medial prefrontal cortex [60] and reduced GABA in occipital cortex [61] in unmedicated patients. Nevertheless, it seems too early to conclude on regional variations in GABA concentrations.

Studies comparing brain metabolism between schizophrenia and BD patients only used single voxel ^{1}H-MRS placed in cortical areas and on medicated patients [6]. The main results showed decreased NAA/Cr in dorsolateral prefrontal cortex greater in schizophrenia than BD and decreased Glu, NAA, and inositol concentrations in the left Heschl's gyrus in BD but not in schizophrenia.

Finally, last but not least, many phosphorus MRS (^{31}P-MRS) studies have been conducted so far in schizophrenia. Yuksel et al. reviewed 52 studies but found heterogenic results [62]. One of the more consistent was a decrease in PDE in chronic patients in subcortical structures. The authors then recommend that future studies should have a more rigorous methodology.

Obsessive-Compulsive Disorder

Obsessive-compulsive disorder (OCD) is a common disease with a lifetime prevalence between 1.2% and 2.4% [63–65]. The age of onset of the disease has a bimodal distribution with a peak around puberty that includes most patients (12–14 years) and a later one (20–22 years) [66]. The symptoms of this disease are heterogeneous and classified in obsession (grouped according to their themes) and compulsion (grouped according to the nature of the behavior) dimensions (DSM-V).

OCD is a chronic disease with usually phases of symptomatic exacerbation interspersed with periods of remission and, in general, progressively worsening symptoms over time [67, 68]. These recurrent obsessions and compulsions are severe enough to cause a loss of time of more than 1 h per day or significantly interfere with the patient's usual activities, work, and family life [63, 69].

Recommended first-line treatment is based on cognitive behavioral therapy (CBT) combined with antidepressant treatments (primarily serotonin reuptake inhibitor SRI antidepressants). However, 40–60% of OCD patients exhibit drug resistance, leaving them with a major handicap in everyday life [70]. It is therefore essential to better understand the anatomical substrates and neurophysiopathologic mechanisms of OCD.

Structural Imaging

Structural anomalies associated with OCD can be sought either by the specific study of a given region (region of interest [ROI]) or by a whole-brain analysis with voxel-based morphometry (VBM).

Volumetric Structural MRI

The results of studies measuring the volume of certain brain regions, between patients with OCD and controls, appeared to be highly discordant from one study to another. A meta-analysis of volumetric MRI studies in OCD has been carried out to provide quantitative and objective information on this seemingly discordant literature [71]. Data were collected from 21 studies that included 371 patients with OCD and 407 controls. Five brain regions were regularly evaluated in OCD: the orbitofrontal cortex (COF), anterior cingulate cortex (CCA), thalamus, putamen, and caudate nucleus.

The results showed a decrease in bilateral COF volume as well as an increase in bilateral thalamus volume in OCD. The left CCA tends to be decreased in volume. There is no significant difference for other brain regions, including the striatum. The heterogeneity of the results for the CCA, probably explained by the great variability of the anatomical definitions used, does not make it possible to clearly conclude in favor of a change in the volume of the CSF in the OCD. In conclusion, this meta-analysis of the volumetric MRI data in the OCD showed that there is a consistency of the results concerning structural anomalies involving COF and thalamus and that it is more difficult to conclude concerning possible anomalies concerning the CCA or striatum.

VBM

The meta-analysis of Radua et al. (2010) demonstrated the existence of a bilateral decrease in the volume of the cortex: orbitofrontal and anterior cingulate, as well as a bilateral increase of that of the caudate nucleus [22]. In addition, the mean volume of the hippocampus and the amygdala is decreased in patients with OCD [72, 73]. Moreover, the severity of the OCD would be related to an increase in the volumes of gray matter at the level of the central gray nuclei. No effect of antidepressant treatments on cerebral volumes has been found [74]. Finally, the meta-analysis of Piras et al. (2013) has found a reduction in the volume of the prefrontal cortex (dorsomedial, ventrolateral, and fronto-polar) and temporo-parieto-occipital associative areas [75]. These broad morphometric changes suggest that a wider network is involved in OCD and this could explain the clinical and pathophysiological heterogeneity.

DTI

Tractography (DTI) is a special MRI technique that determines the trajectory of white matter beams in a noninvasive and in vivo manner, in order to detect the microstructural abnormalities of the latter.

Anomalies of structural connectivity (diffusion tensor) between the basal ganglia, the orbitofrontal cortex, and the cingulate cortex have been demonstrated in several studies and meta-analyses [76, 77]. Two trends are observed:

- Hypo-connectivity, linked to changes in DTI indices in favor of structural alteration, at several structures: cingulate, lower fronto-occipital beam, uncinate beam, and lower and upper longitudinal beam
- Hyper-connectivity, linked to changes in DTI indices in favor of structural reinforcement, at the level of the anterior and posterior arm of the internal capsule, the thalamus, and the knee of the corpus callosum

Those results have been broadly confirmed by Gan et al. lately using a tract-based spatial statistics (TBSS) approach [78]. They found lower fractional anisotropy (FA) values, therefore hypo-connectivity, in the corpus callosum, left anterior corona radiata, left superior corona radiata, and left superior longitudinal fasciculus, and on the other hand, a higher radial diffusivity, therefore hyper-connectivity, in the genu and body of corpus callosum.

Functional Imaging

fMRI provided the most interesting results in terms of activation and functional connectivity. Studies have shown bilateral hyperactivation in patients with OCT in the COF, anterior cingulate cortex (CAB), striatum (caudate nucleus and putamen), and amygdala [76, 79]. Recently, Jung et al. (2013) demonstrated an abnormal modulation of cerebral interactions between the nucleus accumbens (NA) and medial and lateral COF in the resting state and between the NA and the limbic system (especially the amygdala) during a reward task [80]. To this is added a correlation between the severity of the symptoms (evaluated by the Y-BOCS) and the strength of the functional connection between the NA and the orbitofrontal regions.

In addition, it was mainly with fMRI that the dimensional aspect of cortical and parietal activations could be highlighted and the debate on the physiopathological model of OCD [76]. Indeed, in-depth analysis of neuropsychological studies and multimodal neuroimaging revealed the involvement of other regions, notably the dorsolateral prefrontal cortex (DLPFC) and the parietal cortex [76].

Moreover, recent studies, based on the realization of neuropsychological tasks during an fMRI, revealed the existence of correlations between neuropsychological dysfunctions and the clinical symptoms of OCD [79].

Metabolic Imaging

Magnetic resonance spectroscopy (MRS) is a noninvasive technique that can assess biochemical alterations in the brain in vivo [81, 82]. A dramatic increase of metabolic knowledge in psychiatric disorders has led to focus on the potential ability of spectroscopy to provide "in vivo" insights on this topic. In a specific way, proton (^{1}H) MRS can detect several metabolites implicated in neuronal viability and glial functions. First, N-acetylaspartate (NAA), is a quantitative marker of neuronal suffering; then creatine (Cr) is a marker of cellular energy metabolism; myo-ionositol (mI) is a marker for glial density; the glutamatergic complex (Glx) divided into glutamate (excitation neurotransmitter), glutamine (derived from glutamate), and gamma-amino butyric acid (GABA) (inhibitory neurotransmitter); and finally, choline (Cho) is a marker of membrane metabolism. Increased choline levels have been associated with either increased synthesis or degradation of the metabolism of cell membrane in Alzheimer and multiple sclerosis [83, 84].

Considering CSTC network dysfunction, several studies using ^{1}H-MRS have found changes in neural metabolite concentrations among OCD patients in this area.

NAA is considered to be a quantitative marker of neuronal suffering. Even if some of its functions in the central nervous system remain unclear, it seems to be an indirect measure of neuronal integrity and synaptic abundance [85]. Concentration changes of NAA are not specific to a particular disorder and can be found in

Alzheimer's disease [86], Parkinson's disease [87], bipolar disorder [88], and schizophrenia [89]. The NAA rate seems to vary according to the structural and functional status of neuronal cells.

The main results showed increased concentrations of N-acetylaspartate (NAA) in frontal white matter (FWM) [90] and decreased NAA concentrations in ACC [91–94], striatum [95, 96], thalamus [97, 98], and hippocampus [99] in OCD patients. Hatchondo et al. (2017) found a significant decrease of the NAA/Cr ratio only in the striatum, in OCD patients compared to healthy controls ($p = 0.035$) [100].

These metabolite changes could therefore reflect not a definitive neuronal loss but rather a potentially reversible alteration. Therefore, decreased NAA in the CSTC circuit could indicate suffering involving neural structures.

Cho is a membrane turnover marker involved in the metabolism of the phospholipid membrane structure (phosphatidylcholine and sphingomyelin); increased concentrations of Cho may be due to an increased membrane turnover related to a greater need for membrane renewal due to alteration/destruction or membrane proliferation.

Smith et al. [101] and Weber et al. [90] described significantly increased Cho concentrations in left and right medial thalamic and right prefrontal white matter, respectively, but only in pediatric OCD patients. More recently, Fan et al. [102] found higher Cho concentrations in the thalamus in OCD patients, especially compared to Tourette's disorder patients and to a lesser extent compared to healthy controls. Hatchondo et al. [100] showed significantly increased concentrations of Cho between OCD patients and healthy controls in the ACC, the striatum, and the thalamus (Fig. 12.1). This is associated with the facts that Cho/Cr ratio significantly increased in the ACC and that NAA/Cho ratio significantly decreased in the striatum, signaling a metabolic imbalance in these ROIs in OCD patients.

The other ^{1}H-MRS studies that found increased ratios of Cho/Cr have dealt with neurodegenerative diseases such as Alzheimer's disease [103] and Huntington's disease [104]. This suggests a potential link between Cho and neuronal membrane degradation before neuronal loss. Furthermore, multiple sclerosis (MS) patients exhibit increased concentrations of Cho, particularly in the active demyelination plaques [105–107], thereby suggesting an association between the Cho increase and

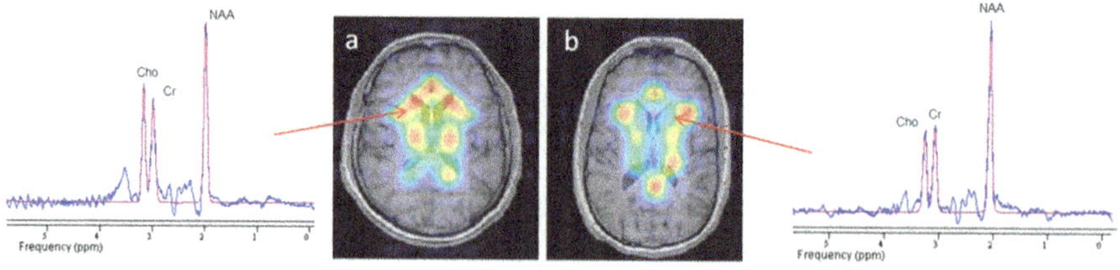

Fig. 12.1 Hatchondo et al. (2017) [100]: Illustration of the increased concentration of choline especially in the striatum between an OCD patient and a healthy control. Respective spectra are shown on the side and respective Cho cartography in the middle. The red color shows the localization of the highest levels of Cho, whereas blue color shows the lowest, respectively, for the OCD patient and the healthy control

the demyelination phenomenon. The Cho increase in OCD may consequently reflect alteration of the myelin and/or cell membrane instability [108]. This interpretation is reinforced by the findings of white matter abnormalities in patients with OCD [77, 109] and by the potential association between this disorder and the genes involved in myelination [110]. Given that Cho is a membrane turnover marker involved in the metabolism of the phospholipid membrane structure (phosphatidyl-choline and sphingomyelin), increased concentrations of Cho may be due to an increased membrane turnover related to a greater need for membrane renewal due to alteration/destruction or membrane proliferation.

Few other studies have yielded similar results. The most interesting one showed NAA/Cr decrease in the right striatum and Cho/Cr increase in the right thalamus, in OCD patients with SRI treatment resistance, similar to the patients in our study [111]. Other studies have found NAA/Cr decrease in OCD patients: in the hippocampus [99] and in the left ACC [93]. Two studies have found Cho increase in pediatric OCD patients, respectively, in medial thalamus and right prefrontal white matter [90, 101]. The low number of concurring studies in the literature can be explained by small sample sizes (<20 patients) [90, 92, 99, 112, 113], less efficient techniques preferably using 1.5 T MRI [93, 114–116], or that are not reproducible because they provide no visual information [96].

Glutamate complex (Glx) variations in the caudate nucleus have been observed after treatment (SSRIs or CBT) [92, 97, 116–118]. More recently, Simpson et al., using high spatial resolution at 3.0 T, found no significant differences in glutamate levels between unmedicated OCD patients and healthy control subjects in three striatal subregions [119].

Nevertheless, the results of those numerous previous studies remain inconsistent [108, 120, 121]. Methodological issues may need to be considered in view of improving data acquisition and post-processing. First, magnetic field strength plays an important role in the quality of MRS data, as the spatial resolution of MR spectra and the signal-to-noise ratio of the metabolites increase linearly with magnetic field strength. Second, as relaxation times (both T1 and T2) and magnetic susceptibility are highly sensitive to the calibration of experimental parameters, it is important to apply a robust protocol and to perform the totality of a study on a high field magnet (≥ 3 T). This is a major issue with regard to MRSI. All of these considerations should be taken into account to allow experimental time to be minimized, a crucial factor in clinical settings. Third, the choice of post-processing, leading to accurate quantification of the data, is of equally paramount importance.

Conclusion

Based on phenomenology and neuropsychology results, structural and functional neuroimaging studies have shown the involvement of the cortico-striato-thalamo-cortical (CSTC) network in OCD including the frontal cortex with the orbitofrontal cortex (OFC), the anterior cingulate cortex (ACC) and the dorsolateral prefrontal cortex (DLPFC), and basal ganglia (striatum and thalamus) [76, 79, 122, 123].

MRI Neuroimaging and Treatments

Neuroimaging, and more specifically MRI neuroimaging, is a very interesting tool to better understand, in vivo, the mechanisms of action of psychiatric treatments (drug or others). The final aim of these studies is to incorporate the use of neuroimaging into the daily clinical practice to evaluate which treatment could be the best for one patient and/or to assess as early as possible the treatment effectiveness at a preclinical stage, before the improvement of symptoms.

For this chapter, we decided to focus our review on functional and metabolic imaging.

Functional Imaging

fMRI studies have been broadly conducted for years to examine the effects of treatments and how the neural networks change after treatment in almost all psychiatric diseases.

In unipolar depression, fMRI studies have shown modification of neural activation correlated to mood improvement after drug treatment. Indeed, Arnone et al. found an abolishment of the abnormal amygdala response to the sad faces task after 8 weeks of citalopram intake (i.e., antidepressant), without altering responses to fearful faces [124]. In the same way, another study found that 2 months of venlafaxine treatment (i.e., antidepressant) induced increased activation of nucleus accumbens and fronto-striatal connectivity, correlated to mood improvement [125].

Moreover, a recent literature review on brain functional effects of psychopharmacological treatment in major depression (MD) found that selective serotonin reuptake inhibitors (SSRI) induced a decreased limbic activation (mainly amygdala) in response to aversive stimuli in healthy and depressed individuals, whereas noradrenalin reuptake inhibitors (NRI) seemed to facilitate processing to positive information (e.g., memory for positive events) and reward [126]. The authors also showed that MD patients specifically had a decreased activation of ACC and that this could be relevant as neural marker of clinical response.

Another review showed that response to transcranial magnetic stimulation (TMS) was consistently predicted by subcallosal cortex connectivity and that electroconvulsive therapy (ECT) treatment induced an increased connectivity between the posterior default mode and left dorsolateral prefrontal cortex in responding patients [127]. ECT also seemed to modulate the level of activity in the left subgenual ACC and its networks that negatively correlated the reduction of depressive symptoms [128].

In bipolar disorder (BD), a very interesting study investigated the effects of 8 weeks of mindfulness-based cognitive therapy (MBCT) in a controlled fMRI study [129]. After MBCT treatment, they found a significant increased activation in the medial prefrontal cortex during the mindfulness task and significant improvements in measures of mindfulness, anxiety, and cognitive tasks performances.

In anxious adolescent with a familial history of bipolar disorder, MBCT has also shown significant effects on modulating the activity of structures involved in interoception and processing of internal stimuli (insula, lentiform nucleus, thalamus, left ACC), correlated to the improvement of these functions [130].

Brooks and Vizueta review on neuroimaging in BD reported that risperidone treatment was associated with increased insula activity during affective task and that lamotrigine treatment induced decreased amygdala activation during negative stimuli in adolescents with bipolar depression [26].

Regarding lithium treatment, a very recent resting-state fMRI study revealed increased connectivity between amygdala and medial orbitofrontal cortex after 8 weeks of treatment, relative to healthy control subjects, with a positive correlation to clinical improvement [131].

In schizophrenia, a review on neuroimaging findings in treatment-resistant schizophrenia patients reported that clozapine decreases prefrontal cortex and basal ganglia activity, while it increases occipital, cingulate, and insular cortices activity [132]. The correlation to clinical improvement has remained inconsistent throughout the studies reviewed. Nevertheless, it seemed that clozapine responders had higher BOLD signal in prefrontal cortex, basal ganglia, and thalamus than non-responders before treatment, and those who had higher metabolic activity in the dorsolateral prefrontal cortex were more likely to improve negative symptoms after clozapine treatment.

Another recent review examined the effect of cognitive remediation therapy on brain functioning through neuroimaging in schizophrenia patients [133]. The authors reported that cognitive remediation improves brain activation, mainly in the prefrontal and thalamic regions.

Studies on antipsychotic treatment in antipsychotic-naïve or antipsychotic-free schizophrenia patients have reported the normalization of BOLD signal in some neural regions (e.g., reduction of impairments in dorsal cortical attention networks, suggesting a beneficial effect on neural systems for attention) but with potential adverse effects as a denormalization of other regions activity, mainly prefrontal functions [134, 135].

Metabolic Imaging

Even if MRS is more and more used to explore in vivo brain metabolism in psychiatric diseases, the number of studies evaluating the effects of treatments is more limited.

In major depression (MD), some studies have reported modification in Cho levels and its ratios after treatment. Indeed, one study found an increase of Cho/Cr ratio in basal ganglia after 8 weeks of fluoxetine treatment, 20 mg/day in responder patients relative to non-responders or placebo [136]. Likewise, another study showed an increase of total choline concentration in the left dorsolateral prefrontal cortex in responders after 10 days of high-frequency (20 Hz) rTMS [137].

Other studies have reported, in responder depressive patients, an increase of Glx, of glutamate (Glu) in the dorsolateral prefrontal cortex after ECT or rTMS treatment [137, 138], and of GABA levels in occipital cortex after ECT or SSRI treatment [139, 140].

Finally, it seems that antidepressant medication induced an increase of NAA levels in medial frontal cortex suggesting a potential neurotrophic effect [141, 142].

In BD, some ^{1}H-MRS studies have focused on the effects of lithium therapy. Bustillo review reported no consistent clues of metabolic changing. Nevertheless, it seems that lithium could induce an increase of NAA levels and a decrease of Glx and myo-inositol levels. Recent studies used innovative approaches to explore the effects of lithium therapy. The first one used ^{7}Li-MRS to evaluate brain lithium levels after 6 weeks of lithium therapy and the correlation to plasma lithium concentrations [143]. They found a significant association between central and peripheral lithium levels only in remitters, suggesting that non-remitters may not transport lithium properly to the brain, which may underlie treatment resistance to lithium in BD. The other study used ^{1}H-MRS to evaluate cingulate cortex (CC) lactate concentrations during acute depressive episodes in BD and the possible effects induced by lithium monotherapy. Their results showed higher lactate levels at baseline (relative to healthy control subjects) which significantly decreased after 6 weeks of lithium treatment (relative to baseline). Lactate levels in cingulate cortex were also associated with family history of mood disorders and plasma lithium levels.

Regarding other pharmaceutical treatments, studies have described changes in Glu concentrations in the left ventrolateral prefrontal cortex related to the degree of symptomatic improvement after divalproex treatment [144], a trend to decreased NAA levels in ACC after quetiapine therapy [145], and significant increase of NAA in cingulate gyrus and decrease of Cr and Cho/Cr in left basal ganglia after lovastatin therapy [146]. These studies have also proposed some metabolic features to predict the treatment response: low Glx levels may predict successful treatment of mania with divalproex, whereas high NAA concentration in right ventrolateral prefrontal cortex may predict successful response to quetiapine [144, 145].

In schizophrenia, Bustillo review reported no consistent results concerning the impact of antipsychotic medications on NAA brain concentration [58]. Furthermore, two literature reviews have shown that medication-naïve or medication-free patients had higher glutamatergic indices (mainly in prefrontal cortex and basal ganglia) at baseline and that antipsychotics may reduce or normalize glutamatergic indices [54, 58]. Again, those results may lead to possible use of noninvasive brain imaging such as ^{1}H-MRS to assess treatment response rather than only rely on clinical assessment.

Finally, a recent review focused on ^{31}P-MRS studies in schizophrenia [62]. This type of MRS sequence is very interesting as it allows in vivo measurement of multiple functionally important phosphorus-containing metabolites divided in two main groups: cell membrane-related phospholipids and energy-related metabolites. Unfortunately, the results remain inconsistent concerning the effects of antipsychotics on phosphorus brain metabolism. Nevertheless, it seems that the strongest pattern is related to phosphodiester (PDE) changes after antipsychotic therapy. Four studies have found similar results as increased PDE levels after antipsychotic treatment (e.g., risperidone, olanzapine), in frontal regions (mainly, dorsolateral prefrontal cortex) [147–150].

In OCD, MRS studies conducted after drug (citalopram) or cognitive behavioral therapy (CBT) treatment have shown normalization of NAA concentrations in the prefrontal cortex, FWM, ACC, and caudate nucleus [91, 151]. These modifications may therefore reflect not definitive neuronal loss but rather a potentially reversible alteration.

References

1. NIMH. Bipolar Disorder [Internet] (2016) Available from: https://www.nimh.nih.gov/health/topics/bipolar-disorder/index.shtml#part_152505
2. Association AP (2013) DSM-5 : diagnostic and statistical manual of mental disorders, 5th edn. American Psychiatric Pub. 1629 p
3. Hajek T, Kopecek M, Kozeny J, Gunde E, Alda M, Höschl C (2009) Amygdala volumes in mood disorders — meta-analysis of magnetic resonance volumetry studies. J Affect Disord 115(3):395–410
4. Bora E, Fornito A, Yücel M, Pantelis C (2010) Voxelwise meta-analysis of gray matter abnormalities in bipolar disorder. Biol Psychiatry 67(11):1097–1105
5. Ellison-Wright I, Bullmore E (2010) Anatomy of bipolar disorder and schizophrenia: a meta-analysis. Schizophr Res 117(1):1):1–1)12
6. Birur B, Kraguljac NV, Shelton RC, Lahti AC (2017) Brain structure, function, and neurochemistry in schizophrenia and bipolar disorder—a systematic review of the magnetic resonance neuroimaging literature. NPJ Schizophr [Internet]. 2017 Apr 3;3. Available from: https://www.ncbi.nlm.nih.gov/pmc/articles/PMC5441538/
7. Lee J, Choi S, Kang J, Won E, Tae W-S, Lee M-S et al (2017) Structural characteristics of the brain reward circuit regions in patients with bipolar I disorder: a voxel-based morphometric study. Psychiatry Res Neuroimaging 269:82–89
8. Adler CM, Holland SK, Schmithorst V, Wilke M, Weiss KL, Pan H et al (2004) Abnormal frontal white matter tracts in bipolar disorder: a diffusion tensor imaging study. Bipolar Disord 6(3):197–203
9. Wang F, Jackowski M, Kalmar JH, Chepenik LG, Tie K, Qiu M et al (2008) Abnormal anterior cingulum integrity in bipolar disorder determined through diffusion tensor imaging. Br J Psychiatry J Ment Sci 193(2):126–129
10. Haznedar MM, Roversi F, Pallanti S, Baldini-Rossi N, Schnur DB, LiCalzi EM et al (2005) Fronto-thalamo-striatal gray and white matter volumes and anisotropy of their connections in bipolar spectrum illnesses. Biol Psychiatry 57(7):733–742
11. Bruno S, Cercignani M, Ron MA (2008) White matter abnormalities in bipolar disorder: a voxel-based diffusion tensor imaging study. Bipolar Disord 10(4):460–468

12. Mahon K, Wu J, Malhotra AK, Burdick KE, DeRosse P, Ardekani BA et al (2009) A voxel-based diffusion tensor imaging study of white matter in bipolar disorder. Neuropsychopharmacol Off Publ Am Coll Neuropsychopharmacol 34(6):1590–1600
13. Sussmann JE, Lymer GKS, McKirdy J, Moorhead TWJ, Maniega SM, Job D et al (2009) White matter abnormalities in bipolar disorder and schizophrenia detected using diffusion tensor magnetic resonance imaging. Bipolar Disord 11(1):11–18
14. Zanetti MV, Jackowski MP, Versace A, Almeida JRC, Hassel S, Duran FLS et al (2009) State-dependent microstructural white matter changes in bipolar I depression. Eur Arch Psychiatry Clin Neurosci 259(6):316–328
15. Barysheva M, Jahanshad N, Foland-Ross L, Altshuler LL, Thompson PM (2013) White matter microstructural abnormalities in bipolar disorder: a whole brain diffusion tensor imaging study. Neuroimage Clin 2(Supplement C):558–568
16. Emsell L, Leemans A, Langan C, Van Hecke W, Barker GJ, McCarthy P et al (2013) Limbic and Callosal white matter changes in euthymic bipolar I disorder: an advanced diffusion magnetic resonance imaging Tractography study. Biol Psychiatry 73(2):194–201
17. Linke J, King AV, Poupon C, Hennerici MG, Gass A, Wessa M (2013) Impaired anatomical connectivity and related executive functions: differentiating vulnerability and disease marker in bipolar disorder. Biol Psychiatry 74(12):908–916
18. Nortje G, Stein DJ, Radua J, Mataix-Cols D, Horn N (2013) Systematic review and voxel-based meta-analysis of diffusion tensor imaging studies in bipolar disorder. J Affect Disord 150(2):192–200
19. Lu LH, Zhou XJ, Keedy SK, Reilly JL, Sweeney JA (2011) White matter microstructure in untreated first episode bipolar disorder with psychosis: comparison with schizophrenia. Bipolar Disord 13(0):604–613
20. Ganzola R, Nickson T, Bastin ME, Giles S, Macdonald A, Sussmann J et al (2017) Longitudinal differences in white matter integrity in youth at high familial risk for bipolar disorder. Bipolar Disord 19(3):158–167
21. Vargas C, López-Jaramillo C, Vieta E (2013) A systematic literature review of resting state network—functional MRI in bipolar disorder. J Affect Disord 150(3):727–735
22. Argyelan M, Ikuta T, DeRosse P, Braga RJ, Burdick KE, John M et al (2014) Resting-state fMRI connectivity impairment in schizophrenia and bipolar disorder. Schizophr Bull 40(1):100–110
23. Blumberg HP, Kaufman J, Martin A, Whiteman R, Zhang JH, Gore JC et al (2003) Amygdala and hippocampal volumes in adolescents and adults with bipolar disorder. Arch Gen Psychiatry 60(12):1201–1208
24. Lawrence NS, Williams AM, Surguladze S, Giampietro V, Brammer MJ, Andrew C et al (2004) Subcortical and ventral prefrontal cortical neural responses to facial expressions distinguish patients with bipolar disorder and major depression. Biol Psychiatry 55(6):578–587
25. Delvecchio G, Fossati P, Boyer P, Brambilla P, Falkai P, Gruber O et al (2012) Common and distinct neural correlates of emotional processing in bipolar disorder and major depressive disorder: a voxel-based meta-analysis of functional magnetic resonance imaging studies. Eur Neuropsychopharmacol 22(2):100–113
26. Brooks JO, Vizueta N (2014) Diagnostic and clinical implications of functional neuroimaging in bipolar disorder. J Psychiatr Res 57(Supplement C):12–25
27. Townsend JD, Torrisi SJ, Lieberman MD, Sugar CA, Bookheimer SY, Altshuler LL (2013) Frontal-amygdala connectivity alterations during emotion down-regulation in bipolar I disorder. Biol Psychiatry 73(2):127–135
28. Cremaschi L, Penzo B, Palazzo M, Dobrea C, Cristoffanini M, Dell'Osso B et al (2013) Assessing working memory via N-back task in euthymic bipolar I disorder patients: a review of functional magnetic resonance imaging studies. Neuropsychobiology 68(2):63–70

29. McKenna BS, Sutherland AN, Legenkaya AP, Eyler LT (2014) Abnormalities of brain response during encoding into verbal working memory among euthymic patients with bipolar disorder. Bipolar Disord 16(3):289–299
30. Dell'Osso B, Cinnante C, Giorgio AD, Cremaschi L, Palazzo MC, Cristoffanini M et al (2015) Altered prefrontal cortex activity during working memory task in bipolar disorder: a functional magnetic resonance imaging study in euthymic bipolar I and II patients. J Affect Disord 184:116–122
31. Chase HW, Nusslock R, Almeida JR, Forbes EE, LaBarbara EJ, Phillips ML (2013) Dissociable patterns of abnormal frontal cortical activation during anticipation of an uncertain reward or loss in bipolar versus major depression. Bipolar Disord 15(8):839–854
32. Urošević S, Luciana M, Jensen JB, Youngstrom EA, Thomas KM (2016) Age associations with neural processing of reward anticipation in adolescents with bipolar disorders. Neuroimage Clin 11:476–485
33. Trost S, Diekhof EK, Zvonik K, Lewandowski M, Usher J, Keil M et al (2014) Disturbed anterior prefrontal control of the mesolimbic reward system and increased impulsivity in bipolar disorder. Neuropsychopharmacology 39(8):1914–1923
34. Dutra SJ, Man V, Kober H, Cunningham WA, Gruber J (2017) Disrupted cortico-limbic connectivity during reward processing in remitted bipolar I disorder. Bipolar Disord 19:661
35. Berghorst LH, Kumar P, Greve DN, Deckersbach T, Ongur D, Dutra S et al (2016) Stress and reward processing in bipolar disorder: an fMRI study. Bipolar Disord 18(7):602–611
36. Kraguljac NV, Reid M, White D, Jones R, den Hollander J, Lowman D et al (2012) Neurometabolites in schizophrenia and bipolar disorder – a systematic review and meta-analysis. Psychiatry Res 203(2–3):111–125
37. Atagün Mİ, Şıkoğlu EM, Can SS, Karakaş-Uğurlu G, Ulusoy-Kaymak S, Çayköylü A et al (2015) Investigation of Heschl's gyrus and planum temporale in patients with schizophrenia and bipolar disorder: a proton magnetic resonance spectroscopy study. Schizophr Res 161(2):202–209
38. Li H, Xu H, Zhang Y, Guan J, Zhang J, Xu C et al (2016) Differential neurometabolite alterations in brains of medication-free individuals with bipolar disorder and those with unipolar depression: a two-dimensional proton magnetic resonance spectroscopy study. Bipolar Disord 18(7):583–590
39. Chitty KM, Lagopoulos J, Lee RSC, Hickie IB, Hermens DF (2013) A systematic review and meta-analysis of proton magnetic resonance spectroscopy and mismatch negativity in bipolar disorder. Eur Neuropsychopharmacol 23(11):1348–1363
40. Kubo H, Nakataki M, Sumitani S, Iga J, Numata S (2017) Kameoka N, et al. 1H-magnetic resonance spectroscopy study of glutamate-related abnormality in bipolar disorder. J Affect Disord 208(Supplement C):139–144
41. Silveira LE, Bond DJ, MacMillan EL, Kozicky J-M, Muralidharan K, Bücker J et al (2017) Hippocampal neurochemical markers in bipolar disorder patients following the first-manic episode: a prospective 12-month proton magnetic resonance spectroscopy study. Aust N Z J Psychiatry 51(1):65–74
42. Shi X-F, Carlson PJ, Sung Y-H, Fiedler KK, Forrest LN, Hellem TL et al (2015) Decreased brain PME/PDE ratio in bipolar disorder: a preliminary 31P magnetic resonance spectroscopy study. Bipolar Disord 17(7):743–752
43. Andres RH, Ducray AD, Schlattner U, Wallimann T, Widmer HR (2008) Functions and effects of creatine in the central nervous system. Brain Res Bull 76(4):329–343
44. NIMH (2016) Schizophrenia [Internet]. Available from: https://www.nimh.nih.gov/health/topics/schizophrenia/index.shtml
45. Honea R, Crow TJ, Passingham D, Mackay CE (2005) Regional deficits in brain volume in schizophrenia: a meta-analysis of voxel-based morphometry studies. Am J Psychiatry 162(12):2233–2245
46. Crow TJ (1997) Is schizophrenia the price that Homo sapiens pays for language? Schizophr Res 28(2–3):127–141

47. Rajarethinam RP, DeQuardo JR, Nalepa R, Tandon R (2000) Superior temporal gyrus in schizophrenia: a volumetric magnetic resonance imaging study. Schizophr Res 41(2):303–312
48. Dietsche B, Kircher T, Falkenberg I (2017) Structural brain changes in schizophrenia at different stages of the illness: a selective review of longitudinal magnetic resonance imaging studies. Aust N Z J Psychiatry 51(5):500–508
49. Sumner PJ, Bell IH, Rossell SL (2017) A systematic review of the structural neuroimaging correlates of thought disorder. Neurosci Biobehav Rev [Internet]. Available from: http://www.sciencedirect.com/science/article/pii/S0149763417301252
50. Wheeler AL, Voineskos AN (2014) A review of structural neuroimaging in schizophrenia: from connectivity to connectomics. Front Hum Neurosci [Internet]. Available from: https://www.ncbi.nlm.nih.gov/pmc/articles/PMC4142355/
51. Yao L, Lui S, Liao Y, Du M-Y, Hu N, Thomas JA et al (2013) White matter deficits in first episode schizophrenia: an activation likelihood estimation meta-analysis. Prog Neuro-Psychopharmacol Biol Psychiatry 45:100–106
52. Goghari VM, Sponheim SR, MacDonald AW (2010) The functional neuroanatomy of symptom dimensions in schizophrenia: a qualitative and quantitative review of a persistent question. Neurosci Biobehav Rev 34(3):468
53. Mwansisya TE, Hu A, Li Y, Chen X, Wu G, Huang X, et al (2017) Task and resting-state fMRI studies in first-episode schizophrenia: A systematic review. Schizophr Res [Internet]. [cited 2017 Nov 17];0(0). Available from: http://www.schres-journal.com/article/S0920-9964(17)30115-9/fulltext
54. Poels EMP, Kegeles LS, Kantrowitz JT, Javitt DC, Lieberman JA, Abi-Dargham A et al (2014) Glutamatergic abnormalities in schizophrenia: a review of proton MRS findings. Schizophr Res 152(0):325–332
55. Egan MF, Goldberg TE, Kolachana BS, Callicott JH, Mazzanti CM, Straub RE et al (2001) Effect of COMT Val108/158 met genotype on frontal lobe function and risk for schizophrenia. Proc Natl Acad Sci U S A 98(12):6917–6922
56. Meyer-Lindenberg A, Kolachana B, Weinberger DR, Buckholtz J, Ding J, Callicott JH et al (2006) Impact of complex genetic variation in COMT on human brain function. Mol Psychiatry 11(9):867
57. Steen RG, Hamer L (2005) Measurement of brain metabolites by 1H magnetic resonance spectroscopy in patients with schizophrenia: a systematic review and meta-analysis. Neuropsychopharmacology 30(11):1949
58. Bustillo JR (2013) Use of proton magnetic resonance spectroscopy in the treatment of psychiatric disorders: a critical update. Dialogues Clin Neurosci 15(3):329–337
59. Taylor SF, Tso IF (2015) GABA abnormalities in schizophrenia: a methodological review of in vivo studies. Schizophr Res 167(0):84–90
60. Kegeles LS, Mao X, Stanford AD, Girgis R, Ojeil N, Xu X et al (2012) Elevated prefrontal cortex γ-aminobutyric acid and glutamate-glutamine levels in schizophrenia measured in vivo with proton magnetic resonance spectroscopy. Arch Gen Psychiatry 69(5):449–459
61. Kelemen O, Kiss I, Benedek G, Kéri S (2013) Perceptual and cognitive effects of antipsychotics in first-episode schizophrenia: The potential impact of GABA concentration in the visual cortex. Prog Neuro-Psychopharmacol Biol Psychiatry 47(Supplement C):13–19
62. Yuksel C, Tegin C, O'Connor L, Du F, Ahat E, Cohen BM et al (2015) Phosphorus magnetic resonance spectroscopy studies in schizophrenia. J Psychiatr Res 68:157–166
63. Karno M, Golding JM, Sorenson SB, Burnam M (1988) THe epidemiology of obsessive-compulsive disorder in five us communities. Arch Gen Psychiatry 45(12):1094–1099
64. Robins LN, Helzer JE, Orvaschel H, Anthony JC, Blazer DG, Burnam A, et al (1985) 8 - The diagnostic interview schedule. In Kessler WWEG, editor. Epidemiologic field methods in psychiatry [Internet]. San Diego: Academic Press; [cited 2014 Oct 6]. pp 143–70. Available from: http://www.sciencedirect.com/science/article/pii/B9780080917986500129

65. Ruscio AM, Stein DJ, Chiu WT, Kessler RC (2010) The epidemiology of obsessive-compulsive disorder in the National Comorbidity Survey Replication. Mol Psychiatry 15(1):53–63
66. Rasmussen SA, Eisen JL (1992) The epidemiology and clinical features of obsessive compulsive disorder. Psychiatr Clin North Am 15(4):743–758
67. Rasmussen SA, Tsuang MT (1984) The epidemiology of obsessive compulsive disorder. J Clin Psychiatry 45(11):450–457
68. Skoog G, Skoog I (1999) A 40-year follow-up of patients with obsessive-compulsive disorder [see comments]. Arch Gen Psychiatry 56(2):121–127
69. Koran LM, Thienemann ML, Davenport R (1996) Quality of life for patients with obsessive-compulsive disorder. Am J Psychiatry 153(6):783–788
70. Jaafari N, Daniel M-L, Lacoste J, Bacconnier M, Belin D, Rotge J-Y (2011) Insight, obsession et vérification dans le trouble obsessionnel-compulsif. Ann Méd Psychol Rev Psychiatr 169(7):453–456
71. Rotge J-Y, Guehl D, Dilharreguy B, Tignol J, Bioulac B, Allard M et al (2009) Meta-analysis of brain volume changes in obsessive-compulsive disorder. Biol Psychiatry 65(1):75–83
72. Atmaca M, Yildirim H, Ozdemir H, Ozler S, Kara B, Ozler Z et al (2008) Hippocampus and amygdalar volumes in patients with refractory obsessive-compulsive disorder. Prog Neuro-Psychopharmacol Biol Psychiatry 32(5):1283–1286
73. Atmaca M (2011) Review of structural neuroimaging in patients with refractory obsessive-compulsive disorder. Neurosci Bull 27(3):215–220
74. Radua J, Mataix-Cols D (2009) Voxel-wise meta-analysis of grey matter changes in obsessive–compulsive disorder. Br J Psychiatry 195(5):393–402
75. Piras F, Piras F, Chiapponi C, Girardi P, Caltagirone C, Spalletta G Widespread structural brain changes in OCD: a systematic review of voxel-based morphometry studies. Cortex [Internet]. [cited 2014 Aug 4]; Available from: http://www.sciencedirect.com/science/article/pii/S0010945213000464
76. Menzies L, Chamberlain SR, Laird AR, Thelen SM, Sahakian BJ, Bullmore ET (2008) Integrating evidence from neuroimaging and neuropsychological studies of obsessive-compulsive disorder: the orbitofronto-striatal model revisited. Neurosci Biobehav Rev 32(3):525–549
77. Piras F, Piras F, Caltagirone C, Spalletta G Brain circuitries of obsessive compulsive disorder: a systematic review and meta-analysis of diffusion tensor imaging studies. Neurosci Biobehav Rev [Internet]. [cited 2013 Nov 8]; Available from: http://www.sciencedirect.com/science/article/pii/S0149763413002327
78. Gan J, Zhong M, Fan J, Liu W, Niu C, Cai S et al (2017) Abnormal white matter structural connectivity in adults with obsessive-compulsive disorder. Transl Psychiatry 7(3):e1062
79. Nakao T, Okada K, Kanba S (2014) Neurobiological model of obsessive-compulsive disorder: evidence from recent neuropsychological and neuroimaging findings. Psychiatry Clin Neurosci 68(8):587–605
80. Jung WH, Kang D-H, Kim E, Shin KS, Jang JH, Kwon JS (2013) Abnormal corticostriatal-limbic functional connectivity in obsessive–compulsive disorder during reward processing and resting-state. Neuroimage Clin 3:27–38
81. Keshavan MS, Stanley JA, Pettegrew JW (2000) Magnetic resonance spectroscopy in schizophrenia: methodological issues and findings--part II. Biol Psychiatry 48(5):369–380
82. Stanley JA (2002) In vivo magnetic resonance spectroscopy and its application to neuropsychiatric disorders. Can J Psychiatry 47(4):315–326
83. McClure RJ, Kanfer JN, Panchalingam K, Klunk WE, Pettegrew JW (1994) Alzheimer's disease: membrane-associated metabolic changes. Ann N Y Acad Sci 747:110–124
84. Tartaglia MC, Narayanan S, De Stefano N, Arnaoutelis R, Antel SB, Francis SJ et al (2002) Choline is increased in pre-lesional normal appearing white matter in multiple sclerosis. J Neurol 249(10):1382–1390

85. Maier M, Ron MA, Barker GJ, Tofts PS (1995) Proton magnetic resonance spectroscopy: an in vivo method of estimating hippocampal neuronal depletion in schizophrenia. Psychol Med 25(6):1201–1209
86. Kantarci K (2013) Magnetic resonance spectroscopy in common dementias. Neuroimaging Clin N Am 23(3):393–406
87. Nie K, Zhang Y, Huang B, Wang L, Zhao J, Huang Z et al (2013) Marked N-acetylaspartate and choline metabolite changes in Parkinson's disease patients with mild cognitive impairment. Parkinsonism Relat Disord 19(3):329–334
88. Frye MA, Thomas MA, Yue K, Binesh N, Davanzo P, Ventura J et al (2007) Reduced concentrations of N-acetylaspartate (NAA) and the NAA–creatine ratio in the basal ganglia in bipolar disorder: a study using 3-tesla proton magnetic resonance spectroscopy. Psychiatry Res Neuroimaging 154(3):259–265
89. Bertolino A, Callicott JH, Mattay VS, Weidenhammer KM, Rakow R, Egan MF et al (2001) The effect of treatment with antipsychotic drugs on brain N-acetylaspartate measures in patients with schizophrenia. Biol Psychiatry 49(1):39–46
90. Weber AM, Soreni N, Stanley JA, Greco A, Mendlowitz S, Szatmari P et al (2014) Proton magnetic resonance spectroscopy of prefrontal white matter in psychotropic naïve children and adolescents with obsessive–compulsive disorder. Psychiatry Res Neuroimaging 222(1–2):67–74
91. Jang J, Kwon J, Jang D, Moon W-J, Lee J-M, Ha T et al (2006) A proton MRSI study of brain N-acetylaspartate level after 12 weeks of citalopram treatment in drug-naive patients with obsessive-compulsive disorder. Am J Psychiatry 163(7):1202–1207
92. O'Neill J, Gorbis E, Feusner JD, Yip JC, Chang S, Maidment KM et al (2013) Effects of intensive cognitive-behavioral therapy on cingulate neurochemistry in obsessive–compulsive disorder. J Psychiatr Res 47(4):494–504
93. Tükel R, Özata B, Öztürk N, Ertekin BA, Ertekin E, Saruhan Direskeneli G (2014) The role of the brain-derived neurotrophic factor SNP rs2883187 in the phenotypic expression of obsessive-compulsive disorder. J Clin Neurosci 21(5):790–793
94. Yücel M, Harrison BJ, Wood SJ, Fornito A, Wellard RM, Pujol J et al (2007) Functional and biochemical alterations of the medial frontal cortex in obsessive-compulsive disorder. Arch Gen Psychiatry 64(8):946
95. Bartha R, Stein MB, Williamson PC, Drost DJ, Neufeld RWJ, Carr TJ et al (1998) A short Echo 1H spectroscopy and volumetric MRI study of the Corpus striatum in patients with obsessive- compulsive disorder and comparison subjects. Am J Psychiatry 155(11):1584–1591
96. Ebert D, Speck O, König A, Berger M, Hennig J, Hohagen F (1997) 1H-magnetic resonance spectroscopy in obsessive-compulsive disorder: evidence for neuronal loss in the cingulate gyrus and the right striatum. Psychiatry Res Neuroimaging 74(3):173–176
97. Fitzgerald KD, Moore GJ, Paulson LA, Stewart CM, Rosenberg DR (2000) Proton spectroscopic imaging of the thalamus in treatment-naive pediatric obsessive–compulsive disorder∗. Biol Psychiatry 47(3):174–182
98. Rosenberg DR, Amponsah A, Sullivan A, MacMillan S, Moore GJ (2001) Increased medial thalamic choline in pediatric obsessive-compulsive disorder as detected by quantitative in vivo spectroscopic imaging. J Child Neurol 16(9):636–641
99. Atmaca M, Yildirim H, Ozdemir H, Koc M, Ozler S, Tezcan E (2009) Neurochemistry of the hippocampus in patients with obsessive–compulsive disorder. Psychiatry Clin Neurosci 63(4):486–490
100. Hatchondo L, Jaafari N, Langbour N, Maillochaud S, Herpe G (2017) Guillevin R, et al. 1H magnetic resonance spectroscopy suggests neural membrane alteration in specific regions involved in obsessive-compulsive disorder. Psychiatry Res Neuroimaging 269:48–53
101. Smith EA, Russell A, Lorch E, Banerjee SP, Rose M, Ivey J et al (2003) Increased medial thalamic choline found in pediatric patients with obsessive-compulsive disorder versus major

depression or healthy control subjects: a magnetic resonance spectroscopy study. Biol Psychiatry 54(12):1399–1405

102. Fan S, Cath DC, van den Heuvel OA, van der Werf YD, Schöls C, Veltman DJ et al (2017) Abnormalities in metabolite concentrations in tourette's disorder and obsessive-compulsive disorder—a proton magnetic resonance spectroscopy study. Psychoneuroendocrinology 77:211–217

103. Meyerhoff DJ, MacKay S, Constans JM, Norman D, Van Dyke C, Fein G et al (1994) Axonal injury and membrane alterations in Alzheimer's disease suggested by in vivo proton magnetic resonance spectroscopic imaging. Ann Neurol 36(1):40–47

104. Jenkins BG, Koroshetz WJ, Beal MF, Rosen BR (1993) Evidence for impairment of energy metabolism in vivo in Huntington's disease using localized 1H NMR spectroscopy. Neurology 43(12):2689–2695

105. Arnold DL, Matthews PM, Francis GS, O'Connor J, Antel JP (1992) Proton magnetic resonance spectroscopic imaging for metabolic characterization of demyelinating plaques. Ann Neurol 31(3):235–241

106. Hattingen E, Magerkurth J, Pilatus U, Hübers A, Wahl M, Ziemann U (2011) Combined 1H and 31P spectroscopy provides new insights into the pathobiochemistry of brain damage in multiple sclerosis. NMR Biomed 24(5):536–546

107. Zaaraoui W, Audoin B, Pelletier J, Cozzone PJ, Ranjeva J-P (2010) Advanced magnetic resonance imaging techniques to better understand multiple sclerosis. Biophys Rev 2(2):83–90

108. Brennan BP, Rauch SL, Jensen JE, Pope HG Jr (2013) A critical review of magnetic resonance spectroscopy studies of obsessive-compulsive disorder. Biol Psychiatry 73(1):24–31

109. Atmaca M, Onalan E, Yildirim H, Yuce H, Koc M, Korkmaz S (2010) The association of myelin oligodendrocyte glycoprotein gene and white matter volume in obsessive–compulsive disorder. J Affect Disord 124(3):309–313

110. Stewart SE, Platko J, Fagerness J, Birns J, Jenike E, Smoller JW et al (2007) A genetic family-based association study of OLIG2 in obsessive-compulsive disorder. Arch Gen Psychiatry 64(2):209–214

111. Mohamed MA, Smith MA, Schlund MW, Nestadt G, Barker PB, Hoehn-Saric R (2007) Proton magnetic resonance spectroscopy in obsessive-compulsive disorder: a pilot investigation comparing treatment responders and non-responders. Psychiatry Res Neuroimaging 156(2):175–179

112. Bédard M-J, Chantal S (2011) Brain magnetic resonance spectroscopy in obsessive–compulsive disorder: the importance of considering subclinical symptoms of anxiety and depression. Psychiatry Res Neuroimaging 192(1):45–54

113. Yücel M, Wood SJ, Wellard RM, Harrison BJ, Fornito A, Pujol J et al (2008) Anterior cingulate glutamate–glutamine levels predict symptom severity in women with obsessive–compulsive disorder. Aust N Z J Psychiatry 42(6):467–477

114. Ohara K, Isoda H, Suzuki Y, Takehara Y, Ochiai M, Takeda H et al (1999) Proton magnetic resonance spectroscopy of lenticular nuclei in obsessive–compulsive disorder. Psychiatry Res Neuroimaging 92(2):83–91

115. Sumitani S, Harada M, Kubo H, Ohmori T (2007) Proton magnetic resonance spectroscopy reveals an abnormality in the anterior cingulate of a subgroup of obsessive–compulsive disorder patients. Psychiatry Res Neuroimaging 154(1):85–92

116. Whiteside SPH, Abramowitz JS, Port JD (2012) Decreased caudate N-acetyl-l-aspartic acid in pediatric obsessive-compulsive disorder and the effects of behavior therapy. Psychiatry Res Neuroimaging 202(1):53–59

117. Bolton J, Moore GJ, MacMillan S, Stewart CM, Rosenberg D (2001) Case study: caudate glutamatergic changes with paroxetine persist after medication discontinuation in pediatric OCD. J Am Acad Child Adolesc Psychiatry 40(8):903–906

118. Moore GJ, MacMaster F, Stewart CM, Rosenberg D (1998) Case study: caudate glutamatergic changes with paroxetine therapy for pediatric obsessive-compulsive disorder. J Am Acad Child Adolesc Psychiatry 37(6):663–667

119. Simpson HB, Kegeles LS, Hunter L, Mao X, Van Meter P, Xu X et al (2015) Assessment of glutamate in striatal subregions in obsessive-compulsive disorder with proton magnetic resonance spectroscopy. Psychiatry Res 232(1):65–70

120. Aoki Y, Aoki A, Suwa H (2012) Reduction of N-acetylaspartate in the medial prefrontal cortex correlated with symptom severity in obsessive-compulsive disorder: meta-analyses of 1H-MRS studies. Transl Psychiatry 2(8):e153

121. O'Neill J, Lai TM, Sheen C, Salgari GC, Ly R, Armstrong C et al (2016) Cingulate and thalamic metabolites in obsessive-compulsive disorder. Psychiatry Res 254:34–40

122. Aouizerate B, Guehl D, Cuny E, Rougier A, Bioulac B, Tignol J et al (2004) Pathophysiology of obsessive–compulsive disorder: A necessary link between phenomenology, neuropsychology, imagery and physiology. Prog Neurobiol 72(3):195–221

123. Pauls DL, Abramovitch A, Rauch SL, Geller DA (2014) Obsessive-compulsive disorder: an integrative genetic and neurobiological perspective. Nat Rev Neurosci 15(6):410–424

124. Arnone D, McKie S, Elliott R, Thomas EJ, Downey D, Juhasz G et al (2012) Increased amygdala responses to sad but not fearful faces in major depression: relation to mood state and pharmacological treatment. Am J Psychiatry 169(8):841–850

125. Heller AS, Johnstone T, Light S, Peterson MJ, Kolden GG, Kalin NH et al (2013) Relationships between changes in sustained Fronto-striatal connectivity and positive affect with antidepressant treatment in major depression. Am J Psychiatry 170(2):197–206

126. Wessa M, Lois G (2015) Brain functional effects of psychopharmacological treatment in major depression: a focus on neural circuitry of affective processing. Curr Neuropharmacol 13(4):466–479

127. Dichter GS, Gibbs D, Smoski MJ (2015) A systematic review of relations between resting-state functional-MRI and treatment response in major depressive disorder. J Affect Disord 172:8–17

128. Liu Y, Du L, Li Y, Liu H, Zhao W, Liu D, et al Antidepressant effects of electroconvulsive therapy correlate with subgenual anterior cingulate activity and connectivity in depression. Medicine (Baltimore) [Internet]. 2015 13 [cited 2017 Dec 17];94(45). Available from: https://www.ncbi.nlm.nih.gov/pmc/articles/PMC4912303/

129. Ives-Deliperi VL, Howells F, Stein DJ, Meintjes EM, Horn N (2013) The effects of mindfulness-based cognitive therapy in patients with bipolar disorder: a controlled functional MRI investigation. J Affect Disord 150(3):1152–1157

130. Strawn JR, Cotton S, Luberto CM, Patino LR, Stahl LA, Weber WA et al (2016) Neural function before and after mindfulness-based cognitive therapy in anxious adolescents at risk for developing bipolar disorder. J Child Adolesc Psychopharmacol 26(4):372–379

131. Altinay M, Karne H, Anand A (2018) Lithium monotherapy associated clinical improvement effects on amygdala-ventromedial prefrontal cortex resting state connectivity in bipolar disorder. J Affect Disord 225:4–12

132. Nakajima S, Takeuchi H, Plitman E, Fervaha G, Gerretsen P, Caravaggio F et al (2015) Neuroimaging findings in treatment-resistant schizophrenia: a systematic review. Schizophr Res 164(0):164–175

133. Penadés R, González-Rodríguez A, Catalán R, Segura B, Bernardo M, Junqué C (2017) Neuroimaging studies of cognitive remediation in schizophrenia: a systematic and critical review. World J Psychiatr 7(1):34–43

134. Abbott CC, Jaramillo A, Wilcox CE, Hamilton DA (2013) Antipsychotic drug effects in schizophrenia: a review of longitudinal fMRI investigations and neural interpretations. Curr Med Chem 20(3):428–437

135. Keedy SK, Reilly JL, Bishop JR, Weiden PJ, Sweeney JA (2015) Impact of antipsychotic treatment on attention and motor learning Systems in First-Episode Schizophrenia. Schizophr Bull 41(2):355–365

136. Sonawalla SB, Renshaw PF, Moore CM, Alpert JE, Nierenberg AA, Rosenbaum JF et al (1999) Compounds containing cytosolic choline in the basal ganglia: a potential biological marker of true drug response to fluoxetine. Am J Psychiatry 156(10):1638–1640

137. Luborzewski A, Schubert F, Seifert F, Danker-Hopfe H, Brakemeier E-L, Schlattmann P et al (2007) Metabolic alterations in the dorsolateral prefrontal cortex after treatment with high-frequency repetitive transcranial magnetic stimulation in patients with unipolar major depression. J Psychiatr Res 41(7):606–615

138. Michael N, Erfurth A, Ohrmann P, Arolt V, Heindel W, Pfleiderer B (2003) Metabolic changes within the left dorsolateral prefrontal cortex occurring with electroconvulsive therapy in patients with treatment resistant unipolar depression. Psychol Med 33(7):1277–1284

139. Sanacora G, Mason GF, Rothman DL, Hyder F, Ciarcia JJ, Ostroff RB et al (2003) Increased cortical GABA concentrations in depressed patients receiving ECT. Am J Psychiatry 160(3):577–579

140. Sanacora G, Fenton LR, Fasula MK, Rothman DL, Levin Y, Krystal JH et al (2006) Cortical γ-aminobutyric acid concentrations in depressed patients receiving cognitive behavioral therapy. Biol Psychiatry 59(3):284–286

141. Gonul AS, Kitis O, Ozan E, Akdeniz F, Eker C, Eker OD et al (2006) The effect of antidepressant treatment on N-acetyl aspartate levels of medial frontal cortex in drug-free depressed patients. Prog Neuro-Psychopharmacol Biol Psychiatry 30(1):120–125

142. Taylor MJ, Godlewska BR, Norbury R, Selvaraj S, Near J, Cowen PJ (2012) Early increase in marker of neuronal integrity with antidepressant treatment of major depression: 1H-magnetic resonance spectroscopy of N-acetyl-aspartate. Int J Neuropsychopharmacol 15(10):1541–1546

143. Machado-Vieira R, Otaduy MC, Zanetti MV, De Sousa RT, Dias VV, Leite CC et al (2016) A selective association between central and peripheral Lithium levels in remitters in bipolar depression: a 3T-(7) li magnetic resonance spectroscopy study. Acta Psychiatr Scand 133(3):214–220

144. Strawn JR, Patel NC, Chu W-J, Lee J-H, Adler CM, Kim MJ et al (2012) Glutamatergic effects of divalproex in manic adolescents: a proton magnetic resonance spectroscopy study. J Am Acad Child Adolesc Psychiatry 51(6):642–651

145. Adler CM, DelBello MP, Weber WA, Jarvis KB, Welge J, Chu W-J et al (2013) Neurochemical effects of quetiapine in patients with bipolar mania: a proton magnetic resonance spectroscopy study. J Clin Psychopharmacol 33(4):528–532

146. Lotfi M, Shafiee S, Ghanizadeh A, Sigaroudi MO, Razeghian L (2017) A magnetic resonance spectroscopy study of lovastatin for treating bipolar mood disorder: a 4-week randomized double-blind, placebo- controlled clinical trial. Recent Patents Inflamm Allergy Drug Discov 10(2):133–141

147. Stanley JA, Williamson PC, Drost DJ, Carr TJ, Rylett RJ, Malla A et al (1995) An in vivo study of the prefrontal cortex of schizophrenic patients at different stages of illness via phosphorus magnetic resonance spectroscopy. Arch Gen Psychiatry 52(5):399–406

148. Volz H-P, Riehemann S, Maurer I, Smesny S, Sommer M, Rzanny R et al (2000) Reduced phosphodiesters and high-energy phosphates in the frontal lobe of schizophrenic patients: a 31P chemical shift spectroscopic-imaging study. Biol Psychiatry 47(11):954–961

149. Smesny S, Langbein K, Rzanny R, Gussew A, Burmeister HP, Reichenbach JR et al (2012) Antipsychotic drug effects on left prefrontal phospholipid metabolism: a follow-up 31P-2D-CSI study of haloperidol and risperidone in acutely ill chronic schizophrenia patients. Schizophr Res 138(2):164–170

150. Nenadic I, Dietzek M, Langbein K, Rzanny R, Gussew A, Reichenbach JR et al (2013) Effects of olanzapine on 31P MRS metabolic markers in schizophrenia. Hum Psychopharmacol Clin Exp 28(1):91–93

151. Whiteside SPH, Abramowitz JS, Port JD (2011) The effect of behavior therapy on caudate N-acetyl-l-aspartic acid in adults with obsessive–compulsive disorder. Psychiatry Res Neuroimaging 201(1):10–16

Chapter 13
The Implications of Brain Plasticity and Task Selectivity for Visual Rehabilitation of Blind and Visually Impaired Individuals

Daniel-Robert Chebat, Benedetta Heimler, Shir Hofsetter, and Amir Amedi

Introduction

The amount of cortical space devoted to each sensory modality differs substantially among phylogenetically different species. By observing the amount of cortical space devoted to vision among primitive mammals, and then for primates and humans, it is possible to infer the evolution of the importance of vision across different taxonomic groups. Pending millions of years of primate and human evolution, the portion of the neocortex devoted to visual processing has expanded substantially in response to environmental demands. To survive as a species, our ancestors had to develop the use of increasingly complex tools; they also had to rely on complex navigational strategies to hunt and gather, find an ancient encampment, or return to an old hunting ground during seasonal migrations. These tasks required the expansion of the visual

D.-R. Chebat (✉)
Visual and Cognitive Neuroscience Laboratory (VCN lab), Department of Behavioral Sciences, Faculty of Social Sciences and Humanities, Ariel University, Ariel, Israel
e-mail: danielc@ariel.ac.il

B. Heimler · S. Hofsetter
Department of Medical Neurobiology, The Institute for Medical Research Israel-Canada, Faculty of Medicine, The Hebrew University of Jerusalem, Jerusalem, Israel

The Edmond and Lily Safra Centre for Brain Sciences (ELSC), The Hebrew University of Jerusalem, Jerusalem, Israel

A. Amedi (✉)
Department of Medical Neurobiology, The Institute for Medical Research Israel-Canada, Faculty of Medicine, The Hebrew University of Jerusalem, Jerusalem, Israel

The Edmond and Lily Safra Centre for Brain Sciences (ELSC), The Hebrew University of Jerusalem, Jerusalem, Israel

Cognitive Sciences Program, The Hebrew University of Jerusalem, Jerusalem, Israel
e-mail: amira@ekmd.huji.ac.il

© Springer International Publishing AG, part of Springer Nature 2018
C. Habas (ed.), *The Neuroimaging of Brain Diseases*, Contemporary Clinical Neuroscience, https://doi.org/10.1007/978-3-319-78926-2_13

cortex to allow for complex navigational behaviors, the focalization of objects in the central visual field, and the accurate judgment of distances with binocular vision.

Because of the high degree of consistency in terms of the loci of sensory specializations in the human brain (visual, somatosensory, auditory, and even olfactory specialized areas), classical theories on brain organization have conceptualized the sensory responsiveness of different brain areas as modality specific. This notion has been strongly put into question by results obtained with sensory-deprived populations and mainly with congenitally blind adults. It has been consistently shown that many specialized regions in the visually deprived occipital cortex maintain their functional specialization (i.e., responded to the same category of stimuli) even though the input was provided through a different sensory modality (for review see [1]). For example, the perception of objects, faces and facial expressions, body postures, motion, and text using the tactile and auditory modalities recruits the same brain areas as in sighted individuals for similar categories (for review see [2–5]; see also section "Functional Brain Plasticity and Task Selectivity" in this review). This led to the notion that the evolutionary pressures mentioned in the previous paragraph forged an incredibly flexible sensory brain capable of adapting to many different classes of stimuli and tasks, regardless of the input sensory modality. Thus, the organization of cortical areas might be better described in terms of a flexible task-based organization, rather than a sensory-dependent classification as was classically conceived (for review: [2, 5–7]).

In this review chapter, we discuss theories of task-specific adaptability, amodal cortical processing, and plasticity using blindness as a working model. Special emphasis will be given to results obtained in this population using sensory substitution devices. Sensory substitution devices are capable of conveying information generally provided in one sensory modality (e.g., vision) into a different sensory modality (e.g., touch; audition) through a specific algorithm. These devices are typically composed of three parts: The first part is a sensor (i.e., a camera, an infrared laser, or a motion sensor) to pick up information from the environment. The second part consists of a computer chip to convert the information from the sensor. The third part is a way to convey the information to the user either through touch (e.g., a tactile grid) or sound (e.g., using an auditory cue, or vibrations). Sensory substitution device information can be interpreted through focused training and practice (for review: [8–10]). One fascinating aspect of this training is the shift in processing of this information in people who are congenitally blind. The information switches from the modality specific area before training to task-specific areas after training. We discuss different approaches to sensory substitution and what these approaches have taught us about brain organization in this review. We also discuss mechanisms of navigation in people who are blind and the neural correlates thereof. We frame the implications of these novel results using sensory substitution to understand brain mechanisms in terms of visual rehabilitation programs for people who are blind.

Artificial Forms of Vision: Sight Restoration and Sensory Substitution

There are two main approaches to rehabilitate visual functions for people who were born blind, or lost their sight later in life, namely, invasive and noninvasive approaches. Invasive approaches require the implantation of an electric grid directly on the nervous surface, thus attempting to bypass the lesioned or missing sense, and stimulate the rest of the intact visual system electronically. This grid is implanted either on the retina, on the optic nerve, or directly on the cellular cortical surface. Currently, these implants do not allow to entirely recover vision, but they provide some visual "qualia" to their users such as *phosphenes*. Phosphenes are the perception of retinotopically organized flashes of light, obtained by stimulating the receptive fields of neurons in the primary visual cortex. The second approach includes all noninvasive methods to convey visual information through other modalities. Here we will focus on one of the most effective forms of noninvasive approaches to visual recovery, namely, sensory substitution devices which capture specific aspects of visual information from different types of sensors or cameras and translate them into either touch or sound. Certain sensory substitution devices code general aspects of visual information (i.e., image based), while others concentrate on translating one specific aspect of the visual information (e.g., distance or depth). While image-based sensory substitution devices generally require many hours of supervised training in order to learn to interpret the conveyed image, other sensory substitution devices are more minimalistic and convey information much more intuitively, thus requiring very little training. The following sections discuss these two approaches in greater detail.

Sight Restoration

Sight restoration attempts to bypass the lesioned, absent, or damaged part of the visual system causing total or partial blindness (for review see [11, 12]) by transmitting visual information provided by a camera via an electric grid innervating visual cells of the retina, optic nerve, or occipital cortex. Some researchers have relied on animal models to identify the most effective ways to improve this approach [13, 14], while others have attempted this technique in humans already [15–18]. One important aspect that needs to be understood is the criteria to help determine who is the ideal candidate to receive these implants, and who might not be able to benefit from these implants [19], which might be influenced by mechanisms of brain plasticity [20].

For retinal, optic nerve, or cortical prostheses, the electric grid is composed of many tiny electrodes capable of directly stimulating neurons, mimicking neuronal action potentials, and aiming at conveying the reconstructed neuronal information to the next visual relay in the visual system. Several human patients, blind or with visual impairments, have been implanted with a grid stimulating the anterior visual pathways [15], optic nerve [18], retina [21], or directly on the visual cortex [16, 17]. The outcomes of these interventions are very variable, but some of these patients report to experience visual *phosphene* sensations (i.e., flashes of lights retinotopically organized), and following extensive training with their new sensory input, some patients were able to recognize also large-scale visual patterns [22].

The most widespread optical implant in humans is by far the Argus II (second sight, Sylmar, CA) [23–25] which is designed specifically to directly stimulate the inner surface of the retina (i.e., epiretinal implant) and is currently approved as a commercial product. Typically, the electric grid is implanted in the worse seeing eye of a blind patient (with some light perception in their good eye). A small video camera is mounted on a pair of eyeglasses that are worn by the subject, or can sometimes be implanted directly in the ocular orbit, and a video processor converts the image into stimulation patterns that are sent to the electrode array on the retina. After extensive training with the device, users have been shown to perform significantly better when the system is on than when it is turned off in a variety of orientation and mobility tasks [26] and were able to learn to locate bright objects on a screen, read very large print, and show improved visual acuity and visual field perimetry [23, 27–31].

In addition, reaching and grasping movements are improved after implantation of the device [32], such as the ability to localize and apprehend objects in space [33]. How do they do this? When sighted subjects attempt to reach for a target, they are often guided by their eye position. When blind people who are implanted with the Argus II attempt to reach for a target, they are guided by the information from the camera, but these participants appear to be still influenced by their eye positions and gaze shifts. Knowledge by the user of eye position seems particularly important for these reaching and grasping movements in order to achieve correct visuomotor coordination with the device [34]. Because misalignment of head and gaze directions are observed in all subjects, it is interesting to note that participants develop compensatory strategies in order to attenuate the impact of eye and head misalignment. Sabbah et al. [34] observe that when using these strategies, implanted patients are able to perform simple visually directed movements toward targets.

A recent study shows that the Argus II retinal prosthesis elicits in its users visual BOLD responses in early visual areas and the lateral geniculate nucleus, areas known to be involved in normal visual processing in sighted participants [35]. Specifically, after prolonged use of the Argus II, these visual responses were enhanced in people implanted with the device, suggesting a degree of cortical plasticity enabling participants to encode and respond to the restored sensory input.

One major limiting factor of this approach is that it can only be applied to a specific population of visually impaired individuals, namely, people with retinitis pigmentosa, i.e., a degenerative eye disease that causes severe vision impairment due to the progressive degeneration of the rods and cones, photoreceptor cells in the retina. Furthermore, despite being a very promising avenue for visual restoration, the results are still rather poor, especially compared with noninvasive approaches of sensory substitution (see next section on sensory substitution devices).

A possible explanation for the reported poor outcomes may be the emergence of adaptive experience-induced plasticity in the visual cortex of blind individuals [5]. It has been proposed that the presence of such plasticity might actually hinder the ability of implanted patients to effectively process visual information ([36, 37], but see [3, 38] for a different conclusion; see also section "Visual Rehabilitation"). This conclusion might be even more problematic when applied to the possibility of implanting congenitally blind people. In this population, early visual areas have been shown to be recruited during tactile and auditory tasks. For instance, transcranial magnetic stimulation (TMS) applied to early visual areas of highly trained congenitally blind braille readers induces somatotopically organized phosphenes in the fingers used to read braille [39], and in the tongue of congenitally blind participants trained to perceive visual information by stimulating the tongue via sensory substitution (see next section on tactile sensory substitution methods) [40]. Thus, it is possible to assume that it is because the primary visual areas in people who are congenitally blind are recruited for tactile and auditory tasks, electrical stimulation of those areas will not elicit *visual* qualia, but rather tactile or auditory ones.

The main challenge with invasive approaches to sight restoration is to effectively reconstruct the visual input, regardless of the type of implant used (retinal, optic nerve, or cortical implants). Indeed, all these devices rely on the assumption that the visuotopic representation of the visual field on the retina, optic nerve, or cortical surface is represented in the same way as the retinal image. However, the way our visual field is represented on the cortical surface is not simply a point-to-point representation. Adjustments must be made in order to allow the signal produced by the implant to reach the visual cortex in the proper arrangement.

Taken together, it appears clear that much more work is needed in order to improve the outcomes of the invasive implants for sight restoration. Given the fast advances in biotechnological methods, these implants as well as other approaches such as gene therapy [41], transplantation of photoreceptors [42], and chemical interventions [43–46] are rapidly improving and might become a successful option in coming years. Thus, there is a real need to formulate sight restoration approaches and be able to predict their success from the available evidence.

Sensory Substitution

Sensory substitution is rooted in the idea that it is possible to convey information from a given modality (e.g., vision) through another modality (such as touch or sound). Today, most sensory substitution devices attempt to restore visual functions, but not exclusively. For example, some devices are intended for the sexual rehabilitation of men with spinal cord injuries. The information from below the waist that can no longer be perceived is relocated to a different part of the skin via a tactile grid to an area of skin that is still sensitive [47]. Other devices are geared toward conveying balance information for people with vestibular loss [48, 49]. Sensory substitution devices can also serve as *sensory augmentation* devices by providing additional information to users [50]. There are many other uses for sensory substitution or augmentation devices, but in this review we will concentrate on devices that attempt to convey visual information for people who are blind via other sensory modalities (typically touch or sound).

At first, the concept of sensory substitution seems very intuitive. Everybody relies on different sensory modalities when visual information is unavailable: we rely on touch when fumbling for the light switch in the dark or when we use our fingers to feel the walls to find the keyhole when attempting to open the door in the dark. Similarly, people who are blind become experts at using tactile and auditory cues to understand object shapes and spatial layouts. A good example of a sensory substitution system widely used by the blind community is the Braille alphabet which codes written text through raised dots to allow reading. Visually impaired people can learn the Braille alphabet via focused training and be able to read texts, an ability that otherwise will be impossible for them. Another strategy often used by blind people to replace the missing visual information includes the use of naturally produced sounds to create echolocation in order to recognize and locate objects. These examples are personal strategies used in order to successfully solve *visual* tasks. Sensory substitution devices extend these naturally occurring strategies and abilities developed by people who are blind by use of predetermined algorithms that people can learn in order to interpret visual information. With the advent of microtechnologies, more complex and sophisticated sensory substitution devices have been conceived, the properties of which will be discussed in the next sections. These sensory substitution devices allow visually impaired individuals to perform many complex visual tasks such as shape recognition and navigation.

Several authors have referred to the experience of sensory substitution devices as an "acquired synesthesia" [51]. Synesthesia is the phenomenological merging of one or more attributes from one sensory modality to another; for example, many synesthetes report sensing specific colors linked to specific musical tones [52–56], or colors being associated with the imagined perception of a geometric form [57]. The experience of sensory substitution devices has been linked to synesthesia because of the phenomenological attribution of visual properties to tactile or auditory stimulation [58, 59]. There are certain differences, however, between synesthesia and sensory substitution perception. Synesthesia is an innate automatic process where the individual cannot help but perceive a merging of the senses, whereas the sensory substitution perceptual process is an active process at first. The user must first learn to interpret the tactile or auditory sensory signal from the sensory substitution devices, and only after training this process becomes automatic. Sensory substitution users who are blind often use terms such as "visualize" or report *seeing* objects or movement during sensory substitution experiments. The substituted *visual* information allows for the active sensing and manipulation of stimuli, for example, during walking and avoiding obstacles [60, 61], finger maze learning [62], recognizing routes [63], navigating a maze [64], identifying the direction of motion [65], locating and recognizing an object [66], recognizing body postures [67], and reading words [68] and numbers [69]. Users who are blind often respond to sensory substitution very enthusiastically and emotionally [70–72]. There are even reports of users which became so emotionally attached to their sensory substitution device-related information that removing access to it resulted in a feeling of loss [9]. In highly trained users, the resulting embodiment of sensory substitution stimulation is perceived as an extension of the senses, not as an outside apparatus [73]. *Visual* sensory substitution devices differ from one another in terms of their respective approaches for capturing, transforming, and sending information [9]. In the next sections, we discuss in details two types of sensory substitution devices: image-based that transforms whole-image visual scenes into tactile or auditory information and minimalistic sensory substitution devices which convey only specific aspects of visual information.

Visual-To-Tactile Image-Based Sensory Substitution Devices

In the late 1960s, Professor Paul Bach-Y-Rita started developing sensory substitution devices capable of substituting a missing or lesioned sense by another sense. In the preface to his now famous book, *Brain Mechanisms in Sensory Substitution* (1972), Paul Bach-y-Rita, who is typically considered the father of sensory substitution as a formulated field, aimed his work at restoring visual functions in blind people [71]. Professor Bach-Y-Rita asks whether the eyes are essential to vision and if the ears are essential to audition. This question might seem absurd, but in reality it is exactly the problem posed by sensory substitution. Can braille reading be qualified as vision? Or is it rather a tactile experience that replaces vision? Bach-Y-Rita explains that the images captured by our pupils never leave the retina. From the retina to the brain, this information travels in the form of electric and chemical impulses, and it is the brain that interprets this information as vision. The perception of an image requires much more from the brain than a simple image analysis. Visual perception is based on memory, learning, and interpretation of contextual factors [74]. It is this phenomenological observation that doubtlessly prompted the idea of a sensory substitution system that replaces the visual input of the eye by tactile stimulation, and phenomenologically, the *visual* presence of sensory substitution information is close to vision (for review see [9]). In 1970, Professor Paul Bach-Y-Rita develops an apparatus capable of transmitting images on the back of users. A camera captures an image that is transmitted to an electrode grid positioned on the back of users to the cutaneous receptors of the skin. Case studies conducted in Bach-Y-Rita's laboratory demonstrate that after extensive training, it is possible to use this device to make judgments of distance, grab objects in motion, and even recognize novel objects. Later, this device was adapted to stimulate tongue somesthesia. The reason for the choice of the tongue rather than the skin of the back is twofold: firstly the tongue is embedded in a wet milieu, making possible to use much safer micro-currents for stimulation, and secondly the tongue is a much more sensitive organ with many more receptive fields (and with higher resolution) than those on the back. Today, the camera is the size of a webcam, and the computer that transforms the image is portable.

Studies show that visual-to-tactile sensory substitution device users can even succeed in perceiving certain functional subjective aspects of vision [75], such as optical illusions ([76]; but see also [77]). The presence of optical illusions in the congenitally blind using sensory substitution devices hints at the fact that, at least from a phenomenological point of view, information translated through sensory substitution devices shares some properties with vision. This phenomenological shift from tactile sensations to *visual*-like percepts has been observed and reported in many different studies using sensory substitution device (for review see [9, 78]).

Visual-To-Auditory Image-Based Sensory Substitution Devices

Echolocation is a great example of auditory information filling in for visual information. A blind person who uses echolocation will send out a sound, tongue clicks or even footsteps, and will listen for the echoes (reverberations) that come back. Reverberations that come back quickly indicate objects that are very close, while later reverberations indicate objects at a distance.

> **Box 13.1**
> Early researchers trying to understand how blind people are able to navigate hypothesized that their other senses must be heightened, allowing them to substitute visual information. It was believed that blind people possess a heightened tactile facial sense enabling them to detect obstacles and discriminate between textures [79]. At the time, researchers hypothesized that congenitally blind people possessed the ability to sense atmospheric pressure changes on their faces, which enabled them to perceive obstacles [79] and even textures [80]. However, it was later discovered that these abilities were not tied to somatosensory abilities at all, but rather their ability to use echolocation [81], which is the ability to sense the environment by hearing echoes.

A few clicks in the room, and a blind person will be able to detect a chair and walk over it and sit down in front of the table. This ability seems to develop early on in the life of certain congenitally blind individuals that are capable of echolocation. Children can use echolocation to detect and avoid obstacles, such as small boxes [82]. Blind people who are apt echolocators report that the quality of the sound is different for different textures, the same way something that is hollow has a different sound from something that is full. This ability requires much training and allows for the collection of spatial information far beyond the reach of even the long cane. This is the same ability used by bats, who also need to operate in the dark. In most cases, blind and visually impaired individuals use echolocation as complementary to tactile and other vestibular/proprioceptive information. This ability to echolocate can be conceived as a form of visual-to-auditory sensory substitution system that requires no external device.

With the development of speakers, researchers were able to produce uniform sounds in order to study human echolocation [83]. Auditory sensory substitution is founded on these studies, which gave birth much later to the idea of sonar systems for the blind capable of sending visual information through sound [84, 85]. These systems use information captured by a camera and transmitted to a computer that transform these images into soundscapes [76]. The image is coded in terms of the amplitude and frequency of the sound. The subject can thus learn to decode an auditory image, to eventually grasp objects and move in space. One such system is the vOICe developed by [86], or the EyeMusic [69], which adds the element of color to the image.

Minimalistic Sensory Substitution Devices

Minimalistic sensory substitution devices attempt to translate one specific aspect of visual information and code it through a different modality. Minimalistic sensory substitution devices can use either touch or sound or the combination of both modalities to convey visuospatial information. For example, translating only distance information, or the indication of the magnetic north, is minimalistic compared to the complex information delivered through image-based sensory substitution devices discussed in the sections above. Minimalistic sensory substitution devices can substitute certain functional aspects of vision in many different tasks and can be very useful for aiding navigation. These devices have the advantage of being extremely intuitive to use since they only code one single aspect of vision; thus they only need very little training.

Recently, there has been an increase in the advent of technological aids (for review see [87]) and sensory substitution devices (for review see [88]) to help blind or visually impaired people to navigate. Research in laboratory settings has demonstrated the potential level of spatial abilities that blind people can display when using sensory substitution devices ([9, 61, 63, 64, 89–93], for review see [9, 78]). With the tongue display unit (TDU), a visuo-tactile image-based sensory substitution device, congenitally blind subjects outperform sighted blindfolded participants in a high-contrast, life-sized obstacle course [61]. Results show that all participants showed greater difficulty in stepping over an obstacle, than going around it [61]. Using a depth-to-audio minimalistic sensory substitution device (the EyeCane), Buchs et al. [60] tested obstacle navigation and reported the same difficulty in stepping over obstacles in their blind subjects. Blindfolded participants using the EyeCane performed better than another blindfolded group using a white cane, demonstrating that sensory substitution devices can be more efficient and can convey a wider variety and more complex information than a white cane alone. Similar results were also obtained using virtual mazes, in which participants outperformed both the no-device and white-cane groups [88]. These studies show it is possible for blind people to learn how to use sensory substitution devices for navigation. In both cases of the TDU and EyeCane, results were similar, but training times were shorter in the case of the EyeCane. For the EyeCane, participants could intuitively use the device with less than 5-min training. The experiment with the TDU included 15–30-min familiarization period before testing.

The EyeCane has been also used in a real life-sized Hebb-Williams maze, and results showed that congenitally blind, late blind, and sighted blindfolded participants could learn to navigate with a statistically similar performance level than sighted visual control (SVC) group [64].

Minimalistic sensory substitution devices can also transfer landmark or positioning information, for example, the FeelSpace [92] or NavBelt [94, 95] indicates the magnetic north from a compass via vibrotactile stimulations on a belt that are updated with the wearer's movements [50]. Researchers found that after several weeks of training, late blind participants improved on a pointing task, maintaining a sense of direction over long distances and finding shortcuts in familiar environments [50]. After participants have learned the novel sensory stimulation code for the FeelSpace device, the acquired new perception is "embodied," which not only helps them select more efficient routes (i.e., shortcuts) during navigation but also creates the feeling of a "new" sense [94].

Minimalistic sensory substitution devices have the objective and advantage of being very easy to learn and require less training to interpret, but other aspects of the visual information must be inferred indirectly. For example, the EyeCane does not directly convey any shape information, but by scanning the environment, it is possible to infer it. The user can use perceived distance information regarding objects to infer shape information or the layout of a room. This demonstrates that it is possible to learn the spatial layout of an environment from even minimalistic sensory substitution devices coding only distance through different sensory modalities, regardless of visual experience.

Functional Brain Plasticity in the Case of Blindness and Sensory Substitution

In the last decades, neuroimaging techniques provided crucial information for unraveling the properties of brain organization in the case of congenital and late sensory deprivation. Imaging techniques applied to the field of sensory deprivation combined with the use of psychophysical tests using sensory substitution devices have led researchers to identify two forms of reorganization in the blind brain which mutually influence each other, namely, amodal task selectivity [2] and experience-induced brain plasticity [6]. In the next few sections, we discuss evidence for task selectivity and also for brain plasticity in the congenitally blind brain. We will further discuss the implications of functional studies as well as of volumetric [39, 96, 97], metabolic [98, 99], and connectivity studies [39, 97, 100] in congenitally blind compared to sighted adults.

Task Selectivity in the Perceptual "Visual" Streams

One of the first indications of task-selective brain organization was demonstrated using fMRI and TMS in sighted participants showing the role of visual imagery in normal tactile detection of orientation selectivity (Zangaladze, et al. 1999). It was found that in normally sighted individuals, application of TMS to the occipital cortex disrupts tactile discrimination of gratings, showing that normal occipital processing is crucial for tactile discrimination. In a later experiment using PET and TMS, visual cortical involvement was found for discrimination of tactile orientation, but not frequency [101]. The region that was found to be most activated during tactile grating orientation was the extrastriate occipitoparietal junction. These important studies on mental imagery during tactile exploration were among the first to suggest the implications of task-selective brain organization.

Neuroimaging studies with expert sensory substitution device users (mainly congenitally blind users) showed that almost all of the known specialized "visual" regions maintained their category selectivity even if the input was conveyed through an atypical sensory modality (i.e., audition; touch). First of all, using a *visual-to-auditory* sensory substitution device, Striem-Amit [66] and colleagues showed that the main broad division of labor in the visual cortex between dorsal (where) and ventral (what) visual pathways [102, 103] could develop in the total absence of vision after few hours of sensory substitution training. Furthermore, specializations were maintained even within specialized regions in both the ventral and dorsal *visual* pathways [104].

Task Selectivity in the Ventral Stream

The ventral visual stream is subdivided into regions specialized for specific categories, such as recognizing simple shapes, faces, body postures, or the shape of words when reading. Several studies showed that after sensory substitution device-related training, congenitally blind expert users consistently activate those specialized regions in a similar way than their fully sighted counterparts for a variety of tasks involving shape recognition of different kinds [105]. For example, congenitally blind sensory substitution device users activate the lateral occipital cortex (LOC/LOtv) to recognize the shape of an object (audition: [106]), namely, the region in the visual cortex that is activated in the sighted brain when performing visual object recognition tasks. In addition, again similarly to the sighted brain when processing the same stimuli categories through vision, congenitally blind sensory substitution device users activate the visual word form area (VWFA) for recognizing letters and to read words (audition: Striem-Amit [68, 107]; touch: [108], the extrastriate body area (EBA) to recognize body shapes (audition: [67]), and the visual number form area to recognize numbers [69].

Task Selectivity Dorsal Stream

The dorsal visual stream can also be subdivided into cortical nodes defined by their specific functional specializations, such as a specialized area for motion processing, for the location of objects, and for spatial relations. Studies using sensory substitution devices have repeatedly demonstrated the preservation of these subdivisions in the dorsal visual stream of the congenitally blind brain [65, 66, 106, 109], for review: [2, 9, 110]. For example, it has been showed that congenitally blind sensory substitution device users activate the right dorsal extrastriate visual cortex when using sounds to localize the position of objects in space, namely, the same region that is activated in the sighted when performing visual localization tasks [111]. Using the Tongue Display Unit, another study showed that area hMT+ was activated in congenitally blind adults when trying to determine the direction of motion through touch [109], or for the direction of motion through sound [112], even without using sensory substitution devices. Dorsal parietal areas such as the dorsal parietal cortex and the precuneus were also observed to be activated in congenitally blind participants recognizing routes using a visual-to-tactile sensory substitution device, a pattern of activation that was similar to the one observed in sighted participants doing the task visually [63].

Taken together, these results on the blind brain processing of sensory substitution information suggest that not only is the occipital cortex of the blind recruited when using sensory substitution devices by an active reinterpretation of the signal but also that the brain treats the afferent information in an amodal, task-dependent way (for review see [3, 5]).

Two Principles Mediating Task Selectivity

The aforementioned evidence of the preservation of the division of labor between the dorsal and ventral streams in people who are congenitally blind had led certain researchers to conceive the brain as an amodal and flexible task-specific machine ([113]; for review see [2, 3]). In other words, it is proposed that sensory brain regions are not tailored to respond to a specific input sensory modality but rather to specific tasks and computations. Confirming evidence of this proposal also comes from studies with sighted sensory substitution device users. In a recent study, for instance, it has been shown that similar to the congenitally blind adults, Braille reading activates the visual word form area in the sighted brain and, furthermore, that TMS pulses applied to this region disrupt such ability ([114]; but see also Hertz and Amedi [115] for some differences in brain organization elicited by sensory substitution device training between sighted and blind sensory substitution device users).

If sensory input does not drive sensory brain organization, what does drive this organization, i.e., the emergence of task selectivity in the human brain? We propose that a combination of two nonmutually exclusive principles may drive the emergence of task-selective brain organization [2, 3, 116]: the first is known as the "biased connectivity principle," which posits that task-specific recruitment draws on preexisting cortical connections linking occipital cortex task-selective regions to the rest of the network processing information for a specific computational task. Evidence in favor of this principle comes again from sensory substitution studies with congenitally blind adults. Using resting-state functional connectivity magnetic imaging, which exploits the assumption that correlations in the activity of different brain regions during resting state (i.e., without an explicit task) reflect functionally relevant correlations in neuronal firing [117, 118], it has been shown that congenitally blind participants have preserved functional-connectivity patterns between specific category-selective "visual" regions and other brain regions relevant for that computation ([67, 69]; see for review Hannagan et al. [2, 3, 119, 120]). For instance, as described above, it has been shown that in congenitally blind participants, the visual number form area (NFA) is recruited in a task-selective manner after a relatively short sensory substitution training period on a number identification task [69]. In the same congenitally blind participants, this recruitment was accompanied by preserved cortical connections between NFA and other essential areas involved in the representation of quantities in the sighted population [69, 121]), and the visual word form area showed preserved connections to fundamental areas for language processing ([122, 123]; see also [68]). Similar results have been recently reported for the extrastriate body area [67]. Activity in this area was elicited in congenitally blind adults by the perception of sensory substitution-presented body shapes, and such activity was accompanied by preserved functional connectivity between this region and other areas considered to be an integral part of the body-image network in the sighted population, such as the posterior superior temporal sulcus and the temporal-parietal junction [67]. Another recent study showed that category preference and functional connectivity yield comparable results in vast areas of the visual cortex of the blind [124].

The second principle is known as the "task-distinctive feature sensitivity principle" and states that task-specific recruitment can emerge from the intrinsic circuitry of the visual cortex which may be tuned to the extraction of the specific but invariant task features of an object [2, 3]. In other words, this extraction is expected to occur independently of translation, rotation, size, distance, or other variations in the stimulus properties and, moreover, independently of the sensory modality through which the information is conveyed [3]. This proposal still leaves several critical questions open, the most crucial ones being related to the generalization of these two principles to the organization of the early sensory cortices and, more practically, the implications of these principles for sensory restoration—could including them in rehabilitative programs help maximize sensory recovery?

Brain Plasticity in the Primary Visual Cortex

The counterpart of task-selective amodality in the brain is the concept of experience-induced brain plasticity in the case of sensory deprivation (Cechetti et al. 2016). This theory proposes that in cases of sensory deprivation, like congenital blindness, primary sensory areas, such as primary visual cortex (V1), are recruited to process information from other modalities in a non-task-selective manner. Indeed, to date there is no conclusive evidence regarding which task-selective computational tasks these cortices should maintain if deprived of their natural input from birth. A few studies on early blind populations have reported recruitment of the deprived V1 by low-level spatially related features (e.g., [39, 125, 126]). For example, electrotactile stimulation of the tongue via a sensory substitution device activates certain regions of the early visual cortex in the congenitally blind [125]. It was demonstrated in this study that the occipital lobe takes charge of the discriminatory function of the image projected unto the tongue. These results are further corroborated when highlighting that TMS of the occipital lobe in sighted participants results in the appearance of visual phosphenes [127]. By contrast, TMS of the occipital lobe in congenitally blind participants trained using the visual-to-tactile TDU creates the sensation of somatotopically organized sensations on the tongue [40]. This same stimulation in sighted subjects does not create tactile sensations on the tongue of sighted participants.

However, such reports of low-level spatially organized recruitment are rare and weaker compared to the accumulating evidence of "task-switching" in V1 toward higher cognitive functions in the case of language and memory tasks [128–131] and in tasks requiring focused attention [120] or executive control [132]. These results are thought to diverge dramatically from the predictions of task-selective brain organization, because such functions do not typically recruit early visual areas in sighted individuals (but see [133]).

For example, Bedny et al. [129] demonstrated that congenitally blind adults, but not late blind children (2012), recruit early occipital areas during language tasks. Amedi and collaborators demonstrated that deprived early visual cortices are activated during episodic retrieval [134], that the strength of activations correlate with superior memory performance [128], and that transcranial magnetic stimulation of this area can also interfere with verbal processing of learned words [135]. Taken together, these studies suggest that the "task-distinctive features sensitivity" principle is not respected in deprived V1. However, a unifying theory attempting to explain comprehensively primary visual cortex recruitment in congenitally blind participants will have to better characterize the properties of perceptual versus higher cognitive functions activations.

At the same time, studies should also investigate to what extent the biased connectivity principle is present within deprived primary sensory cortices. Recent evidence in congenitally blind adults supports this conclusion by showing retained functional connectivity patterns mimicking retinotopic organization, a hallmark of the visual cortex structural architecture [108]. These retained patterns were observed for all three main retinotopic mapping axes: eccentricity (center-periphery), laterality (left-right), and elevation (upper-lower), throughout the early and high-level ventral and dorsal streams. This functional connectivity architecture was also observed in people whose eyes did not fully develop in utero (i.e., without any possible visual experience). Thus this architecture appears to be hardwired and dependent on genetic blueprints, rather than on experience-dependent or even activity-dependent mechanisms [3]. Further supportive findings were reported by other groups for the retained fine-detailed functional connectivity within V1 [136] and for retained visual callosal anatomical connectivity [137]. However, consistent with previous studies [138–142], this retained organization coexisted with some level of divergent organization in the blind [143]. These latter connections varied in accordance with retinotopic division. The blind central V1 showed increased functional connectivity to the left frontal language areas, and their peripheral V1 showed increased functional connectivity to the parieto-frontal attention networks. This might indicate distinctive V1 localizations for the two functional roles generally attributed to the blind V1, namely, higher-order cognitive functions such as language processing [129] and nonvisual spatial (and also nonspatial) attention [144]. Future studies should further clarify the functional meaning of the findings indicating both the retention and divergence of early visual cortices in functional connectivity organization and systematically test its effects on sight restoration outcomes.

Structural Connectivity Studies in Blindness

We showed in the section above that the deprived primary visual cortex can be recruited by tactile and auditory information [63, 125] and even by higher cognitive functions such as complex memory tasks [134]. In this section we discuss differences at the level of the white matter in blind individuals. Early studies on the visual system of the blind used magnetic resonance imaging (MRI) to explore the extent to which the primary afferent visual pathways were altered by blindness. Breitenseher et al. [145] visually detected atrophies, or thinning, at the level of the optic chiasm, optic nerve and tract, and optic radiation [145]. A corresponding reduction in volume of the afferent visual pathway was also found using voxel-based morphometry (VBM) in the congenital and early blind [68, 96, 145, 146]. The reduction in volume of the optic radiation was found to correlate with age at blindness onset as well as blindness duration [146]. Ptito et al. [39] demonstrated changes in white matter volume of regions beyond the optical pathway. These included reduction in the volume of the splenium of the corpus callosum and inferior longitudinal fasciculus, and enlargement of the occipitofrontal fasciculus, the superior longitudinal fasciculus, and the genu of the corpus callosum. Noppeney et al. [96], in a small group of mixed congenital and early blind, reported on increase white matter in the sensorimotor system. The findings of structural modifications in the optic tract and radiation were further corroborated by other MRI modalities such as diffusion tensor imaging (DTI) and tractography [97, 147] that evaluate white matter integrity at the microstructural level. These studies revealed changes in diffusion characteristics and connectivity in the geniculocalcarine tract and ventral splenium of the corpus callosum [97, 147], suggesting neuronal degeneration in these affected tracts [147].

Metabolic Differences in the Congenitally Blind Brain

Positron emission tomography (PET) is one of the very first approaches that was used to investigate differences in brain organization between the blind and the sighted people. Through use of this method, it is possible to measure the regional changes in metabolic glucose activity of cells. These seminal studies demonstrated that people who are congenitally blind possess a supranormal metabolism in visual areas at rest compared to sighted participants [98, 99, 148]. The increase in glucose level is observed only in people who are congenitally blind, whereas people who became blind later in life had a reduction in activity in visual areas during rest compared to sighted participants [149].

This augmented metabolism in visual areas suggests that, despite volumetric reductions of the visual cortex (see following section on *Volumetric Plasticity in the Congenitally Blind Brain*), this cortex is very active in people who are congenitally blind. It is hypothesized that this higher activity is what enables the recruitment of these areas for other modalities ([39, 150]; for review see [9]). In addition, it has been proposed that such increased metabolic activity might reflect the neurochemical changes in the pericalcarine cortex in congenitally blind individuals, such as higher levels of choline and glutamate that have been suggested as possible triggers for plasticity [151].

Volumetric Gray Matter Plasticity in the Congenitally Blind Brain

This section discusses volumetric and morphometric differences that were found at the level of the gray matter in the blind brain, and then those studies that attempt to link structure with function, i.e., that discuss these differences in terms of behavioral abilities of the blind. The brain of congenitally blind people undergoes massive structural and volumetric changes in all of the visual structures [39, 150, 152, 153]. These volumetric changes in the blind depend upon the duration of blindness [153], and hypertrophy of nonvisual frontal and cerebellar areas suggests compensatory adaptations in blindness [152].

Breitenheser et al. [145] and Shimony [97] did not observe structural changes in visual cortex in blind participants. Other studies [39, 96, 146], however, found a reduction of the primary visual areas (V1 or B.A. 17 and 18) by using voxel-based morphometry. Ptito et al. [39] found bilateral reduction of 25% in V1 and 20% in V2. The dorsal lateral geniculate nucleus (LGN), the pulvinar, and the hMT+ were also dramatically reduced [39]. These reductions are comparable to those found in the rodent by Desgent and colleagues (2009).

Perspectives on Brain Organization, Amodality, and Implications for Visual Rehabilitation

The results discussed in previous sections suggest that not only is the occipital cortex of the blind recruited when using sensory substitution devices by an active reinterpretation of the signal (top-down processing) but also that the brain treats the afferent information in an amodal, task-selective way (for review see [2, 3, 5]). These findings force a reevaluation of current models of brain organization according to sensory modality in terms of a novel task-selective brain organization hypothesis. Findings obtained in deprived early visual cortices show that some key aspects of task selectivity are present not only in associative sensory areas but are also present in primary sensory areas. Using a visual-to-auditory sensory substitution device, Striem-Amit and colleagues demonstrated that large-scale retinotopic structural organization is preserved in V1 [143]. Other aspects of task selectivity seem lost however in these primary visual areas in the blind. For example, these areas seem to process higher-order cognitive functions in people who are congenitally blind, not tied to any particular modality (e.g., [129, 135] but see [68, 125]). This suggests that there may be a complex interaction between task selectivity and task-switching plasticity in the brain of people who are blind. These new ideas have implications for strategies used by researchers attempting to restore certain aspects of visual information for people who are blind. In the next two sections, we discuss the implications of these findings for the development of theories on brain organization and for the development of training programs for visual rehabilitation for people who are blind using sensory substitution devices.

Implications for Theories on Brain Reorganization and Amodality

The results cited in this review have many implications for theories on brain organization and reorganization. We put forward the theory that the brain is an amodal task-selective machine, capable of rewiring itself to adapt and maximize its ability to process sensory information. We have shown behavioral, anatomical, and functional evidence for this theory.

Many different studies using sensory substitution devices suggest that congenitally blind participants tend to have better performance than late blind and sighted blindfolded participants in terms of spatial tasks. The recruitment of these highly specialized cortical nodes by task-specific, rather than modality-specific, stimuli enables the blind to the conscious perception of sensory qualia in the same way that sighted people do [4]. We suggest that the human brain has evolved to develop amodal properties ([128]; for review see [2, 3]) that possess an amodal functional organization of the dorsal and ventral streams [154]. We demonstrate that the recruitment of the dorsal [109] and ventral streams [109, 110] selectively treats movement and shape via sensory substitution [155]. These findings of improved spatial performance and preserved selectivity in the dorsal and ventral streams indicate that instead of developing to form modality-specific regions in the brain, the brain has evolved to form highly adaptable task-specific regions that can be modulated according to the sensory input that is available to it. In other words, although vision is usually the most suited sense to accomplish many different tasks in the dorsal and ventral streams, the brain is capable of adapting itself to use other sensory modalities to accomplish these tasks and the regions of the brain that are associated with these tasks are not really interested in what modality the information comes from, but rather what task it is trying to accomplish. Thus, the brain is not compartmentalized according to sensory modality it would seem, rather is a complex task-machine capable of adapting to environmental pressures for survival.

Implications for Visual Rehabilitation

Taken together, the series of results discussed in this review would imply that it is possible to train people who are blind to use existing senses to replace missing visual information, and such information will recruit the corresponding sensory-independent task-specific region in the deprived visual cortex. This would imply that sensory substitution devices can potentially help people who are blind recover certain high-level visual functions (category-specific selectivity of tasks in the ventral and dorsal streams). Furthermore, we have recently proposed that sensory substitution device training on specific stimuli categories might be able to guide the newly restored visual signal to recruit its dedicated brain regions, ultimately maximizing the outcomes of visual recovery [3]. Future work must try to implement a generally agreed-upon training program for sensory substitution use and evaluation (for review on the development of a generally agreed-upon training protocol for sensory substitution: [10, 156]). This implies that training with SSDs can help open up the world for people who are blind in a very meaningful way [88, 157, 158]. For example, it is possible to transfer spatial knowledge using sensory substitution between a virtual reality paradigm and real-world environments for blind individuals [159] or train participants to perform better at navigational tasks [64].

In terms of visual implants, however, we have also reported in this review that early visual cortices seem to be mainly recruited in the blind group by higher-level functions such as language and memory. This recruitment might instead prevent efficient use of implants in terms of visual recovery of functions. Encouraging evidence are coming from animal studies indicating that chemical interventions can release molecular "breaks" of plasticity (involving the balance between inhibition and excitation) and reset juvenile brain plasticity ultimately increasing sensitivity to external inputs (e.g., [44, 45]). Treatment based on this approach is being piloted for amblyopia [43, 46]. If this approach will prove useful, it may be expanded to people recovering from blindness. In this case, the maintenance of the macro-structural organization of V1 (e.g., [108]), along with rejuvenating its ability to wire and refine its connections once visual input is restored, may facilitate a vision efficient takeover of the re-afferented visual cortex. The combination of this approach with the aforementioned training of "visual" categories through sensory substitution devices might become the most promising approach for successful visual restoration. Sensory substitution devices could also be used to help train participants to use their retinal, optic nerve, or cortical implants [5], in a way awakening the congenitally blind brain to new experiences using new senses, for example, using sensory substitution. Indeed, sensory substitution constitutes a new sense because it is not quite auditory or tactile information; it is the convergence of visual information on tactile or auditory cues which enable the perception of newly acquired qualia.

Conclusions

In this chapter we have reported and discussed the most recent advances in sensory substitution techniques and the latest neuroimaging studies in the blind, as well as perspectives on the amodality of the brain. This new evidence of plasticity and task selectivity highlighted the fact that the brain has evolved to be an efficient, ever-adapting, amodal task machine. From an evolutionary point of view, in response to environmental demands, the brain has to adapt to specific sensory tasks, such as recognizing faces, motion, and distance and also develop and be able to use tools of a growing complexity. The use of these tools required the development of the visual cortex for the focalization of objects in the central visual field and for the ability to judge distance with binocular vision. As a general principle of brain organization, it was a predicate for survival to be able to use vision for hunting and gathering, to return to old hunting grounds, to find an ancient encampment, and to accomplish seasonal migrations. All of these different skills depend greatly on vision, and their necessity encouraged the development of specific cortical areas that are visually responsive to many different classes of stimuli.

Behavioral, anatomic, and *f*MRI results that we highlight in this review are supported by an ever-expanding literature on the visual system in the blind. From a behavioral point of view, we show that the perceptual advantages of the blind extend to recognition of routes, navigation, obstacle detection and avoidance, and real and virtual maze solving via sensory substitution. This adds fuel to the amodal task-specific brain organization theory for cognitive, tactile, and auditory tasks. From an anatomical point of view, we demonstrate that this perceptual advantage relies on morphological changes in the blind brain. We are far from having elucidated all the mysteries of the sensory-deprived brain, and we must continue to explore these mechanisms and manifestations of plasticity, reorganization, and task-specific organization in the blind brain.

Sensory substitution offers many promises to study these mechanisms but also to correct certain perceptual deficits in people who are blind, and the possibility to accomplish a panoply of tasks and also help mobility in everyday life. The fact that blind brain can adapt and retain some of the functions of genetically determined nodes in the brain leads to very promising avenues of research. It is important to continue studying this particular brain machine interface. Future studies should further investigate and characterize the interactions and implications of these two forms of brain reorganization in primary and secondary *visual* areas.

References

1. Heimler B, Baruffaldi F, Bonmassar C, Venturini M, Pavani F (2017) Multisensory interference in early deaf adults. J Deaf Stud Deaf Educ 22(4):422–433
2. Amedi A, Hofstetter S, Maidenbaum S, Heimler B (2017) Task selectivity as a comprehensive principle for brain organization. Trends Cogn Sci 21(5):307–310
3. Heimler B, Striem-Amit E, Amedi A (2015) Origins of task-specific sensory-independent organization in the visual and auditory brain: neuroscience evidence, open questions and clinical implications. Curr Opin Neurobiol 35:169–177
4. Kupers R, Ptito M (2011) Insights from darkness: what the study of blindness has taught us about brain structure and function. In: Progress in brain research, vol 192. Elsevier, Amsterdam, pp 17–31
5. Reich L, Maidenbaum S, Amedi A (2012) The brain as a flexible task machine: implications for visual rehabilitation using noninvasive vs. invasive approaches. Curr Opin Neurol 25 (1):86–95
6. Cecchetti L, Kupers R, Ptito M, Pietrini P, Ricciardi E (2016) Are supramodality and cross-modal plasticity the yin and yang of brain development? From blindness to rehabilitation. Front Syst Neurosci 10:89
7. Ptito M, Chebat DR, Kupers R (2008a) The blind get a taste of vision. In: Human haptic perception: basics and applications. Birkhäuser, Basel, pp 481–489
8. Bach-y-Rita P, Kercel SW (2003) Sensory substitution and the human–machine interface. Trends Cogn Sci 7(12):541–546
9. Chebat DR, Harrar V, Kupers R, Maidenbaum S, Amedi A, Ptito M (2018) Sensory substitution and the neural correlates of navigation in blindness. In: Mobility of Visually Impaired People. Springer, Cham, pp 167–200
10. Proulx MJ, Ptito M, Amedi A (2014) Multisensory integration, sensory substitution and visual rehabilitation. Neurosci Biobehav Rev 41:1
11. Fine I, Boynton GM (2015) Pulse trains to percepts: the challenge of creating a perceptually intelligible world with sight recovery technologies. Philos Trans R Soc Lond B Biol Sci 370 (1677):20140208
12. Fine I, Cepko CL, Landy MS (2015) Vision research special issue: sight restoration: prosthetics, optogenetics and gene therapy. Vis Res 111(Pt B):115
13. Arabi K, Sawan MA (1999) Electronic design of a multichannel programmable implant for neuromuscular electrical stimulation. J IEEE Trans Rehabil Eng 7:204–214
14. Sawan M, Hu Y, Coulombe J (2005) Wireless smart implants dedicated to multichannel monitoring and microstimulation. IEEE Circuits Syst Mag 5:21–39
15. Delbeke J, Pins D, Michaux G, Wanet-Defalque MC, Parrini S, Veraart C (2001) Electrical stimulation of anterior visual pathways in retinitis pigmentosa. Invest Ophthalmol Vis Sci 42 (1):291–297
16. Dobelle WH (2000) Artificial vision for the blind by connecting a television camera to the visual cortex. ASAIO J 46(1):3–9
17. Dobelle WH, Quest DO, Antunes JL, Roberts TS, Girvin JP (1979) Artificial vision for the blind by electrical stimulation of the visual cortex. Neurosurgery 5(4):521–527

18. Veraart C, Raftopoulos C, Mortimer JT, Delbeke J, Pins D, Michaux G, Vanlierde A, Parrini S, Wanet-Defalque MC (1998) Visual sensations produced by optic nerve stimulation using an implanted self-sizing spiral cuff electrode. Brain Res 813(1):181–186

19. Merabet LB, Rizzo JF III, Pascual-Leone A, Fernandez E (2007) 'Who is the ideal candidate?': Decisions and issues relating to visual neuroprosthesis development, patient testing and neuroplasticity. J Neural Eng 4(1):S130

20. Neville H, Bavelier D (2002) Human brain plasticity: evidence from sensory deprivation and altered language experience. Prog Brain Res 138:177–188

21. Gekeler F, Messias A, Ottinger M, Bartz-Schmidt KU, Zrenner E (2006) Phosphenes electrically evoked with DTL electrodes: a study in patients with retinitis pigmentosa, glaucoma, and homonymous visual field loss and normal subjects. Invest Ophthalmol Vis Sci 47 (11):4966–4974

22. Veraart C, Wanet-Defalque MC, Gerard B, Vanlierde A, Delbeke J (2003) Pattern recognition with the optic nerve visual prosthesis. Artif Organs 27(11):996–1004

23. Humayun MS, Dorn JD, Da Cruz L, Dagnelie G, Sahel JA, Stanga PE, Ho AC (2012) Interim results from the international trial of second Sight's visual prosthesis. Ophthalmology 119 (4):779–788

24. Luo YHL, da Cruz L (2016) The Argus® II retinal prosthesis system. Prog Retin Eye Res 50:89–107

25. Matet A, Amar N, Mohand-Said S, Sahel JA, Barale PO (2016) Argus II retinal prosthesis implantation with scleral flap and autogenous temporalis fascia as alternative patch graft material: a 4-year follow-up. Clin Ophthalmol (Auckland, NZ) 10:1565

26. Dagnelie G, Christopher P, Arditi A, Cruz L, Duncan JL, Ho AC et al (2017) Performance of real-world functional vision tasks by blind subjects improves after implantation with the Argus® II retinal prosthesis system. Clin Experiment Ophthalmol 45(2):152–159

27. Barry MP, Dagnelie G, Argus IISG (2012) Use of the Argus II retinal prosthesis to improve visual guidance of fine hand movements. Invest Ophthalmol Vis Sci 53(9):5095±101. https://doi.org/10.1167/iovs.12-9536. PMID: 22661464; PubMed Central PMCID: PMCPMC3416020

28. Chader GJ, Weiland J, Humayun MS (2009) Artificial vision: needs, functioning, and testing of a retinal electronic prosthesis. Prog Brain Res 175:317±32. https://doi.org/10.1016/S0079-6123(09)17522-2. PMID: 19660665

29. da Cruz L, Coley BF, Dorn J, Merlini F, Filley E, Christopher P et al (2013) The Argus II epiretinal prosthesis system allows letter and word reading and long-term function in patients with profound vision loss. Br J Ophthalmol 97(5):632±6. https://doi.org/10.1136/bjophthalmol-2012-301525. PMID: 23426738; PubMed Central PMCID: PMCPMC3632967

30. Dorn JD, Ahuja AK, Caspi A, da Cruz L, Dagnelie G, Sahel JA et al (2013) The detection of motion by blind subjects with the Epiretinal 60-electrode (Argus II) retinal prosthesis. JAMA Ophthalmol 131(2):183±9. https://doi.org/10.1001/2013.jamaophthalmol.221. PMID: 23544203; PubMed Central PMCID: PMCPMC3924899

31. Rizzo S, Belting C, Cinelli L, Allegrini L, Genovesi-Ebert F, Barca F et al (2014) The Argus II retinal prosthesis: 12-month outcomes from a single-study center. Am J Ophthalmol 157 (6):1282±90. https://doi.org/10.1016/j.ajo.2014.02.039. PMID: 24560994

32. Kotecha A, Zhong J, Stewart D, da Cruz L (2014) The Argus II prosthesis facilitates reaching and grasping tasks: a case series. BMC Ophthalmol 14(1):71

33. Luo YHL, Zhong JJ, Da Cruz L (2015) The use of Argus® II retinal prosthesis by blind subjects to achieve localisation and prehension of objects in 3-dimensional space. Graefes Arch Clin Exp Ophthalmol 253(11):1907–1914

34. Sabbah N, Authié CN, Sanda N, Mohand-Said S, Sahel JA, Safran AB (2014) Importance of eye position on spatial localization in blind subjects wearing an Argus II retinal prosthesis eye position, localization, and retinal prosthesis. Invest Ophthalmol Vis Sci 55(12):8259–8266

35. Castaldi E, Cicchini GM, Cinelli L, Biagi L, Rizzo S, Morrone MC (2016) Visual BOLD response in late blind subjects with Argus II retinal prosthesis. PLoS Biol 14(10):e1002569

36. Huber E, Webster JM, Brewer AA, MacLeod DI, Wandell BA, Boynton GM, Fine I (2015) A lack of experience-dependent plasticity after more than a decade of recovered sight. Psychol Sci. https://doi.org/10.1177/0956797614563957

37. Sabbah N, Sanda N, Authié CN, Mohand-Saïd S, Sahel JA, Habas C, Amedi A, Safran AB (2017) Reorganization of early visual cortex functional connectivity following selective peripheral and central visual loss. Scientific Reports 7:43223

38. Heimler B, Weisz N, Collignon O (2014) Revisiting the adaptive and maladaptive effects of crossmodal plasticity. Neurosci 283:44–63

39. Ptito M, Fumal A, de Noordhout AM, Schoenen J, Gjedde A, Kupers R (2008) TMS of the occipital cortex induces tactile sensations in the fingers of blind braille readers. Exp Brain Res 184:193–200

40. Kupers R, Fumal A, De Noordhout AM, Gjedde A, Schoenen J, Ptito M (2006) Transcranial magnetic stimulation of the visual cortex induces somatotopically organized qualia in blind subjects. Proc Natl Acad Sci 103(35):13256–13260

41. Busskamp V, Duebel J, Balya D, Fradot M, Viney TJ, Siegert S et al (2010) Genetic reactivation of cone photoreceptors restores visual responses in retinitis pigmentosa. Science 329(5990):413–417

42. Yang Y, Mohand-Said S, Léveillard T, Fontaine V, Simonutti M, Sahel JA (2010) Transplantation of photoreceptor and total neural retina preserves cone function in P23H rhodopsin transgenic rat. PLoS One 5(10):e13469

43. Davis MF, Velez DXF, Guevarra RP, Yang MC, Habeeb M, Carathedathu MC, Gandhi SP (2015) Inhibitory neuron transplantation into adult visual cortex creates a new critical period that rescues impaired vision. Neuron 86(4):1055–1066

44. Deidda G, Allegra M, Cerri C, Naskar S, Bony G, Zunino G, Bozzi Y, Caleo M, Cancedda L (2014) Early depolarizing GABA controls critical-period plasticity in the rat visual cortex. Nat Neurosci 18:87

45. Lunghi C, Emir UE, Morrone MC, Bridge H (2015) Short-term monocular deprivation alters GABA in the adult human visual cortex. Curr Biol 25:1496–1501

46. Sengpiel F (2014) Plasticity of the visual cortex and treatment of amblyopia. Curr Biol 24(18): R936–R940

47. Borisoff JF, Elliott SL, Hocaloski S, Birch GE (2010) The development of a sensory substitution system for the sexual rehabilitation of men with chronic spinal cord injury. J Sex Med 7(11):3647–3658

48. Sadeghi SG, Minor LB, Cullen KE (2012) Neural correlates of sensory substitution in vestibular pathways following complete vestibular loss. J Neurosci 32(42):14685–14695

49. Vuillerme N, Hlavackova P, Franco C, Diot B, Demongeot J, Payan Y (2011) Can an electro-tactile vestibular substitution system improve balance in patients with unilateral vestibular loss under altered somatosensory conditions from the foot and ankle? In: Engineering in medicine and biology society, EMBC, 2011 annual international conference of the IEEE. IEEE, pp 1323–1326

50. Kärcher SM, Fenzlaff S, Hartmann D, Nagel SK, König P (2012) Sensory augmentation for the blind. Front Hum Neurosci 6:37

51. Ward J, Meijer P (2010) Visual experiences in the blind induced by an auditory sensory substitution device. Conscious Cogn 19(1):492–500

52. Fornazzari L, Fischer CE, Ringer L, Schweizer TA (2012) "Blue is music to my ears": multimodal synesthesias after a thalamic stroke. Neurocase 18(4):318–322

53. Ione A, Tyler C (2004) Neuroscience, history and the arts synesthesia: is F-sharp colored violet? J Hist Neurosci 13(1):58–65

54. Marks LE (1975) On colored-hearing synesthesia: cross-modal translations of sensory dimensions. Psychol Bull 82(3):303

55. Tyler CW (2005) Varieties of synesthetic experience. In: Robertson LC, Sagiv N (eds) Synesthesia: Perspectives from Cognitive Neuroscience. Oxford University Press, New York

56. Zamm A, Schlaug G, Eagleman DM, Loui P (2013) Pathways to seeing music: enhanced structural connectivity in colored-music synesthesia. NeuroImage 74:359–366
57. Chiou R, Stelter M, Rich AN (2013) Beyond colour perception: auditory–visual synaesthesia induces experiences of geometric objects in specific locations. Cortex 49(6):1750–1763
58. Proulx MJ (2010) Synthetic synaesthesia and sensory substitution. Conscious Cogn 19 (1):501–503
59. Ward J, Wright T (2014) Sensory substitution as an artificially acquired synaesthesia. Neurosci Biobehav Rev 41:26–35
60. Buchs G, Maidenbaum S, Amedi A (2014) Obstacle identification and avoidance using the 'EyeCane': a tactile sensory substitution device for blind individuals. In: International conference on LBHuman haptic sensing and touch enabled computer applications. Springer, Berlin/Heidelberg, pp 96–103
61. Chebat DR, Schneider FC, Kupers R, Ptito M (2011) Navigation with a sensory substitution device in congenitally blind individuals. Neuroreport 22(7):342–347
62. Gagnon L, Schneider FC, Siebner HR, Paulson OB, Kupers R, Ptito M (2012) Activation of the hippocampal complex during tactile maze solving in congenitally blind subjects. Neuropsychologia 50(7):1663–1671
63. Kupers R, Chebat DR, Madsen KH, Paulson OB, Ptito M (2010) Neural correlates of virtual route recognition in congenital blindness. Proc Natl Acad Sci 107(28):12716–12721
64. Chebat DR, Maidenbaum S, Amedi A (2015) Navigation using sensory substitution in real and virtual mazes. PLoS One 10(6):e0126307
65. Matteau I, Kupers R, Ricciardi E, Pietrini P, Ptito M (2010) Beyond visual, aural and haptic movement perception: hMT+ is activated by electrotactile motion stimulation of the tongue in sighted and in congenitally blind individuals. Brain Res Bull 82(5):264–270
66. Striem-Amit E, Dakwar O, Reich L, Amedi A (2011b) The large-scale organization of "visual" streams emerges without visual experience. Cereb Cortex 22(7):1698–1709
67. Striem-Amit E, Amedi A (2014) Visual cortex extrastriate body-selective area activation in congenitally blind people "seeing" by using sounds. Curr Biol 24(6):687–692
68. Striem-Amit E, Cohen L, Dehaene S, Amedi A (2012) Reading with sounds: sensory substitution selectively activates the visual word form area in the blind. Neuron 76(3):640–652
69. Abboud S, Maidenbaum S, Dehaene S, Amedi A (2015) A number-form area in the blind. Nat Commun 6:6026
70. Bach-y-Rita P (1975) Plastic brain mechanisms in sensory substitution. In: Cerebral localization. Springer, Berlin/Heidelberg, pp 203–216
71. Bach-y-Rita P, Collins CC, Saunders FA, White B, Scadden L (1969) Vision substitution by tactile image projection. Nature 221(5184):963–964
72. Nagel SK, Carl C, Kringe T, Märtin R, König P (2005) Beyond sensory substitution—learning the sixth sense. J Neural Eng 2(4):R13
73. Visell Y (2008) Tactile sensory substitution: models for enaction in HCI. Interacting with Computers 21(1-2):38–53
74. Bach-y-Rita P, Aiello GL (1996) Nerve length and volume in synaptic vs diffusion neurotransmission: a model. Neuroreport 7(9):1502–1504
75. Kacznaarek KA, Bach-Y-Rita P (1995) Tactile displays. In: Virtual environments and advanced interface design, vol 55. New York, Oxford, p 349
76. Renier L, Laloyaux C, Collignon O, Tranduy D, Vanlierde A, Bruyer R, De Volder AG (2005) The Ponzo illusion with auditory substitution of vision in sighted and early-blind subjects. Perception 34(7):857–867
77. Renier L, Bruyer R, De Volder AG (2006) Vertical-horizontal illusion present for sighted but not early blind humans using auditory substitution of vision. Attention. Percept Psychophys 68 (4):535–542
78. Schinazi VR, Thrash T, Chebat DR (2016) Spatial navigation by congenitally blind individuals. Wiley Interdiscip Rev Cogn Sci 7(1):37–58

79. Supa M, Cotzin M, Dallenbach KM (1944) " facial vision": the perception of obstacles by the blind. Am J Psychol 57(2):133–183
80. Cotzin M, Dallenbach KM (1950) "Facial Vision": the role of pitch and loudness in the perception of obstacles by the blind. Am J Psychol 63:485–515
81. Cotzin, Dallenbach (1950) "Facial vision": the role of pitch and loudness in the location of obstacles by the blind. Am J Psychol 63:485
82. Ashmead DH, Hill EW, Talor CR (1989) Obstacle perception by ongenitally blind children. Atten Percept Psychophys 46(5):425–433
83. Wilson JP (1967) Psychoacoustics of obstacle detection using ambient or self-generated noise. In: Animal sonar systems: biology and bionics, vol 1. Laboratoire de Physiologie Acoustique, INRA-CNRZ, Jouy-en-Josas, pp 89–114
84. Bronkhorst AW, Houtgast T (1999) Auditory distance perception in rooms. Nature 397 (6719):517–520
85. Kaye HS (2000) Computer and internet use among people with disabilities. Disability Statistics Report 13
86. Meijer PB (1992) An experimental system for auditory image representations. IEEE Trans Biomed Eng 39(2):112–121
87. Hersh MA, Johnson MA (2008) Disability and assistive technology systems. In: Assistive technology for visually impaired and blind people. Springer, London, pp 1–50
88. Maidenbaum S, Abboud S, Amedi A (2014) Sensory substitution: closing the gap between basic research and widespread practical visual rehabilitation. Neurosci Biobehav Rev 41:3–15
89. Dunai L, Peris-Fajarnés G, Lluna E, Defez B (2013) Sensory navigation device for blind people. J Navig 66(3):349–362
90. Hartcher-O'Brien J, Auvray M, Hayward V (2015) Perception of distance-to-obstacle through timedelayed tactile feedback. In: World Haptics Conference (WHC), 2015 I.E. (pp. 7–12). IEEE
91. Segond H, Weiss D, Sampaio E (2005) Human spatial navigation via a visuo-tactile sensory substitution system. Perception 34(10):1231–1249
92. Shoval S, Borenstein J, Koren Y (1998) Auditory guidance with the navbelt-a computerized travel aid for the blind. IEEE Trans Syst Man Cybern Part C Appl Rev 28(3):459–467
93. Stoll C, Palluel-Germain R, Fristot V, Pellerin D, Alleysson D, Graff C (2015) Navigating from a depth image converted into sound. Appl Bionics Biomech 2015:9
94. Kaspar K, König S, Schwandt J, König P (2014) The experience of new sensorimotor contingencies by sensory augmentation. Conscious Cogn 28:47–63
95. König SU, Schumann F, Keyser J, Goeke C, Krause C, Wache S, Lytochkin A, Ebert M, Brunsch V, Wahn B, Kaspar K, Nagel SK, Meilinger T, Bülthoff H, Wolbers T, Büchel C, König P (2016) Learning new sensorimotor contingencies: effects of long-term use of sensory augmentation on the brain and conscious perception. PLoS One 11(12):e0166647
96. Noppeney U, Friston KJ, Ashburner J, Frackowiak R, Price CJ (2005) Early visual deprivation induces structural plasticity in gray and white matter. Curr Biol 15(13):R488–R490
97. Shimony JS, Burton H, Epstein AA, McLaren DG, Sun SW, Snyder AZ (2006) Diffusion tensor imaging reveals white matter reorganization in early blind humans. Cereb Cortex 16 (11):1653–1661
98. De Volder AG, Bol A, Blin J, Robert A, Arno P, Grandin C, Michel C, Veraart C (1997) Brain energy metabolism in early blind subjects: neural activity in the visual cortex. Brain Res 750 (1):235–244
99. Wanet-Defalque MC, Veraart C, De Volder A, Metz R, Michel C, Dooms G, Goffinet A (1988) High metabolic activity in the visual cortex of early blind human subjects. Brain Res 446(2):369–373
100. Noppeney U (2007) The effects of visual deprivation on functional and structural organization of the human brain. Neurosci Biobehav Rev 31(8):1169–1180
101. Sathian K, Zangaladze A (2002) Feeling with the mind's eye: contribution of visual cortex to tactile perception. Behav Brain Res 135(1):127–132

102. Goodale MA, Milner AD (1992) Separate visual pathways for perception and action. Trends Neurosci 15(1):20–25
103. Milner AD, Goodale MA (2008) Two visual systems re-viewed. Neuropsychologia 46 (3):774–785
104. Mishkin M, Ungerleider LG (1982) Contribution of striate inputs to the visuospatial functions of parieto-preoccipital cortex in monkeys. Behav Brain Res 6(1):57–77
105. Ptito M, Matteau I, Zhi Wang A, Paulson OB, Siebner HR, Kupers R (2012) Crossmodal recruitment of the ventral visual stream in congenital blindness. Neural Plast 2012:304045
106. Amedi A, Stern WM, Camprodon JA, Bermpohl F, Merabet L, Rotman S et al (2007) Shape conveyed by visual-to-auditory sensory substitution activates the lateral occipital complex. Nat Neurosci 10(6):687
107. Reich L, Szwed M, Cohen L, Amedi A (2011) A ventral visual stream reading center independent of visual experience. Curr Biol 21(5):363–368
108. Büchel C, Price C, Frackowiak RS, Friston K (1998) Different activation patterns in the visual cortex of late and congenitally blind subjects. Brain J Neurol 121(3):409–419
109. Ptito M, Matteau I, Gjedde A, Kupers R (2009) Recruitment of the middle temporal area by tactile motion in congenital blindness. Neuroreport 20(6):543–547
110. Matteau I, Schneider F, Kupers R, Ptito M (2006) Tactile motion discrimination through the tongue in blindness: a fMRI study. Neuroimage 36(Suppl 1):211
111. Collignon O, Lassonde M, Lepore F, Bastien D, Veraart C (2007) Functional cerebral reorganization for auditory spatial processing and auditory substitution of vision in early blind subjects. Cereb Cortex 17(2):457–465
112. Bedny M, Konkle T, Pelphrey K, Saxe R, Pascual-Leone A (2010) Sensitive period for a multimodal response in human visual motion area MT/MST. Curr Biol 20(21):1900–1906
113. Amedi A, Stern W, Striem E, Hertz U, Meijer P, Pascual-Leone A (2008) A what/where visual-toauditory sensory substitution fMRI study: Can blind and sighted hear shapes and locations in the visual cortex. In: 31st European Conference on Visual Perception
114. Siuda-Krzywicka K, Bola Ł, Paplińska M, Sumera E, Jednoróg K, Marchewka A, Szwed M (2016) Massive cortical reorganization in sighted braille readers. elife 5:e10762
115. Hertz U, Amedi A (2014) Flexibility and stability in sensory processing revealed using visual-to-auditory sensory substitution. Cereb Cortex 25(8):2049–2064
116. Hannagan T, Amedi A, Cohen L, Dehaene-Lambertz G, Dehaene S (2015) Origins of the specialization for letters and numbers in ventral occipitotemporal cortex. Trends Cogn Sci 19 (7):374–382
117. Damoiseaux JS, Greicius MD (2009) Greater than the sum of its parts: a review of studies combining structural connectivity and resting-state functional connectivity. Brain Struct Funct 213:525–533
118. Fox MD, Raichle ME (2007) Spontaneous fluctuations in brain activity observed with functional magnetic resonance imaging. Nat Rev Neurosci 8:700–711
119. Bi Y, Wang X, Caramazza A (2016) Object domain and modality in the ventral visual pathway. Trends Cogn Sci 20(4):282–290
120. Weaver KE, Stevens AA (2007) Attention and sensory interactions within the occipital cortex in the early blind: an fMRI study. J Cogn Neurosci 19(2):315–330
121. Eger E, Sterzer P, Russ MO, Giraud A-L, Kleinschmidt A (2003) A supramodal number representation in human intraparietal cortex. Neuron 37:719–726
122. Price CJ (2012) A review and synthesis of the first 20 years of PET and fMRI studies of heard speech, spoken language and reading. NeuroImage 62:816
123. Vigneau M, Beaucousin V, Herve P-Y, Duffau H, Crivello F, Houde O, Mazoyer B, Tzourio-Mazoyer N (2006) Meta-analyzing left hemisphere language areas: phonology, semantics, and sentence processing. NeuroImage 30:1414–1432
124. Wang X, Peelen MV, Han Z, He C, Caramazza A, Bi Y (2015) How visual is the visual cortex? Comparing connectional and functional fingerprints between congenitally blind and sighted individuals. J Neurosci 35(36):12545–12559

125. Ptito M, Moesgaard SM, Gjedde A, Kupers R (2005) Cross-modal plasticity revealed by electrotactile stimulation of the tongue in the congenitally blind. Brain 128(3):606–614

126. Thaler L, Arnott SR, Goodale MA (2011) Neural correlates of natural human echolocation in early and late blind echolocation experts. PLoS One 6:e20162

127. Théoret H, Merabet L, Pascual-Leone A (2004) Behavioral and neuroplastic changes in the blind: evidence for functionally relevant cross-modal interactions. J Physiol aris 98 (1):221–233

128. Amedi A, Raz N, Pianka P, Malach R, Zohary E (2003) Early 'visual' cortex activation correlates with superior verbal memory performance in the blind. Nat Neurosci 6:758–766

129. Bedny M, Pascual-Leone A, Dodell-Feder D, Fedorenko E, Saxe R (2011) Language processing in the occipital cortex of congenitally blind adults. Proc Natl Acad Sci 108 (11):4429–4434

130. Burton H, Diamond JB, McDermott KB (2003) Dissociating cortical regions activated by semantic and phonological tasks: a FMRI study in blind and sighted people. J Neurophysiol 90:1965–1982. Epub 2003 Jun 1964

131. Sadato N, Pascual-Leone A, Grafman J, Ibanez V, Deiber MP, Dold G, Hallett M (1996) Activation of the primary visual cortex by braille reading in blind subjects. Nature 380:526–528

132. Strnad L, Peelen MV, Bedny M, Caramazza A (2013) Multivoxel pattern analysis reveals auditory motion information in MT+ of both congenitally blind and sighted individuals. PLoS One 8:e63198

133. Vetter P, Smith FW, Muckli L (2014) Decoding sound and imagery content in early visual cortex. Curr Biol 24:1256–1262

134. Raz N, Amedi A, Zohary E (2005) V1 activation in congenitally blind humans is associated with episodic retrieval. Cereb Cortex 15(9):1459–1468

135. Amedi A, Floel A, Knecht S, Zohary E, Cohen LG (2004) Transcranial magnetic stimulation of the occipital pole interferes with verbal processing in blind subjects. Nat Neurosci 7 (11):1266–1270

136. Butt OH, Benson NC, Datta R, Aguirre GK (2013) The fine-scale functional correlation of striate cortex in sighted and blind people. J Neurosci 33:16209

137. Bock AS, Saenz M, Tungaraza R, Boynton GM, Bridge H, Fine I (2013) Visual callosal topography in the absence of retinal input. NeuroImage 81:325

138. Bock AS, Fine I (2014) Anatomical and functional plasticity in early blind individuals and the mixture of experts architecture. Front Hum Neurosci 8:971

139. Burton H, Snyder AZ, Raichle ME (2014) Resting state functional connectivity in early blind humans. Front Syst Neurosci 8:51

140. Deen B, Saxe R, Bedny M (2015) Occipital cortex of blind individuals is functionally coupled with executive control areas of frontal cortex. J Cogn Neurosci 27(8):1633–1647

141. Liu Y, Yu C, Liang M, Li J, Tian L, Zhou Y, Qin W, Li K, Jiang T (2007) Whole brain functional connectivity in the early blind. Brain 130:2085–2096

142. Watkins KE, Cowey A, Alexander I, Filippini N, Kennedy JM, Smith SM, Ragge N, Bridge H (2012) Language networks in anophthalmia: maintained hierarchy of processing in 'visual' cortex. Brain 135:1566

143. Striem-Amit E, Ovadia-Caro S, Caramazza A, Margulies DS, Villringer A, Amedi A (2015) Functional connectivity of visual cortex in the blind follows retinotopic organization principles. Brain 138:1679–1695. awv083

144. Gougoux F, Zatorre RJ, Lassonde M, Voss P, Lepore F (2005) A functional neuroimaging study of sound localization: visual cortex activity predicts performance in early-blind individuals. PLoS Biol 3:e2

145. Breitenseher M, Uhl F, Prayer Wimberger D, Deecke L, Trattnig S, Kramer J (1998) Morphological dissociation between visual pathways and cortex: MRI of visually-deprived patients with congenital peripheral blindness. Neuroradiology 40(7):424–427

146. Pan WJ, Wu G, Li CX, Lin F, Sun J, Lei H (2007) Progressive atrophy in the optic pathway and visual cortex of early blind Chinese adults: a voxel-based morphometry magnetic resonance imaging study. NeuroImage 37(1):212–220

147. Shu N, Li J, Li K, Yu C, Jiang T (2009) Abnormal diffusion of cerebral white matter in early blindness. Hum Brain Mapp 30(1):220–227

148. Christensen R, Grey M, Ptito M, Kupers R (2009) Resting state brain metabolism and functional connectivity of the occipital cortex in congenital blindness: a combined rTMS and PET-FDG study. NeuroImage 47:S64

149. Veraart C, De Volder AG, Wanet-Defalque MC, Bol A, Michel C, Goffinet AM (1990) Glucose utilization in human visual cortex is abnormally elevated in blindness of early onset but decreased in blindness of late onset. Brain Res 510(1):115–121

150. Kupers R, Ptito M (2014) Compensatory plasticity and cross-modal reorganization following early visual deprivation. Neurosci Biobehav Rev 41:36–52

151. Coullon GS, Emir UE, Fine I, Watkins KE, Bridge H (2015) Neurochemical changes in the pericalcarine cortex in congenital blindness attributable to bilateral anophthalmia. J Neurophysiol 114(3):1725–1733

152. Leporé N, Voss P, Lepore F, Chou YY, Fortin M, Gougoux F, Lee AD, Brun C, Lassonde M, Madsen SK, Toga AW, Toga AW, Thompson PM (2010) Brain structure changes visualized in early-and late-onset blind subjects. NeuroImage 49(1):134–140

153. Maller JJ, Thomson RH, Ng A, Mann C, Eager M, Ackland H et al (2016) Brain morphometry in blind and sighted subjects. J Clin Neurosci 33:89–95

154. Bonino D, Ricciardi E, Sani L et al (2008) Tactile spatial working memory activates the dorsal extrastriate cortical pathway in congenitally blind individuals. Arch Ital Biol 146:133–146

155. Striem-Amit E, Hertz U, Amedi A (2011a) Extensive cochleotopic mapping of human auditory cortical fields obtained with phase-encoding FMRI. PLoS One 6(3):e17832

156. Stronks HC, Mitchell EB, Nau AC, Barnes N (2016) Visual task performance in the blind with the BrainPort V100 vision aid. Expert Rev Med Devices 13(10):919–931

157. Maidenbaum S, Hannasi S, Abboud S, Arbel R, Shipuznikov A, Levy-Tzedek S et al (2012) The EyeCane-Distance information for the blind. In: Journal of molecular neuroscience, vol 48. Humana Press Inc, Totowa, pp S75–S76

158. Maidenbaum S, Levy-Tzedek S, Chebat DR, Amedi A (2013) Increasing accessibility to the blind of virtual environments, using a virtual mobility aid based on the "EyeCane": feasibility study. PLoS One 8(8):e72555

159. Chebat DR, Harrar V, Kupers R, Maidenbaum M, Amedi A, Ptito M (2017) Sensory SUbstitution and the neural correlates of navigation in blindness. In: Mobility in visually impaired people. Springer. In press

Chapter 14
Neuroimaging of Pain

S. Espinoza and C. Habas

fMRI/PET and the "Pain Matrix"

fMRI has enabled to compute activation brain mapping during application of acute pain mechanical, thermal, or chemical stimuli on the body skin, while functional connectivity identifies pain-related and polymodal networks involved in the pain experience. fMRI using acute pain stimulation has enlighted a constant brain network, the so-called pain matrix involved in localization and valuation of pain intensity and identity [1].

The Pain Matrix The PM has been subdivided into sensory-discriminative, fast-conducting lateral (lPM), and medial affective systems (mPM) including mainly the spinothalamic afferents, the ventral and intralaminar thalamus, the somatotopically organized posterior insula, the S1/S2 (parietal operculum) and the midcingulate cortex, and the medial thalamus, the anterior insula, and the anterior cingulate cortex (BA 24), respectively (Fig. 14.1). lPM is particularly involved in accurate spatial localization and in intensity evaluation of the painful stimulus. However, it seems that the insula and the adjacent ventral premotor cortex also belong to a polymodal "how much" module evaluating stimulus magnitude and allowing to determine its saliency afterwards [2]. Within the mPM, ACC (anterior cingulate cortex) represents a multi-integrative hub being preferentially implicated in subjective appreciation of unpleasantness, attention, pain anticipation, and motor response programming [3]. Anatomically, ACC has been proposed to be, at least, divided into an anterior perigenual area (BA 32–25) and a dorsal (BA 32) and ventral (BA 24) midcingulate area [4] coping with unpleasantness, phasic, and sustained pain-related attentional reallocation/motor planning, respectively [5]. However, discrepant data summarized by Peyron et al. [3] suggested a more complex anatomo-functional

S. Espinoza · C. Habas (✉)
Service de NeuroImagerie, Paris, France
e-mail: chabas@15-20.fr

© Springer International Publishing AG, part of Springer Nature 2018
C. Habas (ed.), *The Neuroimaging of Brain Diseases*, Contemporary Clinical
Neuroscience, https://doi.org/10.1007/978-3-319-78926-2_14

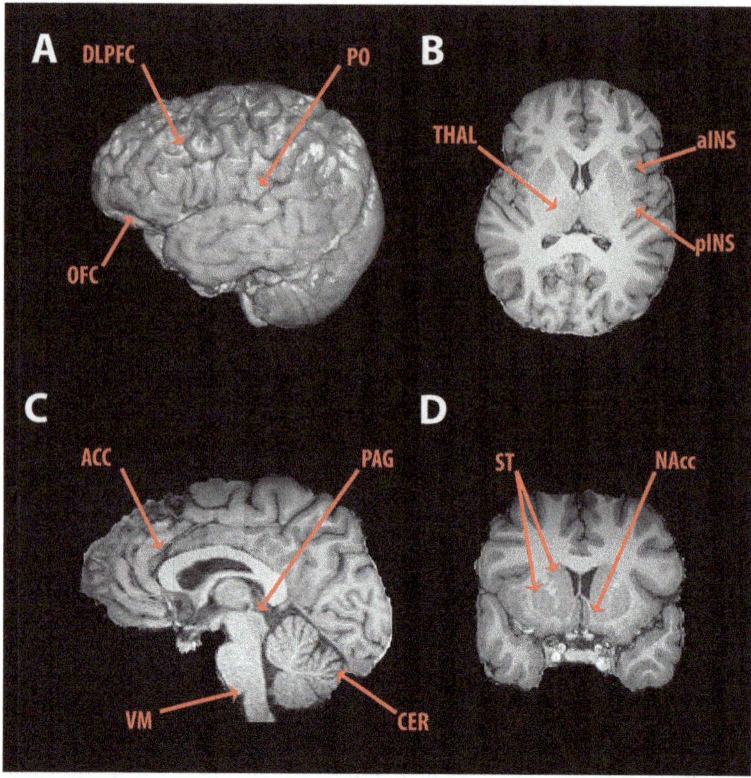

Fig. 14.1 The main brain areas involved in the lateral and medial pain matrix (S1/S2, insula, thalamus, ACC), in prefrontal control, in striatal and cerebellar regulation, and in brainstem pain-gating structures. (**a**) Lateral view of a 3D volume rendering of the brain. (**b**) Axial slice. (**c**) Sagittal slice. (**d**) Coronal slice. Abbreviations: ACC anterior cingulate cortex, CER cerebellum, DLPC dorsolateral prefrontal cortex, a/pINS anterior/posterior insula, Nacc nucleus accumbens septi, OFC orbitofrontal cortex, OP parietal operculum (S1/S2), PAG periaqueductal gray matter (mesencephalon), ST striatum (caudate nucleus and putamen), THAL thalamus, VM ventromedial medulla

parcellation of ACC. Insula links the medial and lateral PM by channeling nociceptive stimuli from its sensory-discriminative caudal zone to its more rostral vegetative and cognitive zone, partly in relation with awareness access [6]. Of importance, functional connectivity between the anterior insula and periaqueductal gray matter (PAG) predicts susceptibility to upcoming pain and varies with personality traits [7].

If the PM represents the first step for processing the nociceptive peripheral afferents and for conscious evaluation of pain magnitude (insula), this circuit appears to be also engaged by other "neutral" somatosensory and teleceptive stimuli in relation to their modality-unspecific saliency [8]. In particular, the anterior insula and midcingulate cortex perform together valuation of the saliency of impending painful stimulus [9]. Therefore, PM is also involved in a bottom-up attentional filtering of distal sensory afferents.

PM-Associated Brain Areas This "core" PM is also but more inconstantly associated with context-dependent activation of other neural systems such as the hippocampus (episodic memory: recall of past pain events or memorization), amygdala (negative mood modulation, threatening evaluation of pain) [10], parietal cortex (recognition of environmental stimuli during painful stimulation) [11], cingulate and frontal motor areas (motor planning), striatum (motor response such as withdrawal) [12], nucleus accumbens septi (cue-based pain prediction, pain relief-related reward) [13, 14], and cerebellum, as well as the red nucleus (modulatory and integrative functions) [15]. Cerebellar activity was found in the ipsilateral anterior vermis of lobule III, lobules III–VI, and deep cerebellar nuclei in relation to high score of pain intensity [16]. The cerebellum and red nucleus were implicated in pain-induced withdrawal [12]. All these systems allow to recognize, to categorize, and to memorize painful stimuli and, then, to generate relevant motor, emotional, and cognitive behavior, and anormal (see below) or epigenetic functional connection between the PM and limbic system, for example, during depression or anxiety, may amplify pain feeling [17]. Consequently, idiosyncratic response to pain turns out to be multifactorial and strongly depends on individual past experience and current mood.

Top-Down Pain Modulation Bantick et al. [18] showed reduced activation of the thalamus, midcingulate cortex, and insula correlated with pain intensity underscoring by volunteers performing an interference task (counting Stroop task) and simultaneously undergoing a painful thermal stimulation. In line with these authors, Valet et al. [19] found that orbitofrontal (OFC) and perigenual cortices (PGC) modulate pain by acting on posterior thalamic and PAG activity during the cognitive distraction task. The dorsolateral prefrontal cortex (DLPFC) also induces decreased perception of pain intensity and unpleasantness by exerting an influence upon the medial thalamus and mesencephalon [20]. Cognition, in terms of belief in self-controlled perception, can also mediate central analgesic by recruiting the right anterolateral prefrontal cortex (ALPC) [21]. Controllable versus uncontrollable painful stimulation entails diminished functional connectivity between the DLPFC and anterior insula [22]. Therefore, pain-related (anticipation, interpretation of noxious stimulation) or pain-unrelated (distraction) cognition can cause analgesia by exerting prefrontal control on PM. Furthermore, besides diencephalo-telencephalic activation, pain-related activation was detected within brainstem structures belonging to the *descending pain pathway* and including PAG, ventral tegmental area (VTA), parabrachial nucleus, raphe magnus, gigantocellular (reticular) nucleus, nucleus tractus solitarii, and rostral ventromedial medulla (RVM) [23, 24]. Strong functional connections were found between the hypothalamus, PAG, and RVM. Fairhurst et al. [23] also observed that anticipatory activation of the entorhinal cortex and VTA, prior to noxious stimulation, was significantly correlated with anterior insula activity and with intensity of perceived pain. Thus, attention to upcoming pain participates in pain modulation. In summary, a central, top-down modulatory system capable of attenuating pain sensation can, consequently, be individualized and, mainly, encompasses ALPFC/DLPFC (cognitive control), OFC, and PGC (emotional control) exerting a differential, direct or indirect, via the insula, action onto the medial and posterior thalamus, as well as the hypothalamus, and onto the brainstem pain-related nuclei, especially PAG. This action could be regarded as a two-level gating control for cortical and amygdalian pain processing, on one side, and for spinal and brainstem (disfacilitated) pain transmission, on the other side.

Studies of placebo-related influence on nociceptive processing (reviewed in [25]) capturing central, belief-based analgesia confirmed downregulation of PM nodes (rostral ACC, insula, and thalamus) and specific activation of DLPFC and OFC, as well as mesencephalic tegmentum, for example, during pain anticipation [26]. Functional connectivity between rostral ACC and PAG plays an important role in placebo analgesia, and PET studies [25] showed the pivotal contribution of endogenous opioid action onto μ-opioid receptors in several brain areas such as the prefrontal cortex, ACC, anterior insula, amygdala, nucleus accumbens septi, and PAG.

Functional connectivity and network level Many brain regions involved in pain processing also take part in well-known canonical networks contributing to the brain resting state, such as the default mode network (DMN), salience network (SN), executive network (EN), or ventral and dorsal attentional networks (vAN and dAN). For instance, the DLFPC/parietal cortex, DLFPC/OFC, insula/ACC, and right anterior insula belong to EN, dAN, SN, and vAN, respectively. Therefore, pain processing may recruit and synchronize, in a context-dependent manner, canonical networks for general purposes such as painful stimulus-oriented attention (AN); filtering of relevant interoceptive, emotional, and vegetative information (SN); selection of pain-related proper response (vAN); and pain awareness and memory (DMN). Moreover, the (right) anterior insula may represent a hub interconnecting and modulating SN, dAN [27], and PM. In this vein, a ASL-based fcMRI [28] demonstrated stronger coherence between the DMN, PGC, and right insula during chronic low back pain and with a positive correlation between DMN insular activity and pain rating. More precisely, Taylor et al. [29] differentiated two distinct networks passing through the insular cortex, an anterior insular PGC/anterior midcingulate cortex and a mid-posterior insular/posterior midcingulate cortex, respectively, implicated in emotion and in environmental monitoring and behavioral response selection. The insula can be, at least, subdivided into three zones contributing to analyzing the painful stimulus and to recruiting canonical networks for further integration and for access to consciousness [30]. First, the posterior insula evaluates pain intensity. Second, based upon the posterior insular output, the anterior insula associated with the SN discriminates saliency of the noxious stimulus. The SN, via ACC node, can engage the limbic system. Third, the right dorsal insula switches activity between competing canonical networks such as DMN and EN. Therefore, some anatomical hubs display a dual function being implicated in nociceptive processing and in affective/cognitive parallel processing, acting as an interface with polymodal canonical and subcortical networks, such as the DLPFC, ACC, and insula. Seminowicz and Moayedi [31] summarized the pivotal role of DLPFC as a pain suppressor, and downregulator of pain catastrophizing (more generally of pain-related emotions), and as a switch, like the right dorsal insula [32], between central executive networks and DMN.

Conclusion Reflecting the multidimensional subjective pain experience, the pain neural system encompasses dynamical synchronization of large-scale networks either (partly) pain-specific such as PM for pain localization and intensity (insula), although PM also contributes to bottom-up attentional selection of salient stimuli, or pain-nonspecific, polymodal networks such as circuits involved in top-down attention, prediction, emotion, memory, associative learning, and autonomic and motor responses. Let us emphasize that some brain areas (amygdala and mid-ACC; [2]) exhibit an anticipatory response to noxious stimuli. Some of the pain-related networks are represented by canonical networks (SN and EN) and include limbic striatal and cerebellar loops. This fluid and plastic pain-related system, especially its fronto-limbic part, can exert anticipatory, facilitatory, or disfacilitatory influences on the "brainstem integrator" (PAG and RVM) which processes and gates nociceptive stimuli [33].

Chronic Pain

Chronic pain, which displays a very heterogeneous clinical presentation, can accompany a large number of diseases such as headaches, unspecific chronic back pain, spinal cord injury, fibromyalgia, irritable bowel syndrome, chronic vulvar pain, complex regional pain syndrome, phantom limb pain, postherpetic neuralgia, menstrual pain, diabetic neuropathy, chronic pancreatitis, temporomandibular disorder, or osteoarthritis (reviewed in [33–35]). This trouble defined as a persistent and daily pain for, at least, 6 months is very frequently associated with the following mood disturbances: general tiredness, anxiety, depression or fear, and hyperalgesia or allodynia. Pain catastrophizing characterized by amplification of experienced pain intensity and inability to suppress pain-oriented attention and pain-related thoughts can also be reported.

From a pathophysiological viewpoint, chronic pain can no longer be regarded (exclusively) as the consequence of a long-lasting sensitization of peripheral nociceptors causing an overstimulation of the PM, for instance [35]. Chronic pain would rather result from long-term neuroplastic changes affecting several associative brain networks likely in relation to reinforcing learning of painful stimuli and pain-related environmental cues [34]. Such changes could yield abnormal functional connectivity between networks and interference with other information processing within the central nervous system.

fMRI complemented by other imaging methods, such as structural MRI, voxel-based morphometry (VBM), measurement of cortical thickness, spectroscopy, tensor diffusion imaging, and PET scan, strives to characterize precisely brain activation pattern associated with chronic pain.

If chronic pain-related activation brain patterns are rather pathology-specific [36], it seems that the prefrontal cortex and anterior insula are observed regardless of the causal disease (reviewed in [37]). For example, studying chronic back pain, Baliki et al. [38] showed that the medial prefrontal cortex (MPFC) and the contiguous rostral ACC activities positively covaried with spontaneous pain intensity, whereas insula activation is related to chronicity. MPFC dysfunction is supposed to ensue from DLPFC hypoactivity/atrophy [39]. Furthermore, the same author also demonstrated two important complementary results. First, pain chronification was accompanied by corticostriatal reorganization [40]: the functional connectivity of the nucleus accumbens septi switched from the insula during subacute pain to MPFC during chronic pain. This reorganization may monitor aberrant reinforcement learning and, thus, leads to erroneous estimation of pain relief. Second, during chronic pain, functional dynamics of DMN was notably altered with lower deactivations in the medial PFC, amygdala, and posterior cingulate cortex when the "rest" task was contrasted with the "attentive" task [41]. DMN dysfunction was extended to osteoarthritis and complex regional pain syndromes, which exhibit diminished coherence between MPFC and posterior cortical areas of DMN but an increased coherence with the insula, correlated with the magnitude of pain, as well as lower phase locking between the DMN and parietal cortex involved in attention [42]. An ASL-based investigation also pointed out the strong coherence between the pregenual ACC (DMN) and the right insula, positively correlated with chronic pain [28]. Kucyi et al. [43] found that "rumination" about patient's own chronic pain was linked to high functional connectivity between MPFC and the (1) posterior cingulate cortex/precuneus, (2) medial thalamus, and (3) midbrain (PAG). Moreover, a frequency shift toward higher spontaneous BOLD frequencies was observed for the anterior DMN and the salience network during the brain resting state of patients affected by chronic pain [44]. Therefore, chronic pain-related brain reorganization can induce strong interference with the general brain dynamics reflected in the correlated-anticorrelated connectivity pattern of canonical networks (DMN, SN, AN), which would explain associated emotional and cognitive impairments. However, some alterations of the power spectral density of low-frequency BOLD signal recorded in S1, DLPFC, supplementary motor area, and especially amygdala were not correlated with anxiety and depression in fibromyalgia [45]. Topological analysis based on mutual partial correlation of spontaneous BOLD fluctuations from various selected brain areas of chronic back pain patients revealed complex modular reorganization within several canonical networks, such as DMN and executive networks, and tighter functional connectivity between emotional, executive, and motor systems [46]. A high connectivity was noticed between ACC and the caudate nucleus.

Downstream, chronic neuropathic pain was accompanied by abnormal functional connectivity between components of the brainstem pain integrator (RVM, ventro-lateral PAG, putative locus coeruleus) [47]. Moreover, stronger coherence was detected between the midbrain and upstream structures such as the hippocampus, nucleus accumbens septi, and ACC. It was postulated that these cortical supratentorial areas might facilitate upward transmission of nociceptive afferents at the mesencephalic level.

It is noteworthy that, if allodynia can be associated with a specific and thus recognizable central signature with altered activation in the prefrontal and insular cortices, thalamus, and striatum [48], it can also result from aberrant functional coherence between PAG and contiguous reticular formation and other pain processing areas [49].

Therefore, chronic pain, at least, stems from central and midbrain changes such as DLPFC dysfunction causing dysregulation of MPFC/ACC involved in pain intensity, pain catastrophizing [50] and emotional content, limbic striatocortical loop passing through the nucleus accumbens septi (erroneous conditioning), canonical networks ("rumination," pain-oriented attention, motivation), and the midbrain pain processor.

Morphological and metabolic MRI techniques also provided important complementary data concerning structural and neurochemical brain reorganization associated with chronic pain. Part of these structural data mirrored fMRI-based functional data. A recent meta-analysis dealing with gray matter alterations (atrophy or hypertrophy) due to several types of chronic pain established that these alterations, often correlated with pain duration, mainly concern the prefrontal (BA 9, 10, 47), cingulate, ventral anterior insular (in relation with SN), inferior parietal and sensorimotor cortices, basal ganglia, thalamus, cerebellum (culmen), and PAG [51]. It is noteworthy that atrophy particularly affects the prefrontal, cingulate, and insular cortices, thalamus, and PAG. Moreover, in a network-oriented analysis, this study showed that gray matter alterations were always observed in canonical networks (DMN, SN, AN) and in the thalamus-basal ganglia circuit, whereas sensory networks are differentially and variably affected by chronic pain pathologies.

Diffusion tensor imaging (DTI) enables to measure the fractional anisotropy (FA) reflecting the tissular microstructural organization and complexity and to reconstruct the cerebral inter-areal tracts within the white matter. Few studies used DTI to identify white matter changes during some pain syndromes [35]. In line with functional and VBM investigations, FA was decreased in the DLPFC, ACC, and insula. Mansour et al. [52] demonstrated that transition from subacute to chronic back pain was predicted by abnormal FA value in correlation with structural connectivity between MPFC and nucleus accumbens septi. Interestingly, anatomoclinical correlations can be established. For example, studying patients with fibromyalgia, Lutz et al. [53] found significant correlations between FA aberrant value in the right superior frontal gyrus, left superior frontal cortex and left ACC, and thalamus and left insula with pain intensity, increased fatigue, and "self-perceived physical impairment," respectively. Tract-based statistics also revealed white matter disorganization. In complex regional pain syndrome [54] and in chronic musculoskeletal pain (lieberma et al. 2014), white matter alterations were found, respectively, in the corpus callosum and the corona radiata, with a significant correlation between abnormal axial diffusivity and motor disability, and in the internal capsule and uncinate fascicle in correlation with pain experience and pain catastrophizing. Therefore, white matter studies can show inter-areal afferentation abnormalities at the origin of clinical impairments.

Few studies resorted to proton and GABA MRI spectroscopy to track chronic pain-related metabolic changes (reviewed in [35]) and coped with a low sample of pain pathologies. The main markers were N-acetyl-aspartate (NAA), glutamine/glutamate (GLX), and GABA. Here again, in agreement with functional and structural results, decreased NAA concentration, a neuronal viability marker, was present in the DLPFC, ACC, and thalamus, while decreased GABA, inhibitory interneuronal marker, was observed in the insula and thalamus.

Conclusion Chronic pain is underlaid by time-dependent, pathology-specific microstructural and metabolic gray matter (atrophy) and microstructural white matter alterations in association with functional and topological reorganization within and between (resting-state) canonical circuits involved in emotion, cognition, memory, anticipation, and learning, such as DMN, and salience, attentional, and executive networks and with brainstem circuits involved in pain modulation and gating. In other words, chronic pain roots in a central, diffuse, and partially reversible or irreversible rewiring affecting not only nociception but other mental processing, especially emotion. More fundamentally, chronic may be conceptualized as "[. . .] the persistence of the memory of pain and/or the inability to extinguish the memory of pain evoked by an initial inciting injury" [36], which entails continuous *aversive emotional associations* with environmental stimuli. Mesolimbic circuit may monitor the pain memory trace reinforcement and the correlative aversive learning, conducting to central rewiring and to downregulation of nociceptive modulation and filtering at the spinal cord and brainstem levels [55]. By many aspects and as noticed earlier, chronic pain impinging persistently on multiple large-scale networks could be compared to neurodegenerative pathology.

Real-Time fMRI-Based Neurofeedback for Pain Treatment

The purpose of real-time neurofeedback (NFB) consists in teaching subjects how to self-regulate activity of a target brain area or neural network, using fMRI [56, 57]. Briefly, inside the MRI scanner, functional images of the brain are acquired and post-processed online, and the resulting activation map is presented back to the subject, who learns to actively modulate the level of activation of the target zone or circuit, using appropriate mental strategies. Usually, feedback signal appears as "thermometer display" figuring this activation. Several sessions may be necessary for the subject to carry out efficiently the mental task. Repeated NFB is thought to induce (long-term) neuroplastic changes in the target zone/network. This method has been successfully applied to downregulation of pain perception by controlling activity within the rostral ACC with trained to NFB, normal subjects and patients suffering from chronic pain managed to control the level of activation in the rostral ACC, and, consequently, downregulate pain perception [58]. NFB targeting the ACC and anterior insula also entailed diminished perception of pain, and these two brain regions were shown to exert their action onto the caudate nucleus and the prefrontal cortex, respectively [59]. Conversely, upregulation of the left posterior insula led to increase in pain unpleasantness [60]. Therefore, several nodes of the pain matrix can be targeted by NFB to potentially self-control acute and chronic pain. However, some of these target regions could be part of a (non-pain-specific) general system of self-regulation, especially the anterior insula and striatum but also the ACC, prefrontal cortex, and temporo-occipital junction [59, 61].

Hypnosis and Meditation Practice

Analgesia can sometimes be obtained by hypnosis and meditation [62]. These two methods can both influence activation of all the components of the pain matrix and of the resting-state networks, such as SN, EN, and DMN. For instance, reduced activity in ACC and in the parieto-insular cortex is correlated with diminished pain unpleasantness and intensity, respectively, during suggestion. But besides of the PM, hypnosis modulates synchronization and amount of activation of SN, DMN, and EN [63]. Therefore, hypnosis should regulate emotional, attentional, and executive processing contributing to pain integration and, in particular, DMN dysfunction observed in chronic pain. Meditator experts exhibited strong SN activation reflecting decreased pain anticipation [64]. It has been assumed that meditation leads to upregulation of sensory processing, through the SN, and downregulation of cognitive processes [33].

Noninvasive Brain Neurostimulation

Medication-resistant chronic pain can benefit from neuromodulation or neurostimulation, transcranial direct current stimulation (tDCS), or repetitive transcranial magnetic stimulation (rTMS), respectively [62, 65]. These methods transiently interfere with central pain processing by targeting specific brain areas belonging to PM or to intrinsic canonical networks. Therefore, the determination of the exact anatomical location of these brain areas is mandatory, what structural and functional imaging enables. Moreover, if repetitive modulation/stimulation is applied to the brain of the patient, neuroplastic long-term neural changes may occur, in correlation with potential clinical improvement, what, here again, can be visualized and followed up with structural imaging, activation brain mapping, and resting-state functional connectivity [66]. Stimulation can be performed offline or online with PET scan or with MRI-compatible stimulator.

Briefly, tDCS entails modulation of neuronal spiking threshold, using cathodal or anodal 1–2 mA current delivered during around 20 min by an electrode positioned over the skull skin toward the targeted brain area. Anodal (versus cathodal) current causes a facilitatory (disfacilitatory) influence on neurons. rTMS consists in transiently and locally generating a strong, fluctuating magnetic field with an external coil, which produces electric current within the underlying cerebral cortex. However, the neuromodulatory effects are not circumscribed at the stimulation point and may thus diffuse to neighboring or more distant regions. A lot of parameters can be tuned to optimize effects of the stimulation, such as placement of electrodes over the scalp, stimulation frequency, synchronization of the stimuli with endogenous brain rhythms, pretreatment stimulation (priming), number of sessions, latency of stimulation-induced analgesia, etc. [67]. Few studies are currently available, sometimes provide discrepant results, and need more randomized controlled protocols [68, 69]. Nevertheless, some meta-analyses suggest, for instance, that (anodal) tDCS stimulation of the motor cortex and DLPFC induce short-term or long-term relief of pain, increasing pain threshold [65, 70]. Immediate effects of tDCS could be mediated by activation of the μ-opioid system [71] and might recruit large network of the cortical areas and cerebellum [72]. Peyron et al. [73] reported in a PET scan study of neuropathic pain that electrical stimulation of motor cortex entailed a poststimulation increased activation of the prefrontal cortex, ACC, and PAG, which might inhibit nociceptive transmission. Moreover, long-lasting effects could be ascribed to long-term depression or potentiation in nodes of the pain-related networks. rTMS applied to the motor cortex, somatosensory, and DLPFC causes pathology- and frequency-dependent effects, but some alleviation has been observed during chronic pain [65, 74, 75], especially centrally originated pain. Finally, rTMS would exert a global pain-relief effect likely in relation with emotional valuation of pain [76]. In this vein, in fibromyalgia, tDCS and rTMS caused reduced pain sensation and depression improvement when stimulating the motor cortex or DLPFC [77].

Conclusion

Advanced neuroimaging has deciphered in human and in vivo the complex and plastic network organization underlying pain perception and pain affective, cognitive, and motor integration. Stimulation functional imaging has identified brain areas involved in pain characterization, in the pain matrix, and in central top-down and midbrain pain regulation. Functional connectivity has shown that pain also recruits canonical networks implicated in its polymodal integration and that several areas of the pain matrix, such as the insula and ACC, constitute hubs of dynamically interconnecting circuits such as DMN, SN, and EN. Besides BOLD fMRI, ASL-based fMRI also allows for quantifying cerebral blood flow and for identifying brain activation pattern caused by spontaneous pain [78, 79]. Furthermore, structural and functional imaging have been applied to understand neural reshaping accompanying chronic pain. Structural MRI and spectroscopy found alterations in tissular architecture (usually atrophy) and metabolism (loss of neurons and interneurons) within pain-related brain regions. Molecular imaging, such as PET scan with specific radiotracers, can also furnish important data about neurotransmitters implicated in pain processes, for instance, μ-opioids and dopamine in pain relief. Diffusion imaging and tractography pointed out white matter fascicular alterations conducting to partial deafferentation. These abnormalities entail aberrant functional pattern with hypoactivation of central (prefrontal cortex) and midbrain/spinal cord pain-inhibiting systems and progressive recruitment especially of limbic systems implicated in catastrophizing, emotion, and aberrant aversive learning. Functional connectivity characterizes short- and long-term disorganization and topological rewiring of the intrinsic resting-state connectivity inducing interferences with global cognition. Notably, functional connectivity could predict transition from subacute pain to chronic pain. Finally, the better understanding of pathology- and subject-specific, pain-related dysfunctional networks would allow to select accurately brain areas to be targeted by online fMRI-based neurofeedback and by noninvasive neuromodulation or neurostimulation methods, which constitute a promising tool to alleviate pain symptoms, even if thorough studies are still required to appraise the real efficacy of these new treatments. Development of machine learning and of multivariate statistical analyses based on clinical and/or neuroimaging data would offer a complementary device to identify multi-network impairments and pathology classification and predict symptom evolution [35]. Undoubtedly, advanced neuroimaging and statistical post-processing provide a penetrating view of the dynamics of normal and pathological pain-related neural networks.

References

1. Garcia L, Peyron R (2013) Pain matrices and neuropathic pain matrices : a review. Pain., elsevier 154(Suppl 1):S29–S43
2. Baliki MN, Geha PY, Apkarian AV (2009) Parsing pain perception between nociceptive representation and magnitude estimation. J Neurophysiol 101(2):875–887
3. Peyron R, Laurent B, Garcia-Larrea L (2000) Functional imaging of brain responses to pain. A review and meta-analysis. Neurophysiol Clin 30:263–288
4. Vogt BA, Derbyshire S, Jones AKP (1996) Pain processing in four regions of human cingulate cortex localized with coregistered PET and fMRI imaging. Eur J Neurosci 8:1461–1473
5. Peyron R, Garcia-Larrea L, Gregoire MC, Costes N, Converts P, Lavenne F, Maugière F, Michel D, Laurent B (1999) Haemodynamic brain responses to acute pain in humans : sensory and attentional networks. Brain 122(9):1765–1780
6. Craig AD (2009) How do you feel-now ? The anterior insula and human awareness. Nat Rev Neurosci 10(1):59–70
7. Ploner M, Lee MC, Wiech K, Bingel U, Tracey I (2010) Prestimulus functional connectivity determines pain perception in humans. PNAS 107(1):355–360
8. Mouraux A, Diukova A, Lee MC, Wise RG, Lannetti GD (2011) A multisensory investigation of the functional significance of the « pain matrix». NeuroImage 54:2237–2249
9. Wiech K, Lin C-S, Brodersen KH, Bingel U, Ploner M, Tracey I (2010) Anterior insula integrates information about salience into perceptual decisions of pain. J Neurosci 30 (48):16324–16331
10. Neugebauer V, Li W, Bird GC, Han JS (2004) The amygdala and persistent pain. Neuroscientist 10:221–234
11. Benuzzi F, Lui F, Duzzi D, Nichelli PF, Porro CA (2008) Does it look painful or disgusting ? Ask your parietal and cingulate cortex. J Neurosci 28(4):923–931

12. Bingel U, Quante M, Knab R, Bromm B, Weiller C, Büchel C (2002) Subcortical structures involved in pain processing : evidence from single-trial fMRI. Pain 1(2):313–321

13. Atlas LY, Bolger N, Lindquist MA, Wager T (2010) Brain mediators of predictive cue effects on perceived pain. J Neurosci 30(39):12964–12977

14. Baliki MN, Geha PY, Fileds HL, Apkarian AV (2010) Predicting value of pain and analgesia : nucleus accumbens response to noxious stimuli changes in the presence of chronic pain. Neuron:66. https://doi.org/10.1371/journal.pone.0106133

15. Moulton EA, Schmahmann JD, Becerra L, Borsook D (2010) The cerebellum and pain : passive integrator or active participator. Brain Res Rev 65(1):14–27

16. Helmchen C, Mohr C, Erdmann C, Petersen D, Nitschke MF (2003) Differential cerebellar activation related to perceived pain intensity during noxious thermal stimulation in humans : a functional magnetic resonance imaging study. Neurosci Lett 335(3):202–206

17. Mclver TA, Kornelsen J, Stroman PW (2017) Diversity in the emotional modulation of pain perception : an account of individual variability. Eur J Pain. Version of Record online: 20 SEP 2017. https://doi.org/10.1002/ejp.1122

18. Bantick SJ, Wise RG, Ploghaus A, Clare S, Smith SM, Tracey I (2002) Imaging how attention modulates pain in humans using functional MRI. Brain 125(2):310–319

19. Valet M, Sprenger T, Boeker H, Willoch F, Rummeny E, Conrad B, Erhard P, Tolle TR (2004) Distraction modulates connectivity of the cingulo-frontal cortex and the midbrain during pain-an fMRI analysis. Pain 109(3):399–408

20. Lorenz J, Minoshima S, Casey KL (2003) Keeping pain out of mind : the role of the dorsolateral prefrontal cortex in pain modulation. Brain 126(5):1079–1091

21. Wiech K, Kalisch R, Weiskopf N, Pleger B, Stephan KE, Dolan RJ (2006) Anterolateral prefrontal cortex mediates the analgesic effect of expected and perceived control over pain. J Neurosci 26(44):11506–11509

22. Bräscher A-K, Becker S, Hoeppllo M-E, Schweinhardt P (2016) Different brain circuitries mediating controllable and uncontrollable pain. J Neurosci 36(18):5013–5025

23. Fairhurst M, Wiech K, Duncley P, Tracey I (2007) Anticipatory brainstem activity predicts neural processing of pain in humans. Pain 128(1–2):101–110

24. Stroman PW, Khan HS, Bosma RL, Cotoi AI, Leung RL, Cadotte DW, Fehlings MG (2016) Changes in pain processing in the spinal cord and brainstem after injury characterized by functional magnetic resonance imaging. J Neurotrauma 33:1450–1460

25. Tracey I (2010) Getting the pain you expect : mechanisms of placebo, nocebo and reappraisal effects in human. Nat Med 16(11):1277–1283

26. Wagner TD, Rilling JK, Smith EE, Sokolik A, Casey KL, Davidson RJ, Kosslyn SM, Rose RM, Cohen JD (2004) Placebo-induced changes in fMRI in the anticipation and experience of pain. Science 303:1163–1166

27. Eckert MA, Menon V, Walczak A, Ahlstrom J, Denslow S, Horwitz A, Dubno JR (2009) At the heart of the ventral attention system : the right anterior insula. Hum Brain Mapp 30 (8):2530–2541

28. Loggia ML, Kim J, Gollub RL, Vangel MG, Kirsch J, Wasan AD, Napadow V (2013) Default mode network connectivity encodes clinical pain : an arterial spin labeling study. Pain 154 (1):24–33

29. Taylor KS, Seminowicz DA, Davis KD (2009) Two systems of resting state connectivity between the insula and cingulate cortex. Hum Brain Mapp 30:2731–2745

30. Uddin LQ (2014) Salience processing and insular cortical function and dysfunction. Nat Neurosci 16(1):55–61

31. Seminowicz DA, Moayedi M (2017) The dorsolateral prefrontal cortex in acute and chronic pain. J Pain 18(9):1027–1035

32. Menon V, Uddin LQ (2010) Saliency, switching, attention and control : a network model of insula function. Brain Struct Funct 214(5–6):655–667

33. Lee MC, Tracey I (2013) Imaging pain : a potent means for investigating pain mechanisms in patients. Br J Anaesth 111(1):64–72

34. Akaparian AV, Hashmi JA, Baliki MN (2011) Pain and the brain : specificity of the brain in clinical chronic pain. Pain 152(3 Suppl):S49–S64
35. Schmidt-Wilcke T (2015) Neuroimaging of chronic pain. Best Pract Res Clin Rheumatol 29:29–41
36. Apkarian AV, Baliki MN, Geha PY (2009) Towards a theory of chronic pain. Prog Neurobiol 87(2):81–97
37. Tracey I, Mantyh PW (2007) The cerebral signature for pain perception and its modulation. Neuron 55(3):377–391
38. Baliki MN, Chialvo DR, Geha PY, Levy RM, Harden LR, Parrish TB, Apkarian AV (2006) Chronic pain and emotional brain : specific brain activity associated with spontaneous fluctuations of intensity of chronic back pain. J Neurosci 22(47):12165–12173
39. Apkarian AV, Sosa Y, Sonty S, Levy RM, Harden RN, Parrish TB, Gitelman DR (2004) Chronic back pain associated with decreased prefrontal and thalamic grey matter density. J Neurosci 24:10410–10415
40. Baliki MN, Petre B, Torbey S, Herrmann KM, Huang L, Schnitzer TJ, Fields HL, Apkarian AV (2012) Corticostriatal functional connectivity predicts transition to chronic back pain. Nat Neurosci 15(8):1117–1119
41. Baliki MN, Geha PY, Apkarian VA, Chialvo DR (2008) Beyond feeling : chronic pain hurts the brain, disrupting the default-mode network dynamics. J Neurosci 28(6):1398–1403
42. Baliki MN, Mansour AR, Baria AT, Apkarian AV (2014) Functional reorganization of the default mode network across chronic pain conditions. PLoS. https://doi.org/10.1371/journal.pone.0106133
43. Kucyi A, Moayedi M, Weissman-Fogel I, Goldberg MB, Freeman BV, Tenenbaum HC, Davis KD (2014) Enhanced medial prefrontal-default mode network functional connectivity in chronic pain and its association within pain rumination. J Neurosci 34(11):3969–3975
44. Otti A, Guendel H, Wohlschläger A, Zimmer C, Noll-Hussong M (2013) Frequency shifts in the anterior default mode network and the salience network in chronic pain. BMC Psychiatry 13:84
45. Kim J-Y, Kim S-H, Seo J, Kim S-H, Han SW, Nam EJ, Kim S-K, Lee HJ, Lee S-J, Kim Y-T, Chang Y (2013) Increased power spectral density in resting-state pain related brain networks in fibromyalgia. Pain 154:1792–1797
46. Balenzuela P, Chernomoretz A, Fraiman D, Cifre I, Sitges C, Ontoya P, Chialvo DR (2010) Modular organization of brain resting state networks in chronic back pain patients. Front Neuroinform 1:116
47. Mills EP, DiPietro F, Alshelh Z, Peck CC, Murray GM, Vickers ER, Henderson LA (2017) Brainstem pain control circuitry connectivity in chronic neuropathic pain. J Neurosci:1647–1617. https://doi.org/10.1523/JNEUROSCI.1647-17.2017
48. Lorenz J, Cross DJ, Minoshima S, Morrow TJ, Paulson PE, Casey JL (2002) A unique representation of heat allodynia in the human brain. Neuron 35(2):383–393
49. Schwedt TJ, Larson-Prior L, Coalson RS, Nolan T, Mar S, Ances BM, Benzinger T, Schlaggar BL (2014) Allodynia and descending pain modulation in migraine : a resting state functional connectivity analysis. Pain Med 15(1):154–165
50. Seminowicz DA, Davis KD (2005) Cortical responses to pain in healthy individuals depends on pain catastrophizing. Pain 120:297–306
51. Cauda F, Palermo S, Costa T, Torta R, Duca S, Vercelli U, Geminiani G, Torta DME (2014) Gray matter alterations in chronic pain : a network-oriented meta-analytic approach. Neuroimage Clin 4:676–686
52. Mansour A, Baliki MN, Huang L, Torbey S, Herrmann K, Schnitzer TJ, Apkarian AV (2013) Brain white matter structural properties predict transition to chronic pain. Pain 154(10):2160–2168
53. Lutz J, Jäger L, de Quervain D, Krauseneck T, Padberg F, Wichnalek M, Beyer A, Stahl R, Zirngibl B, Reiser M, Schelling G (2008) White and grey matter abnormalities in the brain of patients with fibromyalgia : a diffusion-tensor and volumetric imaging study. Arthritis Rheum 58(12):3960–3969

54. Hotta J, Zhou G, Harno H, Forss N, Hari R (2017) Complex regional pain syndrome : the matter of white matter ? Brain Behav 7(5):e00647. https://doi.org/10.1002/brb3.647. eCollection 2017 May

55. Farmer MA, Baliki MN, Apkarian AV (2012) A dynamical network perspective of chronic pain. Neurosci Lett 520(2):197–203

56. Chapin H, Bagarinao E, Mackey S (2012) Real-time applied to pain management. Neurosci Lett 520(2):174–181

57. Sulzer J, Haller S, Scharnowski F, Weiskopf N, Birbaumer N, Blefari ML, Bruehl AB, Cohen LG, deCharms RC, Gassert R, Goebel R, Herwig U, LaConte S, Linden D, Luft A, Seifritz E, Sitaram R (2013) Real-time fMRI neurofeedback : progress and challenges. NeuroImage 73:386–399

58. deCharms RC, Maeda F, Glover GH, Ludlow D, Pauly JM, Soneji D, Gabrieli JDE, Mackey SC (2005) Control over brain activation and pain learned by using real-time functional MRI. PNAS 102(51):18627–18631

59. Emmert K, Breimhorst M, Bauermann T, Birklein F, Van De ville D, Haller S (2014 .; 8 article) Comparison of anterior cingulate vs insular cortex as targets for real-time fMRI regulation during pain stimulation. Front Behav Neurosci 350:1–13

60. Rance M, Ruttorf M, Nees F, Schad LR, Flor H (2014) Real time fMRI feedback of the anterior cingulate and posterior insular cortex in the processing of pain. Hum Brain Mapp 35 (12):5784–5798

61. Emmert K, Kopel R, Sulzer J, Brühl AB, Berman BD, Linden DEJ, Horovitz SG, Caria A, Frank S, Johnston LZL, Paret C, Robineau F, Veit R, Bartsch A, Beckmann CF, Van De Ville D, Haller S (2016) Meta-analysis of real-time fMRI neurofeedback studies using individual participant data : how is brain regulation mediated ? NeuroImage 124:806–812

62. Jensen MP, Day MA, Miro J (2014) Neuromodulatory treatments for chronic pain : efficacy and mechanisms. Nat Rev Neurol 10:167–178

63. Landry M, Lifshitz M, Raz A (2017) Brain correlates of hypnosis : a systematic review and meta-analytic exploration. Neurosci Biobehav Rev 81:75–98

64. Lutz A, McFarlin DR, Perlman DM, Salomons TV, Davidson RJ (2013) Altered anterior insula activation during anticipation and experience of painful stimuli in expert meditators. NeuroImage 64:538–546

65. Fregni F, freedman S, Pascual-Leone A (2007) Recent advances in the treatment of chronic pain with non-invasive brain stimulation techniques. Lancet Neurol 6:188–191

66. Siebner HR, Bergmann TO, Nestmann S, Massimini M et al (2009) Consensus paper : combining transcranial stimulation with neuroimaging. Brain Stimul 2:58–80

67. Klooster DCW, de Louw AJA, Aldenkamp AP, Besseling RMH, Mestrom RMC, Carette S, Zinger S, bergmans JWM, Mess WH, Vonck K, Carrette E, Breuer LEM, Bernas A, Tijuis AG, Boon P (2016) Technical aspects of neurostimulation : focus on equipment, electric field modeling and stimulation protocols. Neurosci Biobehav Rev 65:113–141

68. Luedtke K, Rushton A, Wright C, Geiss B, Juergens TP, May A (2012) Transcranial direct current stimulation for the reduction of clinical and experimentally induced pain : a systematic review and meta-analysis. Clin J Pain 28(5):452–461

69. O'Connell NE, Wand BM, Marston L, Spencer S, Desouza LH (2011) Non-invasive brain stimulation techniques for a chronic pain. A report of a Cochrane systematic review and meta-analysis. Eur J Phys Rehabil Med 47(2):309–326

70. Vaseghi B, Zoghi M, Jaberzadeh S (2014) Does anodal transcranial direct current stimulation modulate sensory perception and pain? A metaanalysis study. Clinical Neurophysiol 125 (9):1847–1858

71. DosSantos MF, Love TM, Martikainen IK, Nascimento TD, Fregni F, Cummiford C, Deboer MD, Zubieta J-K, DaSilva AFM (2012) Immediate effects of tDCS on the μ-opioid system of a chronic pain patient. Front Psych 3:93

72. Lang N, Siebner HR, Ward NS, Lee L, Nitsche MA, Paulus W, Rothwell JC, Lemon RN, rackowiak RS (2005) How does trans-cranial DC stimulation of the primary motor cortex alter regional neuronal activity in the human brain ? Eur J Neurosci 22(2):495–504
73. Peyron R, Faillenot I, Mertens P, Laurent B, Garcia-Larrea L (2007) Motor cortex stimulation in neuropathic pain. Correlations between analgesic effect and hemodynamic changes in the brain. A PET study. NeuroImage 34(1):310–321
74. Jin Y, Xing G, Li G, Wang A, Feng S, Tang Q, Liao X, Guo Z, McClure MA, Mu Q (2015) High frequency repetitive transcranial magnetic stimulation therapy for chronic neuropathic pain : a meta-analysis. Pain Physician 18(6):E1029–E1046
75. Lefaucheur J-P, Antal A, Ahdab R, Ciampi de Andrade D, Fregni F, Khedr EM, Nitsche M, Paulus W (2008) The use of repetitive transcranial magnetic stimulation (rTMS) and transcranial direct current stimulation (tDCS) to relieve pain. Brain Stimul 1:337–344
76. André-Obadia N, Mertens P, Gueguen A, Peyron R, Garcia-Larrea L (2008) Pain relief by rTMS. Differential effect of current flow but no specific action on pain subtypes. Neurology 71 (11):833
77. Hou WH, Wang TY, Kang JH (2016) The effects of add-on non-invasive brain stimulation in fibromyalgia : a meta-analysis and meta-regression of randomized controlled trials. Rheumatology (Oxford) 55(8):1507–1517
78. Maleki N, Brawn J, Barmetter G, Borsook D, Becerra L (2013) Pain responses measured with arterial spin labelling. NMR Biomed 26(6):664–673
79. Segerdahl AR, Mezue M, Okell TW, Farrar JT, Tracey I (2015) The dorsal posterior insula subserves a fundamental role in human pain. Nat Neurosci 18(4):499–500

Index

A

Aβ-amyloid deposition in PD, 154, 155
Aβ-amyloid PET imaging, 153, 154
Acetylcholine, 148, 149
Acquired synesthesia, 300
Acquisition parameters, 7
Aerobic glycolysis, 27
Alzheimer's disease (AD), 60
    amnestic MCI, 253–254
    atypical, 254–255
    connectivity-based models, 256–257
    group of phenotypes, 253
    neurodegenerative disease, 253
    preclinical and at-risk populations, 255–256
Amide proton transfer (APT), 98
Amnestic mild cognitive impairment (aMCI), 60
Amodality, 311–312
Amplitude of low-frequency fluctuations (ALFF), 37
Amygdala, 269
Anaesthetic agent, 67
Analgesia, 331
Anisotropy of diffusion, 15
Antenna, 5
Anterior cingulate cortex (ACC), 269, 323, 324, 326, 328–332
Anterior hippocampus-dorsal attention network connectivity, 66
Anterior insula, 325
Anterior limbic regions, 269
Anterior medial prefrontal cortex (aMPFC), 47
Anterior-posterior segmentation concept, 66
Anterior temporal lobectomy (ATL), 61
Anterolateral prefrontal cortex (ALPC), 325

Anticipation phases, 272
Antiepileptic drugs treatment, 110
Antiparallel spins, 2
Anti-Warburg effect, 97
Anxiety scores, 88
Apparent diffusion coefficients (ADCs), 14, 219
Apraxia of speech, 258
Arcuate fasciculus and language tracts, 134
Argus II, 298
Arterial spin labeling (ASL), 31, 32, 35, 114, 115, 118, 119, 123, 125, 127, 135, 326, 328, 332
Artifacts
    diffusion-weighted MRI, 18
Artificial neural networks (ANNs), 48–50
Artificial neuron, 48
Ataxia with oculomotor apraxia type 2 (AOA2), 226
Ataxias
    clinical features overlapping, 216
    microstructural alterations, 215
    neurochemical compounds, brain, 215
    signal abnormalities, 215
    structural, neurochemical and functional alterations, 216
Ataxia-telangiectasia (AT), 221
Auditory network, 39
Autism Diagnostic Interview-Revised (ADI-R), 236
Autism Diagnostic Observation Schedule-Generic (ADOS-G), 236
Autism spectrum disorder (ASD), 45, 234–238
    brain regions mediation, 240–241
    DSM-5 classification, 233

© Springer International Publishing AG, part of Springer Nature 2018
C. Habas (ed.), *The Neuroimaging of Brain Diseases*, Contemporary Clinical Neuroscience, https://doi.org/10.1007/978-3-319-78926-2

CPSIA information can be obtained
at www.ICGtesting.com
Printed in the USA
LVHW082003110119
603608LV00001B/13/P